IS GOD A REALITY?

IS GOD A REALITY?

A Scientific Investigation

Carmel Paul Attard

IS GOD A REALITY?
A SCIENTIFIC INVESTIGATION

iUniverse books may be ordered through booksellers or by contacting:

iUniverse
1663 Liberty Drive
Bloomington, IN 47403
www.iuniverse.com
1-800-Authors (1-800-288-4677)

Because of the dynamic nature of the Internet, any web addresses or links contained in this book may have changed since publication and may no longer be valid. The views expressed in this work are solely those of the author and do not necessarily reflect the views of the publisher, and the publisher hereby disclaims any responsibility for them.

Any people depicted in stock imagery provided by Thinkstock are models, and such images are being used for illustrative purposes only. Certain stock imagery © Thinkstock.

ISBN: 978-1-5320-1222-8 (sc)
ISBN: 978-1-5320-1221-1 (e)

Library of Congress Control Number: 2016919775

Print information available on the last page.

iUniverse rev. date: 02/07/2017

To my three female generations:
Mary, Lara, and Amelia;
With love

CONTENTS

ACKNOWLEDGEMENTS

I am deeply indebted to my wife, Mary, who tolerated my countless hours of writing and supported me in difficult decisions. I am eternally grateful to my work-colleague Usman Hameed for acting as my audience-of-one while I was writing the first version of my manuscript and for encouraging me all along the way. My heartfelt gratitude goes to my good friends Liborio (Leo) and Diane Panetta for inspiring me on how to refine my research and to expound scientific concepts. My appreciation also goes to my work-colleagues: Veton Visoka, Alex Solomon, Florian Tiganila, Harold Ramhit, John Fletcher, and Bonneville Minott who acted as my human sounding-boards and helped me submit related material to iUniverse. Finally, special thanks go to my friend of over fifty years, George Agius, for critiquing my manuscript from a scientific perspective. I acknowledge that without the overwhelmingly positive reaction and encouragement of the people mentioned above, this book would never have happened.

PREFACE

Ever since I was a young lad, in my early teens, I was fascinated by ocean surfers. I thought there was nothing more beautiful than a short movie strip, in slow motion, of a surfer riding an ocean wave: a large overhanging wave continuously threatening to engulf the surfer but never actually succeeding. Later, as a young man in my early twenties, this fascination made me come up with my own maxim that I have adopted in my studies, work, and family life ever since. It goes, "If you ride the wave, the wave will carry you; if you don't, the wave will crush you." I started to wonder what would be absolutely crucial for a surfer to be able to ride a wave. I am no surfer, so I could be wrong in what I'm about to say: I came to my own conclusion that the key element is *balance*.

I have never been much of an athlete in my youth: I may have made the science soccer team in university; I was decent at swimming; I could passably ride a bike—but that's about all. Consequently, I was never the kind of guy that was very popular with the girls. Instead, I preferred board games, like chess and checkers, and card games, like bridge and euchre. Unlike my father, who was known all over his village for his ability to roller-skate, I couldn't control my body's balance for beans. I never ever managed to learn how to roller-skate, and learning to bike-ride was a real struggle for me. But when it came to balanced thinking, I was at my best. Of course, this natural disposition helped me tremendously in my scientific studies and in my work.

I do not claim to be overly knowledgeable in any particular area of science: I only have a Bachelor of Science degree in physics and mathematics. However, in addition to my balanced way of thinking, I also have a knack for integrating all sorts of information from the many disciplines of science. In the industry, "riding the waves" is synonymous to "delivering the goods": employers only like to pay for results, not effort. So I also tend to possess a results-oriented rather than an effort-oriented personality.

In research, however, one can only go by one's belief, bias, and gut feeling. I do not want to knock research down; we do need research. Sometimes, effort is all these people have to show for decades of work; but "no" is also an answer even in science. I'd rather have a scientist's honestly admitting defeat in a particular field of research than one's giving me a misleading, wishy-washy report in conformance with mainstream scientific paradigms in order to justify one's time. If decades of intense research have not yielded the predicted or desired result, I believe one should admit the remote possibility, at least, that one might have been wrong; or that one might have been looking in the wrong direction, and that perhaps it might be time to look in the opposite direction. I do not want to go into details here; the reader will encounter such instances in the course of this book and see what I mean.

When it comes to science, the evidence collected may be interpreted in many different ways: it is not as easy as it might seem to an outsider. Depending on one's bias, any two people can come to diametrically opposite conclusions. This observation taught me to respect the others' opinion, even if I strongly disagree with them; it made me adopt another maxim by an unknown author: "Always keep your words soft and sweet, just in case you have to eat them." The whole scientific community has been wrong before, big time. To give a couple of simple examples: for centuries, people thought the earth was flat until they discovered that it was actually round (spherical); early scientists thought the sun moved round the earth until they discovered that it was exactly the other way around. I mean, how much more wrong can a whole community, including experts, be? So to be absolutely certain that one's opinion is the right one is, in my opinion, tantamount to ignorance.

This book has been written in this disposition; even if sometimes the words I use may sound too strong in someone's opinion, or if I seem to criticize someone's opinion excessively. I apologize beforehand if I ever seem to cross the line of courteous discussion.

Lee Smolin, in his book *The Trouble with Physics*, makes an interesting observation regarding the attitude of most scientists. Everybody knows that swans are either white or black; but a philosopher he was acquainted with imagines that a red swan suddenly appears on the earth. He quotes the philosopher [1] as saying that normally scientists will not look at the new evidence right in front of their eyes; instead, they will look for the person who painted it red: simply because that is in line with what they already believe. [2] In other words, scientists follow established paradigms. Admittedly, as Lee Smolin also points out, past experience has shown that first reactions, cutting corners and gut feelings, have more often than not resulted in positive results; however, this was not always the case. There is no standard method one can adopt in science to arrive at the truth, or rather, *reality*. I therefore ask the reader to, at times, set aside the assumptions of mainstream science; I ask one to simply examine the evidence I present with as little prejudice as can be mastered and try to come to one's own conclusions. I am not asking scientists to give up their opinion; but only to shed their prejudices for a little while, as I believe I did in coming to the conclusions presented in this book.

In my mid-twenties, a couple of years at my first job back home in Malta, I was required to commute to the smaller island of Gozo. So I had to share a hotel room with a workmate. This workmate is a real person; I'm not making him up, so I do not want to mention him by name. He was an interesting personality: he was a happy-go-lucky kind of guy; I was more of the rigid, pensive type in my youth. What was very revealing to me during this time of my life was the fact that he and I never agreed about anything: I mean *anything*. This made

---

[1]   Lakatos, *Proofs and Refutations*

[2]   Smolin, *The Trouble with Physics*, p. 297

me think: how could an intelligent person like my workmate disagree with me about everything? He has a university degree, like me; he has a well-paying job, like me; he has a beautiful and wonderful girlfriend, like me; why do we think so differently? This period of my life made me respect other people's opinion which, I'm afraid, I didn't before then. We always ended our conversation saying, "You and I never agree about anything; let's agree to disagree." And this is what I'd like to ask from readers who might disagree with my conclusions in this book: let's agree to disagree!

INTRODUCTION

Before I start this book, I would like to apologize to my readers for my personal biases, which will most probably become obvious while reading. As far as is humanly possible, I shall do my utmost to set them all aside, and I shall try to be as fair as I can to those with whom I disagree.

This book deals, mainly, with *scientific* evidence in support of the existence or non-existence of a supreme being whom we normally call "God". I firmly believe that if God exists he would be on the side of the *truth*. Such a God, especially if he is also omnipotent, does not need to be protected by anyone's manipulation of evidence: he can hold his own under any circumstances. If he did need the truth to be thus twisted in his favor, then he would not be on the side of truth. This intuitive rational principle is what inspired me to write this book; I shall, therefore, try to honestly search for the truth. I have no personal agenda: that is, proving or disproving God's existence. I shall present the evidence I researched, as I see it, and let one reach one's own conclusions.

Basically, I believe that whether God exists or not *is* a scientific question because it boils down to knowledge about reality: if science says God does not exist while religion says he does, then one of them must be wrong. I believe that we cannot separate science from religion any longer. I think it is time for scientists and religious scholars to stop putting each other down, throwing monkey wrenches into each other's cogwheels, manipulating evidence, denying or failing to mention facts that compromise their position, etc. and to start working together to come up with some serious conclusions.

A good start is to respect each other; I believe that without honestly and seriously considering another's point of view one simply cannot arrive at the ultimate truth. A tunnel vision, shouting the other side down, or ridiculing the other side is not going to convince anyone on either side. We are all supposed to be looking for the same truth; so it will be to the advantage of both sides if all information were shared in a congenial manner.

Too many times in the past, have we attributed to God's direct action certain natural phenomena we did not understand at the time; only to find out later that there was a scientific explanation for them, after all. Two obvious examples that easily come to mind are our planetary system and evolution.

The basic question I shall address in this book is: "Does God exist?" Is he a reality, or did we just make him up—a figment of our imagination? Although God is, supposedly, a spiritual (non-physical) being, I shall use physical evidence, universal experiential evidence, empirical (data-based) information, and reason (logic) to investigate whether he exists or not. If one tries to tackle the question of God's existence supernaturally, or even philosophically, the final conclusion would be left hanging in the clouds, so to speak: no one would be convinced by it. Science is a search for the truth about reality, so science is a good source in our investigation; hence I shall frequently be using scientific discoveries to reach my conclusions. Science has a lot to say about reality, namely: matter, energy, the universe, space, time, life, evolution, the brain, free will, etc. and God, supposedly, has a finger in all of these things, too.

In this book, I shall not use quotes from any holy scriptures, such as the bible, to prove any point: they are obviously biased in God's favor assuming God's existence. Furthermore, I believe that a book like the bible is a moral book, not a scientific book: so it has very little to tell us about reality as such. I might quote the bible as a historical or an ancient document; but I shall never even hint that it might be directly inspired by God, or that it might be infallible, or that it might be mysteriously superior to any other good book.

To create more of an impact, I shall be using a number of concepts, ideas, and quotes from authorities and experts in their particular field of science; this way my conclusions will be more convincing and obviously bear more weight. I think I am quite safe in stating that no single person can possibly be an expert in everything, especially in a field as vast as science; this is why I quote experts in the field, not to plagiarize their works or their ideas. Most quotes are clearly indented for better identification and, naturally, credited to their authors in footnotes. To the best of my ability, I also referenced words, phrases, sentences, and paragraphs expressing the various authors' concepts and ideas in the footnotes.

In the last couple of centuries, science has been explaining many natural phenomena that were formerly attributed to divine action: this resulted in God's being pushed further and further into a corner, so to speak. Nowadays, anything that the scientists cannot yet explain they call a "god of the gaps": this means that one day they will be able to explain it materialistically. This attitude is, no doubt, the result of science's past successes. However, I think it shows an unhealthy and biased predisposition which is very similar to a religion, or even a blind faith: it describes an attitude which scientists themselves abhor. Good science should keep an open mind and follow an unbiased path that evidence and reason lead it to.

There is obviously an inherent difficulty producing enough reliable and independently verifiable material evidence for an immaterial realm. Nevertheless, it seems to me to be absolutely necessary to do so if modern scientists are to be convinced of the existence of any non-physical entities. I shall try to accomplish such a feat in this book; however, the evidence will be, more often than not, circumstantial rather than direct—this is where our reason comes in.

Whether God, the soul, the afterlife, or any other kind of spiritual entity exists are all significant questions about the reality out there. If there is no God, soul, or afterlife, so be it! But if there is, no amount of shouting the other side down, or name-calling, is going to change that reality. Taking for granted the inexistence of anything supernatural is not going to change the facts, either: there is the possibility that one may end up with a nasty surprise one cannot remedy in the afterlife; that is, if it actually is the reality out there, of course.

Basically I am a scientist, and as such I believe in physical and empirical evidence combined with the power of mathematics and reason to reach conclusions. However, I do not believe that a purely mathematical model that "works" is enough to explain reality (as some very renowned scientists actually do believe): it must, at least, also have some subtle form of connection to reality. Ideally the model should be verifiable in the laboratory or through observations in our universe as true or false, eventually, given the right instrumentation. To me, mathematics and reason are only tools for science, but I also admit that they can direct us into looking for the hidden secrets of reality. Science, ultimately, boils down to the formulation of observed natural phenomena in mathematical terms. Mathematical formulation, while admitting that it usually does not represent reality perfectly, is essential to show relationships (at least to the first order of magnitude) between different parameters. Otherwise our science will sound something like: "an increase in this parameter will result in an increase in that parameter" or "an increase in this parameter will result in a decrease in that parameter". Our five senses are often very misleading in reaching the right conclusions; however, science has shown, in past experience, that with the aid of proper instrumentation this hurdle can be overcome.

For the benefit of the non-science-oriented reader, at times I introduce certain intricate scientific concepts gradually, so that such readers may understand more easily. The science presented may not be very accurate at first blush, but eventually, later on in the chapter or the book, I usually delve deeper into the concept in question to give the reader the current scientific understanding of that particular concept. In such cases, where I do not use the exact term as an introduction, I use a simpler term in quotation marks or brackets.

Before starting to read this book, I suggest the non-science-oriented readers' having a quick glance at the three appendices on "Mathematics", "Chemistry", and "Physics", at the end of the book, and referring to them occasionally when further clarity is deemed required.

CHAPTER 1

MATTER AND ENERGY

There is no denying, as obvious physical evidence, the existence of an abundance of matter in our universe: our earth, the moon, the planets, the sun, the stars, the galaxies, etc. How did all this matter, the whole universe, come about? In 1714, polymath and philosopher Gottfried Leibniz asked, "Why is there something rather than nothing?" [3]

As a general principle of science, matter and energy cannot be created or destroyed; it is similar to balancing a ledger in accounting: to the penny. Scientists still haven't got a clue as to how it could all have come about; yet it is all out there in full view, more matter and energy than one could ever imagine in the mind-boggling immensity of space in our universe. Scientists claim that they haven't got an explanation for the existence of matter in our universe yet, but that they eventually will. They claim that the origin of matter and energy in our universe is a "god of the gaps": this means that currently people may be attributing their origin to God, but that science will eventually find the answer to this question. In my opinion, science should keep on looking for natural explanations—there's nothing wrong with that!

While this may be the case, given the successes science has had in the past explaining many physical phenomena that were previously attributed to divine action, let us look at the evidence we currently have. If we, unquestioningly, assume that science will find the answer to the origin of matter and energy without doubting the scientists'

---

[3]   Leibniz, *The Principles of Nature and of Grace, Based on Reason.*

claims of probable future success, we would be behaving in a similar fashion to those religious people who quote the bible in order to prove the creation of the universe and the earth by God.

Antimatter

Antimatter is not normally found on earth because it would be instantly destroyed by any ordinary matter with which it comes in contact; it would produce a flash and an explosive release of energy. But we know that, just like matter, antimatter exists too; because scientists have made very small quantities of it. To confine it in a cylindrical container, say, antimatter must never touch the walls of the container or any other ordinary matter; so it is kept rotating inside the cylindrical container using electric and magnetic fields in a near-perfect vacuum.

Matter and antimatter particles look identical; the only difference is that they are constructed, internally, as mirror images of each other: they have opposite charge, currents, forces, fields, and spins inside them. Matter and antimatter are what we might call forms of "trapped" energy: according to theoretical physicist Albert Einstein's most famous equation $E=mc^2$ (where "E" stands for "energy", "m" stands for the "mass" or chunk of matter, and "c" stands for the "speed" of light in a vacuum). When a matter particle and a corresponding antimatter particle are combined together, they will both be *annihilated*: that is, no matter at all is left after they make contact with each other. This is the technical term used; however, a tremendous amount of pure energy is released according to the equation $E=2mc^2$: that is, double the energy of the matter or antimatter particle is released because both matter and antimatter particles are annihilated. So it's not that they become "nothingness" and cease to exist altogether; they are converted into energy: in other words, their trapped energy is released.

Conversely, to produce matter and antimatter out of nothingness, so to speak, the same minimum amount of energy ($2mc^2$) is required.

This is a tremendous amount of energy; it is comparable to that released in the atom bomb or the hydrogen bomb. Some extra energy is moreover required in order to give the matter and antimatter particles some additional motion to enable them to separate from each other. Otherwise, if they don't move apart after being created, they would instantaneously recombine and revert to no matter at all: they will just give back the energy initially available.

Energy is not nothingness; yes, it is a non-material substance (that is, it has no mass), however it is a *physical* reality in science that can change into various forms, namely: light, electricity, heat, moving or lifting an object, etc. However, it cannot be created from nothingness or destroyed into nothingness. Whatever energy is in the universe keeps circling it: albeit changing from one form to another or into matter. Following is a quote from particle physicist Frank Close's book *Antimatter* to help us understand a little better:

> There's matter, like the electron; antimatter, like the positron [positive electron]; and there are things that are neither matter nor antimatter. The most familiar example of something that is beyond substance [non-matter] is electromagnetic radiation. All electromagnetic radiation from gamma rays through X-rays and ultra-violet to visible light, infra-red, and radio waves consist of photons [packets] of different energies. Matter and antimatter can cancel one another out, their annihilation leaving non-substance in the form of photons; if the conditions are right this sequence can happen in reverse where photons turn into pieces of matter and antimatter. Pure energy ... is non-substance; it can change from one form to another, such as electrical, chemical, or motion, and it can transubstantiate into matter and antimatter. [4]

Some scientists, for some reason, give ordinary folk the impression that it is easy to produce matter and antimatter pairs from pure energy or even nothingness: that it occurs naturally or

---

[4] Close, *Antimatter*, p. 64.

spontaneously. But in actual fact it is far from easy to construct matter particles from pure energy, never mind nothingness. Notice the phrase "if the conditions are right" in the above quote. In order to spontaneously produce an electron-positron pair, both of which are of extremely small mass, the surrounding temperature has to be raised above six billion (6×10^9) degrees Celsius. This is the temperature at which the average *photons* (energy packets) from *black body* (background) radiation possess sufficient "punch" that it will happen spontaneously. [5] In practice, of course, it is impossible to raise the temperature of any container to such a high temperature to be able to perform the experiment because everything will melt and vaporize at such temperatures. Now, an electron's mass is only about one-two-thousandth (1/2000) of the mass of a proton; so to produce a proton-antiproton pair spontaneously, the surrounding temperature has to be raised much higher: above ten trillion (10^{13}) degrees Celsius. That is the temperature at which the average photons have enough energy to make it happen spontaneously. [6]

It took humanity close to a century, starting prior to 1930, to develop particle accelerators culminating in the large hadron collider in Geneva, Switzerland, and produce the first Higgs boson, which is an elementary particle of matter one cannot see even with a microscope. It is about one-hundredth (1/100) of the size of a proton and one hundred twenty-five (125) times the mass of a proton. The Higgs boson seems to be a necessary "stepping stone" in producing other material particles. In his book *A Brief History of Science*, science historian Thomas Crump writes:

> In the 1960s, the British Physicist, Peter Higgs (1929-) suggested that the whole of space was permeated by the eponymous [named after] 'Higgs Field' (comparable to the electromagnetic field), with the property that all particles,

---

[5] Ryden, "The First Three Minutes": http://www.astronomy.ohio-state. edu/~ryden/ast162_10/notes44.html, para. 2 (23 July 2015).

[6] Ryden, "The First Three Minutes": http://www.astronomy.ohio-state. edu/~ryden/ast162_10/notes44.html, para. 2 (23 July 2015).

as they pass through it, acquired mass. The principle of wave-particle duality [explained presently], fundamental to quantum physics [explained presently], then requires a particle, the 'Higgs boson' to be the agent of interaction. [7]

The concept of *wave-particle duality* and the *quantum physics theory* are discussed in greater detail in chapter eight dealing with the "Mind and Soul". Very briefly for now, wave-particle duality is the fact that, in our universe, sometimes conglomerates of what we know to be particles exhibit wave-like characteristics; vice versa, waves exhibit the general characteristics belonging to particles: thus making it very hard to decide what "things", in nature, really are intrinsically. Quantum physics is miniature scale physics: the phenomena observed at the level of atoms, protons, electrons, and photons.

The odds of a Higgs boson materializing in the large hadron collider was one Higgs boson in about ten billion (10^{10}) collisions. Every collision consisted of two *beams*, each consisting of over three hundred trillion ($\sim$3.23x10^{14}) protons, smashing against each other at speeds almost equal to the speed of light—talk about the right conditions. So even to produce an extremely small particle of matter from pure energy—in this case energy from motion—is very demanding. Note that the Higgs boson is a necessary stepping stone prior to forming other material particles: without it, there would be no matter in our universe.

Dark Matter and Dark Energy

Let us now change gear and look at another aspect of our universe.

When we look at the sky, at very far away distances with a telescope, we can only see the stars: that is, the self-illuminated objects. Stars are similar to our sun, but they look very small because they are extremely far away. The non-self-illuminated objects like our earth, the moon, and the planets do reflect some light from stars

---

[7]   Crump, *A Brief History of Science*, pp. 350–1.

close by: however, they are not bright enough to be seen from very far away distances.

Besides this non-self-illuminated matter, which naturally consists of ordinary matter we are familiar with, there is another type of matter which is invisible even if one shines a light directly on it; scientists have termed this type of matter *dark matter*. However, since all matter gives rise to a gravitational field in its surrounding, dark matter still produces a gravitational effect.

The obvious question here is: if we cannot see dark matter by any means, how do scientists know that it exists? Cosmologists proposed dark matter's existence after astronomers measured the speed of stars orbiting the edge of galaxies. It turns out that the gravitational attraction required to achieve the observed speeds of peripheral stars exceeds, by far, what stars, planets, and moons inside the galaxy could conceivably exert. Here's a quote from particle physicist and mathematician Simon Singh's book *Big Bang* to this effect. (I shall be using concepts and ideas from Singh's book extensively in this and the next chapter.)

> [S]tars orbiting the periphery of galaxies have tremendous speeds, yet the gravitational pull of all the stars closer to the heart of the galaxy is not enough to prevent these peripheral stars from flying off into the cosmos. Therefore, cosmologists believe that there must be vast quantities of dark matter in a galaxy, namely matter that does not shine but which exerts enough of a gravitational pull to keep the stars in their orbits. [8]

We all know about gravity, and we also probably know that when the American astronauts landed on the moon, they felt much lighter than on the earth: they only felt about one-sixth of their normal weight. This is because the *mass* (size) of the moon is much smaller than that of the earth. Size is directly related to mass, the quantity of matter: the bigger the size of something the more matter it contains if it is

---

[8]   Singh, *Big Bang*, p. 479.

made of the same kind of material. Mass is a measure of the quantity of matter, but it also takes into account its *density* (compactness or heaviness of a given size): it is synonymous to weight on the surface of a given planet, like earth. Now, the astronauts' mass (their bodies) did not change when they went to the moon; so the pull of gravity towards the ground also depends on the mass of the planet or celestial body (earth or moon) on which one stands.

Moreover, gravity decreases the further one moves away from a planet or celestial body; in fact, the pull of gravity from the earth did not affect the astronauts much while they were far away on the moon. On the surface of the moon, there was still some pull of gravity exerted by the earth, but it affected the astronauts very minimally because they were significantly far away from the earth: they were practically only affected by the pull of gravity exerted by the moon because they were much closer to it.

Suppose, now, that one goes on the highest mountain on earth and shoots a cannonball from the top of that mountain; it will definitely fall to the ground some distance away. But the stronger is one's cannon the greater will be the horizontal distance covered by the projectile before it lands on the ground. Conceivably (not practically) we can imagine an extremely powerful cannon that can shoot a projectile that does not fall to the ground at all, but it keeps going outside the earth's gravitational influence into outer space. Somewhere in between these two extremes is a happy medium: a speed of fire where the projectile keeps moving round the earth without ever falling to the ground: that is, it keeps the same height of the mountain above the ground.

This may seem strange to some readers, but there are currently over two thousand two hundred (2200) satellites orbiting the earth this way; the only difference is that they are higher up than any mountain. Furthermore, the moon is a satellite of the earth; it keeps happily rotating round the earth: it never falls down to earth. The planets too are satellites of the sun. The speed of the moon, its distance from the earth, and the mass of the earth keep it in space without ever wandering too far away from earth. So, if we know the speed and distance of a satellite from the center of its orbit, we can

7

determine the mass (or equivalent mass distribution) keeping it in orbit: even though it might be far out of our reach. Believe it or not, we can "weigh" the moon, the sun, and the planets this way.

Before we move on to the next piece of scientific evidence, for the benefit of those readers who are not scientifically inclined, I would like to answer the question: how do scientists measure the speeds of these stars at such distances? I shall here explain how relative velocities of stars are determined using *spectroscopic* methods.

When one swims gently towards a point where a pebble was recently dropped in a quiet pool of water, the crests and troughs seem to occur more often than if one did not swim towards that point at all—assuming the original ripples are undisturbed by the gentle swimming. Conversely, the crests and troughs are encountered less often if one gently swims away from that point. The same phenomenon would be observed if we imagine the point where the pebble was dropped (the wave-source) approaching or receding from us, instead of us moving towards or away from the source. Consequently, when an object emitting waves moves towards an observer, the observer perceives a decrease in wavelength: which in practice results in a higher frequency (or pitch). Conversely, when the wave-emitter moves away from the observer, the observer perceives an increase in wavelength: that is, a lower frequency (or pitch).

We experience this in everyday life; when an ambulance approaches us, we hear its siren emitting a higher pitch than its normal (non-moving) pitch; as it passes beside us we hear the real pitch (only instantaneously, though); finally as it moves away from us we hear a lower pitch than normal. The higher the speed of the wave source, relative to the observer, the more noticeable is the perceived change in pitch; hence this change in pitch can serve as a measure of relative speed. Incidentally, the same principle is used by radar speed gun equipment traffic police use to measure our driving speed.

In physics, this phenomenon is known as the *Doppler Effect*, named after its discoverer mathematician and physicist Christian Doppler. Light also consists of waves; so this effect is also observable from illuminated objects, like the stars, if they are moving very fast (that is, with a speed comparable to the speed of light) relative to us.

Most of us know that if the white light, emitted by the sun, is passed through a prism, its various colors will be separated by the prism: it will look like a continuous *spectrum* (span) of colors from red to violet like the colors of a rainbow. However, if one examines it carefully, very fine dark lines can be observed amid the colors. These lines are known as *absorption spectra*. They are caused by the gases in the sun's atmosphere. This is what is happening. All visible light frequencies (colors) are generated and emitted by the sun. The gases in the sun's atmosphere sort of "resonate" with particular light frequencies emitted from the sun: by this I mean that these particular light frequencies match the exact *quantum* (amount or packet) of energy required to extract or change the energy levels (orbits) of electrons in the atoms of the sun's atmospheric gases. These gases are therefore being energized by those particular light frequencies emitted from the sun's surface: these particular light frequencies (colors) are therefore missing from a prism spectrum because they are being absorbed by these gases, and they are thus prevented from reaching us.

Now, every chemical element has a unique, characteristic, *emission spectrum* when heated by a flame. It is the exact negative of the absorption spectrum discussed above: it consists of a few colored lines interspersed on a dark background. This is like a "fingerprint", so to speak, of that particular element. The element emits these light frequencies during transitions to lower energy states: that is, some of its atoms cool down instantaneously inside the flame, and some electrons are captured by its atoms or move to a lower orbit.

Incidentally, this is how the element helium was first discovered. "Helios" in Greek means "sun"; so it was given the sun's name. From the above phenomenon, scientists concluded that there existed a yet undiscovered element in the sun's atmosphere: from its absorption spectrum—even though nobody ever went there. The atmosphere of most stars consists mainly of hydrogen and helium; they are the source and the byproduct, respectively, of the energy emitted by the stars: this process is called *nuclear fusion*—the same phenomenon that produces the energy of a hydrogen bomb.

When the absorption spectrum of helium, say, from almost any star is compared to that from the sun (which is practically stationary relative to us) we notice that it is the same "fingerprint"; except that it is completely shifted towards the longer (red) wavelength side of the light spectrum. This means that the light emitted from the star has become longer in wavelength; which, in turn, implies (from our discussion above) that the star is receding from us. The amount of this shift towards the red side of the spectrum can therefore serve as a measure of the relative velocity of separation between us (earth) and that star. In astronomy literature the Doppler Effect is commonly called the *Doppler shift* and sometimes even *redshift*. If a star exhibits a *blueshift*, it means it is approaching us; but this is very rare, and we shall presently see why.

Both hydrogen and helium absorption spectra (fingerprints) from practically all stars are red-shifted, when compared to those of our sun's absorption spectra. It does not matter which direction of the universe we set our telescope to, the result is always the same: the absorption spectra are red-shifted. This means that practically all stars are moving away from us; our sun is an exception because it remains at approximately the same distance from us—our earth simply orbits round the sun. Moreover, it was discovered that the more distant stars are from us the faster they are moving away from us. This omnidirectional redshift phenomenon does *not* imply that we are at the center of the universe (although it is tempting to think so), but it is evidence that the whole of space in our universe is expanding (we shall explain this further in the next chapter). Not only that, as time passes the universe is actually expanding at a faster and faster rate.

However, someone might ask: how can scientists determine the distance of small specs of light like the stars? Distances of stars from earth are determined by observing what astronomers call *Type 1a supernovae*. Unlike stars, supernovae are very, very bright. According to Wikipedia, a supernova is the explosion of a star at the end of its life, after it is done delivering energy, namely, light and heat. For several weeks, or even a few months, it is so bright that it outshines an entire galaxy: that is, the equivalent brightness of some one hundred billion (10^{11}) stars. It radiates, in this short time, the equivalent amount of energy that an ordinary star emits over its entire life of some ten billion (10^{10}) years—then suddenly it fades away. [9]

Supernovae, being so bright, can therefore be seen even from very remote galaxies. A Type 1a supernova is something like a *standard candle*; it produces a characteristic (known) brightness versus time variation: in other words, it follows a known mathematical equation. Hence, its real intensity at any time may be calculated using this mathematical equation; in a way, therefore, this compensates for the reduction of light intensity as a result of distance from it. From its relative intensity, measured at a particular time along the above relationship, its distance from earth can be calculated using the *inverse square law*, which is explained presently.

Light intensity drops according to the inverse square law; which means that at a distance of two miles the light intensity reduces to one-quarter ($1/4=1/2^2$) of its intensity at one mile away; at a distance of three miles its intensity reduces to one-ninth ($1/9=1/3^2$); at a distance of four miles its intensity reduces to one-sixteenth ($1/16=1/4^2$); and so on. The following extract from Singh's book *Big Bang* explains how distances and recessional speeds of stars from our earth are measured:

> Type 1a supernovae also have the advantage of having a telltale brightness variation [with elapsed time] that can be used to gauge their distances and thus the distances to the galaxies that contain them. And, by using spectroscopy [redshift], it is possible to measure their recessional velocity. As astronomers studied more and more Type 1a supernovae, the measurements seemed to be implying that the universe was actually expanding at an ever increasing rate. ... The repulsive driving force of this

---

[9]   https://en.wikipedia.org/wiki/Supernova (26 July 2015)

> runaway universe is still a mystery, and has been labeled
> *dark energy.* [10]

The fact that our universe's expansion rate seems to be accelerating led cosmologists to propose the existence of what scientists have termed *dark energy*; the properties of this dark energy are still very nebulous.

Finally, according to Anil Ananthaswamy, a former software engineer who became a consulting editor for *New Scientist* in London, England, we know so little about our universe. He ends his book *The Edge of Physics* with the following statement:

> [E]xperiments tell us that the universe is composed of dark energy (~73 percent), dark matter (~23 percent) and normal matter (~4 percent). [11]

It estimated that there are about a hundred billion (10^{11}) visible galaxies like the Milky Way in which our solar system happens to be located, including a total of about seventy sextillion (7×10^{22}) visible stars like our sun in the universe. And this represents only about 4% of our universe. Wow!

Two obvious scientific questions follow from the above considerations. (1) Where did this initial, humungous amount of matter, or energy, in the universe come from? Recall that energy is not nothingness; it is a physical entity that cannot be created out of nothing: it can only easily change form, and it cannot be annihilated either. (2) How did the right conditions required to produce all the matter in the universe come about to start the process?

The mainstream scientific answer to these questions is that our universe was always there: that our universe is *eternal.* In all fairness, in my opinion, it is easier to imagine a simple thing like matter to have no beginning, rather than to conjure up a possibly complex supernatural being (like God) creating the universe. On the other

---

[10] Singh, *Big Bang*, pp. 480–1 (emphasis in original).
[11] Ananthaswamy, *The Edge of Physics*, p. 286

hand, if it could be proved beyond any reasonable doubt that there was really just matter (no God) I think most of us would still be uneasy about it, and we would still wonder, deep down, why there is something rather than nothing. Nothing is much simpler than something; but then if there were nothing, of course, we would not be here to wonder. Anyway, let us tread the path taken by science, assuming that the mainstream scientific consensus is right, and continue examining the evidence to see where it takes us.

CHAPTER 2

BIG BANG THEORY

In 1917, theoretical physicist Albert Einstein applied his *general relativity theory* equations to the entire universe as one integrated system. The numbers showed that the universe possessed a serious inherent instability: that it would not remain in equilibrium (stable), and it would implode (collapse). His gravity formula spelled disaster: every object would be pulled towards every other object. The attraction might start very slowly at first, but eventually everything would end up in an ominous crunch: just as falling objects pick up speed with time. At that time the scientific consensus was that our universe was a *static* universe. By definition, a static universe is one that is both temporally and spatially infinite: thus its space is not expanding or contracting. To work around this problem he used a mathematical trick—adding a "fudge-factor"—to balance his equations of motion. He introduced an *anti-gravity* factor, equivalent to a repulsive force, which he called the *cosmological constant.* [12]

Some people may find it strange that an intelligent person like Einstein, in conjunction with the majority of the scientific community of his time, would accept the notion of an eternal universe: meaning that it never started. Why would the obvious question: "where did our universe come from?" not occur to them? It is not the case that it did not occur to them; they went a step further. If one answers: "God created our universe," then the next obvious scientific question would be: "and where did God come from?" If one answers: "God always

---

[12] Singh, *Big Bang*, pp. 146–9.

existed: he is the *first-cause*", then they would answer: "why can't our universe be the ultimate first-cause?" One must admit that they had a strong point: it is mentally easier (simpler) to accept inanimate matter and energy (the universe) as the first-cause rather than a supposedly omnipotent being with all of God's other attributes.

Anyway, in 1922, physicist and mathematician Alexander Friedman suggested that a continuously expanding universe would not require a cosmological constant (it could be set to *zero* in Einstein's equations of motion) and still the universe would not collapse. (We shall revisit this cosmological constant in chapter four on the "Fine-Tuning of the Universe"; for the time being in this chapter I want to keep things simple.) In Friedman's model of the universe, the initial expansion rate and the mass density of the universe would act against each other producing: (a) a universe that would continue expanding forever; (b) one that would (asymptotically) tend to, but never reach, a fixed size; or (c) one that would eventually collapse after a period of expansion. [13]

In 1927, a Roman Catholic priest, astronomer, and physicist Georges Lemaitre also thought of the expanding universe concept independently from Friedman. Lemaitre was an interesting personality; his general philosophy was, "There are two ways of arriving at the truth; I decided to follow them both." He was, of course, referring to religion and science; truly an inspiring personality, in my opinion. Obviously, he did not believe in "NOMA"; nor does self-declared atheist and evolutionary biologist Richard Dawkins, [14] and neither do I. NOMA is an acronym for "Non-Overlapping Magisteria": meaning that the spiritual (or supernatural) and material realms do not overlap at all, and one cannot prove or disprove the other. No doubt, Lemaitre was an excellent scientist; one who did not allow his background in theology influence his scientific research and conclusions: one who followed the evidence wherever it took him. Lemaitre went one step further than Friedman. He reasoned that if the universe was actually

---

[13]   Singh, *Big Bang*, pp. 151–2.
[14]   Dawkins, *The God Delusion*, pp. 82 & 85

expanding, it must have been smaller the day before, even smaller the previous month or year, and much smaller centuries earlier. And logically, if one goes back far enough in time, the entirety of space must have been compacted into a tiny region, or maybe even an atom.

This conclusion, of course, may have been inspired by his belief in the creation of the universe by God; this is probably a case of arriving at the whole truth by using both science and religion. In principle, however, any conclusions arrived at must converge—they cannot be contradictory—because there is only one truth. Now, if the universe was really expanding, one must admit that a continually expanding universe, combined with winding the clock backwards, logically indicates that most probably the universe had a *beginning*: it is almost unavoidable. [15]

This was all theoretical at the time; until in 1929, astronomer Edwin Hubble was able to measure the receding velocity of galaxies when he observed a longer wavelength, or redshift, in their hydrogen and helium absorption spectra—by using the spectroscopic method described in the previous chapter. Moreover, as already mentioned, Hubble discovered that the further away from us these galaxies were the faster they were receding from us. Scientists agreed that this receding velocity was actually due to *space expansion*: giving us the same impression as receding speed. This model of a still expanding universe corresponds to the first, case (a), of Friedman's three possible models of the universe described above.

It is important to understand here that the galaxies are not travelling *through* space at a relatively higher speed than we are, but that the *whole* of space in the universe is increasing. As an analogy, one can imagine equally spaced dots on a balloon that is being inflated: the dots move apart as it is inflated, but they are not actually moving across the surface of the balloon. This is why galaxies, generally, do not collide with one another (just as the dots on the surface of an expanding balloon never collide) even though they are "travelling" at tremendous speeds. Actually, although they

---

[15] Singh, *Big Bang*, pp. 156–8.

are picking up speed, they are moving further and further apart from each other, and so there is lesser and lesser chance of their colliding over time.

This is the reason why, whichever direction one looks to outer space, the galaxies are *all* receding: similar to looking in any direction at the spots on the surface of an expanding balloon. We are *not* at the center of the universe (as one might wrongly conclude from this last observation); just as any spot on the surface of the expanding balloon cannot be said to be at the center of the surface of the balloon: every single spot experiences all the other spots receding from it. Moreover, this is why we cannot tell "where" the Big Bang happened (in space), any more than we can tell where the balloon started expanding when we started inflating it: it started to expand everywhere. The Big Bang happened in the *entire* space of the universe, not in a particular spot in the universe (like an explosion): similar to the balloon starting to grow as a whole.

This concept is not so easy to understand in the case of the universe because, unlike the expansion of the balloon, which is a surface (two-dimensional) expansion, the expansion of the universe is a volume (three-dimensional) expansion. To appreciate the latter case, we have to think in an extra (fourth) spatial dimension: which is totally foreign to our mind. This is where an area of mathematics (*matrix algebra*) comes in as a handy tool for scientists.

It is also important to appreciate that the universe is finite: space is finite; it is not infinite as one might imagine. There is *no* "space beyond space"; the expression is utterly meaningless: just like the phrase "north of the North Pole" is meaningless. [16]

In his book *Big Bang*, Singh quotes from the *Independent* newspaper of 24th April 1992, describing the first moments of the universe according to the Big Bang Theory:

> First … "all matter and energy were condensed to a point" and then there was an almighty Big Bang. … [It] was not an explosion *in* space but an explosion *of* space.

---

[16] Singh, *Big Bang*, pp. 270–2.

> Similarly the Big Bang was not an explosion *in* time, but
> an explosion *of* time. Both space and time were created at
> the moment of the Big Bang. [17]

The beginning of time in our universe corresponds to the start of the balloon's inflation in the analogy described above. There was no time, as we know it, before the universe's time started: just as there was no inflation before we started to blow up the balloon. This may seem strange, but according to Einstein's general relativity theory there is an intimate interconnection between space and time: they are interconvertible—similar to the interconnection between mass and energy. For example, black holes are so massive and their space is so warped as a consequence, that it seems time stops inside them; too bad that gravity would crush us in there—we would never get old.

Cosmic Microwave Background

Further evidence supporting the Big Bang theory is the *Cosmic Microwave Background* (CMB) radiation. The universe happens to be full of a feeble electromagnetic radiation (similar to that of light) with a wavelength of about one millimeter—that is, in the radio-wave region. Intensity measurement of this radiation, from any location in space and in any direction, is roughly the same. It comes from every direction in space because it happened uniformly everywhere in the universe, as was explained above. [18]

The source (cause) of this phenomenon happened after the universe had existed for about three hundred thousand (3×10^5) years and had cooled down somewhat: below three thousand (3×10^3) degrees Celsius. At this temperature, electrons were trapped by protons and neutrons in atomic nuclei thus forming proper atoms. Here is Singh's account, from his book *Big Bang*, of the origin of this cosmic microwave background radiation:

---

[17] Singh, *Big Bang*, p. 472 (emphasis in original).
[18] Singh, *Big Bang*, p. 333.

The universe continued to expand and cool. It now consisted of simple nuclei [hydrogen, ^{1}H, and helium, ^{4}He], energetic electrons and vast amounts of light, with everything scattering off everything else. After roughly 300,000 years, the temperature of the universe had cooled down sufficiently to allow electrons to slow down, latch onto the nuclei and form fully fledged atoms. This effectively prevented any further scattering of light, which ever since has been sailing through the universe largely unhindered. This light has become known as the cosmic microwave background ... radiation, a sort of luminous echo of the Big Bang. [19]

Recall that energy of a particular wavelength is required to extract an electron from an atom, and that this gives rise to a black line in the absorption spectrum of a given element. Conversely, if an electron is trapped by an atomic nucleus, the same wavelength is emitted: thus forming a bright line in the emission spectrum of that element.

In his book *Big Bang*, Singh continues to explain the evolution of the cosmic microwave background radiation to the present day:

As the universe expanded with time, the wavelength of that primordial light would have been stretched as space itself has been stretched. The light had a wavelength of roughly one-thousandth of a millimeter [10^{-6}m] when it originally emerged from the cosmic fog when the universe was 300,000 years old, but according to the Big Bang model the universe has since expanded by roughly a factor of a thousand. Therefore those light waves should now have a wavelength of roughly 1 millimeter [10^{-3}m], which would place them in the radio region of the electromagnetic spectrum. [20]

Imagine a picture of a light wave in space with its crests and troughs equally spaced. Now imagine this picture is blown up (magnified)

---

[19]   Singh, *Big Bang*, p. 473.
[20]   Singh, *Big Bang*, p. 430.

a thousand times. Obviously, the distance between its crests and troughs would increase a thousand times. Equivalently, because of the expansion of space itself, the original light wave it is not the same light wave it was any longer: now it is not a light wave at all because its wavelength has increased tremendously. It cannot be seen by the eye, either, because the eye is equipped to process only very short wavelengths. It can only be detected by instruments designed to receive radio waves; that is, wavelengths ranging from about one millimeter (10^{-3}m) to about one hundred kilometers (10^{5}m).

Hydrogen to Helium Ratio

Further evidence supporting the Big Bang theory is the ratio (proportion) of the number of hydrogen atoms to the number of helium atoms currently present in the universe.

This ratio was largely determined (fixed) within the first few minutes of the universe's existence. Sophisticated calculations involving the universe's expansion rate, mass density, temperature, and nuclei sizes led scientists to estimate an approximately ten to one (10:1) ratio of hydrogen nuclei to helium nuclei in the first three hundred seconds (5 minutes) of the universe. This is consistent with what we see today. Stars do fuse hydrogen nuclei into helium nuclei $[4(^{1}\text{H}) \rightarrow 2(^{2}\text{H}) \rightarrow {}^{4}\text{He}]$ as their energy source; however, the rate of this nuclear reaction is extremely slow (as we shall see in the next chapter), and it can only account for a very small fraction of the helium known to exist in the universe today. [21] Here is Singh's account of the first few minutes of the universe, again taken from his book *Big Bang*:

> Within a second the super-hot universe expanded and cooled dramatically, its temperature falling from a few trillion [~10^{13}] to a few billion [~10^{10}] degrees. The universe contained mainly protons, neutrons and electrons, all

---

[21] Singh, *Big Bang*, pp. 313–8.

bathed in a sea of light. The protons, equivalent to hydrogen [^{1}H] nuclei, reacted with other particles in the next few minutes to form light nuclei such as helium [^{4}He]. The ratio of hydrogen to helium in the universe was largely fixed within these few minutes, and is consistent with what we see today. [22]

Suffice it to say, at this point, that the Big Bang theory is nowadays accepted by most scientists: it states that the universe started from a very, very small size, possibly the *Planck length* (1.62×10^{-35} meter), if not from a point. According to quantum physics (we shall talk in more detail about this theory in chapter eight on the "Mind and Soul"), space cannot be divided any smaller than the Plank length: just as matter cannot be split smaller and smaller indefinitely.

The Big Bang indicates a beginning (of growth) of our universe: this is, perhaps, consistent with the idea of a moment of *creation* of the universe. If nothing else, it certainly throws a monkey wrench into the cogwheels of a static universe or even an eternal universe. Moreover, something must have triggered this growth.

I am not concluding, from the Big Bang theory, that the universe must have been created by God; however, it is corroborating scientific evidence making it plausible. On the other hand, it certainly goes contrary to a static (eternal, infinite, and non-expanding) universe hypothesis.

Let us simply continue examining the scientific evidence and see where we end up. Still assuming, for the moment, that the universe is eternal, one does not have to explain the origin of matter and energy: so an *eternally oscillating universe* is the only scenario that also fits the evidence from the Big Bang theory. An eternally oscillating universe is one that expands, contracts, and expands again indefinitely. Before we can discuss this hypothesis, we need to understand, in the next chapter, what the *Second Law of Thermodynamics* is all about; afterwards, comes a section on the "Eternally Oscillating Universe".

---

[22]   Singh, *Big Bang*, pp. 472–3.

CHAPTER 3

SECOND LAW OF THERMODYNAMICS

Thermodynamics is the study of how heat (thermo-) produces motion (dynamics). The best way to understand the second law of thermodynamics is to realize a few experiences from everyday life.

After poking burning wood in a fireplace and removing the poker from the fireplace, we notice that the hot end of the poker starts to cool down while the cold end (the handle side) starts to warm up. The cold end of the poker never becomes colder: both sides of the poker eventually equalize to an intermediate (warm) temperature. So if a hot object and a cold object are placed in contact with each other, so that heat can flow freely from one to the other, it is impossible for the hot object to get hotter while the cold object gets colder as a result of this contact. In other words, heat flows freely from hot to cold; on the other hand, heat never flows spontaneously from cold to hot. So far, I think, is obvious. (Please note that "cold" does not exist: it is just the absence of heat, and heat is a form of energy; so only heat flows inside an object, not cold.)

Ice cubes in our drinks, at room temperature (say 20C), always melt; it never happens the other way around: ice cubes are never formed by themselves in our drinks. If one shows a movie strip, in reverse mode, of melting ice, a glass shattering on falling to the ground, or a bursting balloon, even a seven year old child will realize that it is reversed. A clock spring unwinds, and the clock eventually stops; someone has to rewind the spring before the clock can be restarted. I could give a hundred examples, but I think that it is obvious to everyone that most, if not all, everyday phenomena in the universe are *irreversible*. In actual fact, as far as we know, this is the only thing that actually defines the *arrow of time* in our universe. Things, sort of, deteriorate spontaneously if there is no external agent acting on them: if they are left alone, so to speak.

Heat Engine

A heat engine is one that converts heat energy into mechanical energy, say, movement. For a heat engine to work, one *absolutely* requires a difference in temperature: one part of the heat engine must be hotter than another part; this way mechanical energy (motion) can be imparted to the colder part. If they are at the same temperature, absolutely nothing will happen: no heat will flow in either direction; therefore, no energy will be available to do mechanical work on either side.

To understand heat engines it is probably best to think of a locomotive (a train engine), where the term heat engine probably came from. Coal burns; it heats up water, which turns into steam; it pushes a piston, which cranks the wheels; making the locomotive move. Similarly, in an automobile, gasoline is burnt in the combustion chamber; it produces a significant amount of heat expanding (exploding) the gases; it pushes the pistons, which turn the crank shaft and flywheel; the transmission then relays this rotational energy to the wheels; making the automobile move.

Remember then: a temperature difference is absolutely necessary in a heat engine; the higher the temperature difference of the parts (or sub-systems), the higher will be the engine's efficiency. With no temperature difference, a heat engine will remain dead and no mechanical action can be extracted from its *intrinsic* heat: no matter how hot or sophisticated it might be.

Entropy

We already saw, in the introduction to this chapter, that there is an innate irreversibility in almost everything in our universe. This phenomenon of irreversibility of any system is measured in physics by a quantity called *entropy*. It can also be thought of as a measure of the *inability* of a given system to do useful work; by useful work we mean any of the following: making an object start moving, increasing its speed, lifting it up against gravity, making it rotate (spin), extending a spring, compressing air in a cylinder, etc. This type of energy is usually recoverable from *inside* the system itself. Heat energy is not recoverable from inside the system: it is only recoverable from *outside* the system, and only if the other system is at a lower temperature. This is why heat is often considered as the cheapest form of energy.

Notice the word "inability" in describing entropy: the higher the entropy of an *isolated* (or a *closed*) system, the less useful work it can do, and vice-versa. By definition, an isolated (or a closed) system is one that cannot gain or lose any heat from its surroundings: it is, ideally, completely insulated from the rest of the universe. In an isolated system, the higher is the temperature difference between its sub-systems (for example, the sun and the earth—considering them separated from the rest of the universe) the lower is the whole system's entropy, and therefore the more useful work it can perform. Conversely, the lower is the temperature difference of its sub-systems the higher is the whole system's entropy and the less useful work it can perform. When an isolated system reaches thermal equilibrium its entropy is the maximum that system can ever reach.

Let us now examine a simple, everyday experience in some detail. Suppose we have a "system" consisting of (several) ice cubes in a (little) glass of water placed in a (closed) room at normal temperature and left alone. The system initially consists of three sub-systems: (1) ice, at -10C, say; (2) cool tap water, at +10C, say; and (3) air in the room, at +20C, say. We also imagine the system to be isolated (no heat transfer) from the rest of the world (the door of the room is shut). All three subsystems start to find their equilibrium temperatures simultaneously, but for simplicity's sake I shall separate the phenomena that are taking place. (1) First, the water in the glass at 10C starts to warm up the ice from -10C to -5C, say; meanwhile, the water itself cools down from 10C to 0C (the melting point of ice). (2) Second, the air in the room warms up the ice from -5C to 0C (the melting point of ice); during this time, the water remains at 0C. Theoretically, the air temperature cools down a little also, but it is hardly noticeable; however, for simplicity here, let us suppose it cools down from 20C to 19.9C, say. (3) Third, the air in the room starts to melt the ice (that is, change it from solid to liquid); this happens at the same temperature in the glass, 0C (the melting point of ice). However, heat (called *latent* heat) is still required to undo the crystalline structure of the ice and transform it into the liquid structure of water (this process is termed *change of phase*). Meanwhile, the air in the room cools down a little further, for simplicity, from 19.9C to 19.8C, say. (4) Finally, the air in the room keeps pumping heat into the water (there is no more ice now) raising its temperature from 0C to 19.7C, say: the final temperature of the air in the room; since the air will cool down a little bit further (that is, from 19.8C to 19.7C, say).

The physics term, entropy, of the initial system (consisting of ice, water, and air, described in the last paragraph) is a measure of how far the temperature equalization process has progressed. As mentioned above, this equalization process is irreversible. There is no way the ice can reform from the final system (consisting of water and air both at the final temperature of 19.7C): that is, assuming it remains isolated (insulated) from the rest of the world. The isolated system has now increased by the maximum amount of entropy it can possibly change: from here on nothing else will ever happen to it—assuming it remains isolated, of course.

Here's how Wikipedia explains the entropy changes involved in the system described above. The entropy of a system decreases when heat is expelled from it; conversely, the entropy of a system increases when heat is injected into it: that is, when it simply raises its temperature. Since some of the heat of the room (considered as one of the subsystems) is lost (given) to the ice-water (considered as another subsystem), the former's entropy decreases. Naturally, the entropy of the other subsystem (the ice-water) increases, since heat is injected into it. However, the entropy gain by the ice-water subsystem is greater than the entropy loss by the room (the other subsystem). The net effect is that the system as a whole (that is, room and ice-water together) experiences a net gain in entropy. This is always the result: whenever there is heat transfer from one subsystem to another the system as a whole increases in entropy. Thus the entropy of a system at any instant of time is a measure of the temperature equalization process reached at that instant. When (full) thermal equilibrium is reached and there is no more heat transfer, the entropy of that system has reached its maximum. [23]

However, suppose that heat is injected by a first system that manages to heat a sub-system (locally) of a second system: thus raising its temperature higher than another sub-system in the second system; that is, disturbing the thermal equilibrium of the second system. Then this temperature difference in the second system can, temporarily, be used as a heat engine to produce mechanical work inside the second system. In this case, at any instant, the entropy gained by the second system is always greater than the entropy lost by the first system: so the entropy of both systems as a whole, considered as one super-system, always increases. When the second system comes to thermal equilibrium again (and therefore cannot be used as a heat engine any longer) the gain in entropy by the second system is even higher than its initial gain: it is, therefore, always higher than the loss of entropy from the first system. Consequently, for the two systems combined, there is always a net gain in entropy. Wikipedia explains the possibility of a local (sub-system) decrease in the entropy as follows:

> [L]ocally, entropy can be lowered by external action, e.g., solar heating action, and ... this applies to machines, such as a refrigerator, where the entropy in the cold

---

[23] https://en.wikipedia.org/wiki/Entropy (24 May 2015)

chamber is being reduced, to growing crystals, and to living organisms. This local increase in order [entropy decrease—see below] is, however, only possible at the expense of an entropy increase in the surroundings; here more disorder [entropy—see below] must be created. [24]

Physicists often use "entropy" and "disorder" synonymously; we shall refine this relationship later on in this chapter in the section entitled "Order and Disorder".

Thermodynamic Equilibrium

Strictly speaking, we can only define the entropy of a system, as a whole, if it is in complete mutual *thermodynamic equilibrium*. [25] Such systems must be simultaneously, mutually in equilibrium in all respects; there must be no net transfer of any form of energy, namely: matter, heat, mechanical, chemical, radiation, etc., from one sub-system to another. [26] Besides temperature, the pressure and volume of a system are also important in reaching thermodynamic equilibrium, particularly in the case of gases.

The entropy of a system in thermodynamic equilibrium may be defined as the amount of mechanical energy that has been converted into thermal energy: it is therefore a measure of the amount of thermal energy in a system that is unavailable for doing mechanical work. It is given by the equation S=Q/T, where "S" is the "entropy" of the system, "Q" is its "heat" energy, and "T" is its *absolute temperature*".

The absolute, or Kelvin, scale of temperature uses degrees Kelvin (symbol "K") instead of degrees Celsius or degrees Fahrenheit. It is roughly equal to the Celsius temperature plus 273 degrees (more exactly 273.15C). Thus, for example, 0C is equivalent to about 273K, and 100C is about 373K. Anything at about -273C (273 degrees below

---

[24] https://en.wikipedia.org/wiki/Entropy_(order_and_disorder) (24 May 2015)
[25] https://en.wikipedia.org/wiki/Entropy (24 May 2015)
[26] https://en.wikipedia.org/wiki/Thermodynamic_Equilibrium (27 May 2015)

zero Celsius) cannot be cooled any further: that is why we call it absolute temperature.

The thermal, or heat, energy "Q" in the equation for entropy given above may be thought of as the total amount of heat required to raise the temperature of the system in question from absolute zero (0K) to its current temperature "T" in the equation. We may also think of the entropy of a system as the average amount of heat energy required to raise its temperature by one degree Celsius (or Kelvin) from absolute zero (0K) temperature to its current temperature (TK). Entropy is strictly a function of the "state" of a system; this means that it is independent of the method used to get to a particular temperature: it is what it is.

If a system has not yet reached a state of thermodynamic equilibrium, for example, the ice-water-air system described above, then its initial entropy will be given by the equation $S = Q_i/T_i + Q_w/T_w + Q_a/T_a$, where the subscripts' meanings are: "i" for "ice", "w" for "water", and "a" for "air".

Second Law of Thermodynamics

So what does all this business of heat engines and entropy have to do with the second law of thermodynamics? What does it say, anyway? The second law of thermodynamics states:

> [F]or an *isolated* system, the *total entropy* remains *constant* in time if *all* processes occurring within the system are *reversible*; on the other hand, the *total entropy* of an isolated system increases with time if *any* process within the system is *irreversible*. [27]

An irreversible process is one in which *some* of the total energy of a system is converted into heat: that is, molecular energy, or random movement of its molecules, the cheapest form of energy.

---

[27]   Weidner & Sells; *Elementary Classical Physics (Volume 1)*, p. 603 (emphasis in original)

Most processes in everyday life are irreversible. In fact, I cannot think of even one really reversible process; the closest example I can think of is a pendulum. When a pendulum bob is at the left-most position, say, all its energy is potential energy: that is, it possesses energy by virtue of its position—height. When the bob is at the center of its swing, all its potential energy is converted to kinetic energy: that is, now it possesses energy by virtue of its motion—speed. In intermediate positions, the total amount of the bob's energy remains (approximately) the same, additively, but it is now partly potential (height) and partly kinetic (speed). Of course, because of air friction, the pendulum will lose height and eventually stop if undisturbed: showing that it is still an irreversible process. It is only approximately reversible: considering the fact that it will keep swinging for a significant amount of time. If the pendulum is, in addition, placed in a vacuum, it will swing for a much longer time: showing it to be a better approximation of a reversible process. But, even in a vacuum, the pendulum will still lose height and eventually stop: this is because of very slight friction at the supporting knife edge. In practice, there are probably no reversible processes in everyday life—none. There may, however, be reversible processes at the atomic or molecular levels; but we won't go into that, yet.

So for an isolated system (one that does not interact with another in any way, like the whole universe, supposedly), its entropy will consistently increase or, at best, remain unchanged: it never decreases. Now, as we just saw, all actual systems undergo irreversible processes, one time or another; so the total entropy of any real system *always* increases with time. The universe, according to scientists, has existed for about fourteen billion (1.4×10^{10}) years; all this time it has been "unwinding" itself. The obvious scientific question from the above considerations is: how did the universe start in this "wound-up" state?

Order and Disorder

Entropy is sometimes also described or thought of as a measure of the disorder of a system; and so the second law of thermodynamics is sometimes stated in another way:

> An isolated system, free of external influence, will, if it is initially in a state of relative order, always pass into states of relative disorder, until it eventually reaches a state of maximum disorder. [28]

However, we have to be careful how we interpret the words "order" and "disorder" in this context.

Imagine a pack of playing cards initially arranged in order: ace to king, starting with the club suit, then the diamond, heart, and finally the spade suit. If the pack is shuffled once, a certain amount of disorder is introduced into the pack, but there is still some order left in it. Every subsequent shuffle introduces more and more disorder to the pack, until eventually further shuffles make no real difference: the pack is, to all intents and purposes, in a state of maximum disorder. There is no real hope that further shuffles will restore the initial order; someone with *intelligence* must intervene to restore the initial order. In this case, one can see only *some* resemblance to the ice-water-air system described above. This kind of irreversible process resembles in some ways what is happening to the universe over time: but not exactly, if you think about it.

In the ice-water-air example above, we initially had three sub-systems, which corresponded to three different physical states of its reality: (1) a solid, ice at -10C, (2) a liquid, water at 10C, and (3) a gas, air at 20C. Eventually, the system finally settled down to: (1) a liquid, water at 19.7C, and (2) a gas, air also at 19.7C. The three initial states were shuffled, stirred, or mixed and matched, so to speak, to produce two states at the same temperature. Because the final two states are in thermal equilibrium with each other, they cannot do any mechanical

---

[28]   Weidner & Sells; *Elementary Classical Physics (Volume 1)*, p. 540.

work on each other any longer; hence, they can be considered as one state from a thermodynamic perspective. So from a higher order of three different temperature states (sub-systems) we ended up with a lower order of one indistinguishable temperature state, so to speak: that is, from order we ended up with disorder—a mix-and-match.

This is the sense in which the words "order" and "disorder" are to be understood in thermodynamic systems: as an increase or a decrease, respectively, of the different thermodynamic states (thermal, radiative, phase, mechanical, material, chemical, etc.) of the whole system. When all the different states in a system reach thermodynamic equilibrium, there is no net transfer of energy between the sub-systems any longer; that is, barring an external intervention (trigger). This is not exactly the same sense as the card-shuffling example given above; it is basically an equalization of the different physical states. Is there more order in the un-equalized states? The way I would describe it is as follows: the initial (three) states probably require more effort to set-up than the final state where everything is at the same (equilibrium) temperature. No doubt, one needs much more effort to set up the cards in order than to shuffle them—so there is some conceptual similarity.

Now, if the entropy of our universe is continually increasing with time, then it had to be always lower in the past fourteen billion odd years. Achieving this extraordinarily gradual evolution of the universe for such a length of time—that is, preventing things from precipitating abruptly like the explosion of a bomb—must have required quite a bit of ingenuity in setting-up the universe initially.

For example, the reason why things are so controlled here on earth is because of the great distance between us and the sun, our main source of energy. Its relatively high temperature keeps supplying us with enough energy and of the right type for the sustenance of life. This energy is given to us in relatively small doses, because of the relatively slow rate of hydrogen fusion stars—compared to helium fusion stars—as we shall see in the next chapter. All our energy sources: nuclear, hydro, fuels, coal, etc., are still ultimately traceable to the stars' or our sun's action, one way or another.

Scientific Hypothesis for the Universe's Origin

The scientific explanation that is supposed to debunk the concept of a universe starting from a higher thermodynamic order (set-up) is that it arose from total chaos. Here's the relevant hypothesis that particle physicist Victor Stenger gives in us his book *God and the Folly of Faith*:

> We can think of the visible universe as a sphere. For any spherical body, its maximum entropy is that of a black hole [explained presently] of the same size. Thus, as the universe expands, its maximum entropy increases, leaving room inside for order to form. In fact observations are consistent with a model in which 13.7 billion years ago, the universe was confined to a region so small that it was equivalent to a black hole and possessed maximum entropy. That is our universe was in a state of complete chaos in which local order did not form until it started to expand. [29]

However, it is hard to see how the many and various physical states of the sub-systems existing in our universe arose from total chaos—presumably in thermal equilibrium, at least initially. I shall comment further on Stenger' statement right after I explain the properties of a *black hole* in some detail.

Black Hole

A black hole is a location in space where gravity is so strong that neither matter nor radiation (like light) can escape from it. A sufficiently dense mass (matter) will deform the space nearby and give rise to such a region. The boundary inside which no escape is possible (not even light) is called an *event horizon*: it also marks the point of no return for anything approaching it from outside. A black

---

[29] Stenger, *God and the Folly of Faith*, pp. 177–8.

hole acts like an ideal *black body*; it absorbs all the light that hits its event horizon and reflects none: consequently, it is called black, because, from a distance it (obviously) appears black.

Theoretically, event horizons emit radiation, with the same spectrum as a black body at a temperature inversely proportional to its mass: that is, $T=\kappa/m$, where "T" is its absolute temperature, "κ" is a fixed number, and "m" is the mass of the black hole. This temperature is very small: of the order of billionths (10^{-9}) of a degree Kelvin (or Celsius) for black holes having the mass of a star; thus, it is practically impossible to actually measure it, so this concept is more or less theoretical.

Black holes form when very massive stars "die" and implode under their own gravity. After a black hole is formed, it may continue to increase in size by engulfing other stars or merging with other black holes close by. If this continues to happen, they may result in supermassive black holes: the kind of which are widely believed to be located in the centers of most galaxies. [30] The only thing stopping their size continuing to increase indefinitely is simply enough distance from other masses.

Objections to the Scientific Hypothesis for the Universe's Origin

I must admit I could not comprehend what Stenger was trying to convey in the above quote. I honestly cannot understand this *scientific* explanation, despite my trying my utmost to see the concept, at least; maybe it is above my ability to understand it. At first I was intimidated: that is, I dared not criticize a particle physicist; I only have a Bachelor of Science degree in physics and mathematics. However, the book referred to above is not a peer scientific paper; it is a book written for the general public, so I decided to ask some questions in the name of the general public (and in my name). (1) Why would the universe suddenly decide to start expanding? I would imagine the humungous gravity of a black hole will prevent it from

---

[30]   https://en.wikipedia.org/wiki/Black_hole (2 August 2015)

doing so. (2) Why would the universe not stay in a state of minimum energy if it is already there? (3) Where would it get the energy to increase its minimum possible energy from? Total chaos is a state of minimum energy and maximum entropy possible: that of a black hole in thermal equilibrium. (4) Why would it suddenly decide to break the second law of thermodynamics and decrease its entropy? This never happens in our universe: as we have seen, the entropy of an isolated system can never decrease.

I can understand the concept that *if* space manages to expand, and there is matter inside that space, the potential energy (the energy the matter inside that space possesses by virtue of its relative positions) increases: just as the higher (further away) something moves above the earth's surface the higher will be its potential energy. Therefore, I can understand that an increase in the potential energy of a system produces a decrease of entropy. However, as we have seen above, for something in thermal equilibrium to start lowering its entropy, an external intervention (trigger) is required; otherwise nothing new will happen to it: it just sits there. I'll let the reader decide for oneself whether to accept or reject Stenger's hypothesis.

Eternally Oscillating Universe

As promised, we are now in a position to examine the hypothesis of an eternally oscillating universe. Recall that an eternally oscillating universe is one that expands, contracts, and expands again, indefinitely. It is intended to be an alternate model to the Big Bang theory: but without the necessity to explain the origin of matter and energy since it is, supposedly, assumed to be eternal.

However, according to calculations, matter in the universe would have to be ten times what it actually is for the universe to be able to re-contract after reaching maximum size. In his article "Cosmos and Creator", religion-science researcher and analytical philosopher William Craig writes:

For its part, the eternally oscillating model contradicts observational cosmology, which indicates that the universe would have to be ten times denser than it is to allow the universe to re-contract after reaching a maximum distension. Even if the universe did re-contract, however, no known physics could reverse such a Big Crunch and cause the universe to bounce back to a new expansion. Finally, if the universe oscillated, [the second law of] thermodynamics mandates that entropy increases from cycle to cycle. Each successive cycle expands more slowly, with a larger [maximum] radius. Thus, as one traces the cycles back in time [which is supposedly eternally long], they become progressively smaller terminating in a first [possibly a point] cycle and an absolute beginning. [31]

So, ultimately, we are back where we started, namely, that the universe had a beginning—from a very small size, too.

Now scientific evidence is showing that our universe is actually expanding at a faster and faster rate; why would it, all of a sudden, start contracting? If anything, it should be slowing down for it to be able to start contracting at some point in the future: in other words, the current evidence seems to be pointing the other way.

But even if it did: that is, gravity somehow manages to overcome the expansion rate; as self-declared atheist Victor Stenger graciously admits in his book *God and the Folly of Faith*:

This scenario is not to be confused with older proposals about an "oscillating universe" in which our expansion is followed by a contraction and then by another expansion, ad infinitum. That proposal fails because of the second law of thermodynamics. The direction of increase in entropy of the universe does not reverse during the contracting phase but keeps increasing and eventually hits the limit

---

[31]   Craig, "Cosmos and Creator" p. 19.

of total chaos long before the universe has collapsed to
Planck dimensions [smallest size physically possible]. [32]

This means that the universe cannot even complete a first contraction; let alone expand a second time: never mind an infinite number of times.

It follows that what was said above in the previous chapter on the "Big Bang Theory" still holds: namely, that our universe had, what looks like, a beginning. Now, if something has a beginning, the obvious question is: what caused this beginning; what caused it to start happening; what physical explanation can there be behind this trigger point?

---

[32]   Stenger, *God and the Folly of Faith*, pp. 188–9.

CHAPTER 4

FINE-TUNING OF THE UNIVERSE

A very interesting question is whether there is any purpose to the universe, or life on earth for that matter. It becomes much more intriguing when one considers what physics has to say regarding the ultimate fates of the earth and the universe.

Probably, the end of the earth will come when it is absorbed by the sun in about seven and one-half billion (7.5×10^9) years from now: after the sun has become a *red giant* and expanded enough so that its atmosphere reaches the earth's orbit. [33] A red giant is a star at a late stage of its life cycle when its outer atmosphere inflates and becomes rarified. Its radius becomes immense and increases to hundreds of times what it was during its normal energy-producing cycle. Its surface temperature reduces significantly: to anywhere between two thousand five hundred Kelvin (2500K) and five thousand Kelvin (5000K), compared to the sun's current 5778K; but, of course, much too hot for any life, as we know it (carbon based), to survive. [34]

Now, in case you thought life could survive by moving to another planet in the universe, physics also tells us that the ultimate fate of the universe is probably a *heat death*. What does this mean? The second law of thermodynamics (see the previous chapter) predicts that the universe will eventually "wind down" completely and approach the absolute zero of temperature (that is, 0K or about -273C): which is the state of maximum entropy. At this point all energy will be

---

[33] https://en.wikipedia.org/wiki/Future_of_the_Earth (16 April 2016)
[34] https://en.wikipedia.org/wiki/Red_giant (16 April 2016)

evenly distributed, and there will be no free energy left to sustain any processes that consume energy: especially life, of course. So the whole universe is just a ride: a long roller-coaster ride.

It is depressing to realize that all the achievements of humanity through the ages will eventually, literally, end up in flames: nothing will be left. Unless, of course, there is an afterlife; but the majority of scientists think that this is wishful thinking of humanity. We shall talk more about the possibility of an afterlife in chapter eight entitled "Mind and Soul" which deals with our mind, or thinking ability, and our self, or consciousness.

On the other hand, philosophically, can we really say that there is no purpose to life because the ultimate end is bleak? We know that every single one of us will eventually end up dead in a relatively short time, yet most of us strive hard all our lives to enable our loved ones to live a better life. Can we call such a life without purpose? Ultimately, is a roller-coaster ride without purpose? Isn't it supposed to be fun? But anyway, the question is a very valid one: is there a purpose or aim to the universe? Was life, especially intelligent life— that is, us humans—part of this plan?

It is worth noting here, before proceeding with our discussion, that the relatively recent discovery of the universe's increasing expansion rate has cast some doubt on the heat death hypothesis and the applicability to or validity of any simple thermodynamic model of the universe as a whole. There is no doubt that entropy does keep on increasing in the model of an expanding universe; however, this expansion has the opposite effect: it produces cooling. Expansion also increases the potential energy of all matter in the universe by moving them apart. The bottom line is that, in the case of the universe, the maximum possible entropy seems to be rising much more rapidly than heat death is approaching: thus moving it further away from heat death as time passes. [35] Let us now continue our discussion.

---

[35]   https://en.wikipedia.org/wiki/Entropy (3August 2015)

Discoverability

We may not have been placed at the center of the universe; we may even seem very insignificant in the grand scheme of things, but the way the secrets of nature are discoverable by us, it seems at least, that it was made for intelligent life to find. In journalist Lee Strobel's book *The Case for a Creator* physicist and philosopher Robin Collins is quoted as saying:

> "In physics, we see ... something that I call 'discoverability'. By that, I mean that the laws of nature seem to have been carefully arranged so that they can be discovered by beings with our level of intelligence." [36]

(I shall be using concepts and ideas from Strobel's book extensively in this and the next chapter.)

My logical conclusion on this subject is that if we can prove, beyond any reasonable doubt through scientific methods, that the universe and life were designed, rather than their being the result of chance or some other natural process—like physical laws, chemical laws, biological laws, natural selection, etc.—then we can conclude that there probably is an intelligent designer. Moreover, if this is actually the case, then this designer probably had a purpose for the universe; although it may be one that we don't yet understand, or that we may not understand in our lifetime.

Theory versus Hypothesis

Although most people unfamiliar with science may think otherwise, in science a *theory* is *not* certainly true; it is only *probably* true because there is a lot of corroborating evidence supporting it. When the overwhelming majority of the scientific community agrees on a particular proposition of reality, it is called a theory. Sometimes, in view of new developments, a theory falls out of favor

---

[36]   Strobel, *The Case for a Creator*, p. 147

with the scientific community and is relegated to a *hypothesis*. This has happened several times historically: we have already mentioned the *flat earth* and the *geocentric* (the sun orbiting around the earth) hypotheses. A hypothesis is less certain than a theory: there is some corroborating evidence in support of a hypothesis, but not enough to be adopted by the greater majority of the scientific community. This is usually the case at the beginning of a scientific proposal or, as mentioned above, when it falls out of favor because of newer developments. Science does not follow a *dogmatic* policy. Science does retract previous declarations when it finds errors in its previous conclusions.

The general public may also think that this only happened in ancient times when science was still not so sophisticated. However, it seems to be happening even in very modern times, for example, *string theory*; as theoretical physicist Lee Smolin shows in his book *The Trouble with Physics*, even though the majority of the scientific community believed in it and worked on it for the last half century. We shall encounter other modern theories that have been relegated to hypotheses in the course of this book, but I do not want to let the cat out of the bag yet.

When it comes to investigating the existence or non-existence of God as a scientific hypothesis or theory, we therefore have to use the same scientific standards. We should not use a different standard for proving the existence of a particular supernatural phenomenon if we honestly consider it a scientific question. We cannot expect *absolute* certainty only in the case of proving supernatural phenomena because there is no such thing in science: every theory represents a certain degree of probability of its being true, but it can also be proved false anytime. If we were to use the standard of absolute certainty in scientific theories, unlike popular belief, we might as well scrap the whole idea of science—believe it or not!

Design, Chance, and Natural Laws

Was the universe designed? Did it just happen by chance? Or was it the inevitable result of the natural laws of physics, chemistry, natural selection, etc.? Is there a purpose to the universe? Was life on earth planned at the beginning of the universe? If so, was intelligent life, like ours, part of this plan? In this case, discoverability would make sense; if not, why bother? This chapter will try to answer these questions. In similar fashion, later chapters will deal with the question of design, chance, or natural occurrence in life's emergence (the living cell) and evolution.

Please note that I am not assuming anything at this point; I have only been talking philosophically since the beginning of this chapter: I was only saying "if". My sole purpose here is to enable the reader to appreciate what is elucidated in this and later chapters. I am only trying to deal with one issue at a time for clarity's sake; one will have to reach one's own conclusions after reading the chapters referred to above.

In Strobel's book *The Case for a Creator*, Collins continues to introduce the concept of fine-tuning in our universe:

> "When scientists talk about the fine-tuning of the universe they're generally referring to the extraordinary balance of the fundamental laws and parameters of physics, and the initial conditions of the universe." [37]

To investigate the validity of Collins' statement of "extraordinary balance" in our universe, this chapter is divided into various sections concerning: (1) the development of the universe, (2) the pre-requisites for the emergence of life to be possible, and (3) the current actual everyday operation of the universe—that is, the fundamental laws and parameters of physics.

---

[37]   Strobel, *The Case for a Creator*, p. 130

Probability

For the benefit of the non-mathematically-oriented readers, I think we need a primer on *probability* before we embark on discussing odds; thus, what is said in this and the following chapters may be better appreciated.

Imagine a glass jar containing ten (10) same sized marbles: nine (9) black and one white. You are blindfolded; the jar is shaken; you are asked to pick the white marble with your first pick. You might say to yourself, "I'll try, but I really don't like the odds against me." You probably weren't successful in picking the white marble with your first try; but let us suppose, by some fluke, you did; you would feel quite proud of yourself.

You are then asked to proceed to the second part of the experiment. The glass jar now contains one hundred (100) same sized marbles: ninety-nine (99) black and one white. You know the drill: you are blindfolded; the jar is shaken; and you are asked to pick the white marble with your first pick. This is where your jaw drops; the first feat was difficult enough; this is ten times harder. You do it, with a half heart, and you probably won't pick the white marble. But if you do, you will be exhilarated, and you will start jumping up and down for joy.

Next you are taken near a barrel, containing one thousand (1,000) marbles: nine hundred and ninety-nine (999) black and one white. Again: you are blindfolded; the barrel is thoroughly stirred; and you are asked to pick, from anywhere you like, the only white marble with your first pick. Do you think this is possible? By now, I think you get my drift.

Every time we increase a zero (1,000 to 10,000 to 100,000 to 1,000,000, etc.) in the number of marbles in the container, it becomes ten times harder. Now, our mind fails to comprehend the difficulty of picking one marble in a million $(1,000,000=10^6)$, or a billion $(1,000,000,000=10^9)$, or a trillion $(1,000,000,000,000=10^{12})$; we cannot even imagine it. It is practically impossible to pick, blindfolded, a single white marble out of a pool of a trillion black marbles; yet, a trillion has only 12 zeros. Try to remember this; we shall come back to it later.

Odds of Starry Universe

There is a very fine line dividing our cosmos' expanding too rapidly—preventing it from forming any galaxies, stars, and planets—from gravitational implosion, a crunch. The laws of physics and chemistry have to be such as to allow the formation of stars rather than black holes. Stars, like our sun, allow for the gradual release of energy: enabling life, for example, to thrive. Black holes devour and destroy everything around them.

One may have already appreciated the necessity of fine-tuning during the initial five minutes after the Big Bang: when we mentioned the universe's rate of expansion, mass density, temperature, and particle sizes in the section on the "Hydrogen to Helium Ratio". All these fine-tuned conditions had to co-exist at the beginning of the Big Bang in order to result in our universe. Calculations by cosmologists have shown that without this fine-tuning, in practically all probability our universe would have ended up in a bunch of black holes scattered around in empty space. One may easily appreciate that *chaos theory* is a very complex subject. I think I would be quite safe in saying that there is no one on earth that has figured out exactly how the stock market is going behave next. Or why, without there being an accident, the traffic gets jammed and hardly moves at all while a mile down the road it flows freely. If they did figure it out, they haven't done anything to fix Toronto's Highway #401 during rush hour for decades. So I shall be rather superficial about chaos and confine myself to quoting the final figures by experts in the field.

In his article "The Prerequisites of Life in Our Universe" philosopher John Leslie gives the following quote from theoretical physicist, cosmologist, and astrobiologist Paul Davies' book *Other Worlds*: [38]

---

[38]   Leslie, "The Prerequisites of Life in Our Universe", p. 232.

> "[I]f the primeval material was churned about at random it would have been overwhelmingly more probable for it to have produced black holes than stars ... the odds against a starry cosmos [would be] one followed by a thousand billion billion [10^{21}] zeros, at least." [i.e., the odds are 1 against $_{10}10^{21}$]. [39]

(I shall be using concepts and ideas from Leslie's article extensively in this and the next chapter.)

Recall that odds of one in a trillion are practically impossible to achieve, and that a trillion only has twelve zeros; moreover, every time one zero is added to the total number of marbles the feat becomes ten times more difficult. A number such as the above will probably mean very little to the ordinary non-scientifically-oriented reader; so, I would like to give a rough idea of how improbable, even impossible, these odds are.

The flipping and landing of a coin are not exactly totally random events: there are physical laws governing its final position; however, to all intents and purposes, we may assume it to be a close approximation to a completely random event—we are also all quite familiar with it. Imagine someone flipping a *fair* coin once every second: with no breaks for sleep, meals, or going to the washroom for the whole time the universe has existed—that is, about fourteen billion (1.4×10^{10}) years—and consistently getting heads, not a single tails in between, for all this time. That's not even close to the odds described above: it will require the time of over seven thousand five hundred (7500) similar universes to complete the necessary number of coin flips; all the time, of course, obtaining only heads—exclusively. Now, if this actually happened, I ask the reader: Do you think that the coin would have been a fair coin? Or that it was rigged—designed? Do you think mere chance could ever achieve something like this?

---

[39]   Davies, *Other Worlds*, pp. 160–1 & pp. 168–9.

Odds of Life-Sustaining Universe

As if this was not enough of a feat, mathematical physicist Roger Penrose came up with an even more mind-boggling result of a calculation along the same lines; assuming there were no unknown physical phenomena, ensuring a friction-less (smooth) beginning of the universe, he calculated the odds against selecting our universe—culminating in life on earth—from all the possible universes. [40] He writes, "the accuracy ... [was] at least of the order of one part in $_{10}10^{123}$". [41] This was "the precision needed to set the universe on its course" [42] [43] so that it eventually produced a world like ours. It is not even possible to write down, in full, *just the zeros* of this number on a piece of paper, since we would require more zeros than the number of elementary particles—that is, protons, neutrons and electrons—in the whole universe, estimated at about 10^{80}. Our whole universe would still fall far too short; [44] we would need about 10^{43} similar universes just to write down the zeros: assuming every elementary particle represents a zero. Now, recall that with every zero added the difficulty of success increases ten times. These are truly unimaginable odds against a world like ours ever happening by chance alone; and yet, here we are: no one can deny that!

For the benefit of anyone who may be curious enough, actually about 10^{17} [$(6+6+6) \times (100) \times (10 \times 10^{-3} \times 0.1 \times 10^{-3})/(2 \times 70 \times 10^{-12})^2 = 9.2 \times 10^{16}$] elementary particles are required for every zero of 10mm circumference, 0.1mm width, and 100 carbon atoms thickness (similar to the zeros printed in this paragraph)—that is, about 5.1×10^{15} carbon atoms. So about 10^{60} universes similar to ours, made only of carbon atoms, would actually be required to write down just the

[40]   Leslie, "The Prerequisites of Life in Our Universe", p. 232.
[41]   Isham, Penrose, & Sciama, *Quantum Gravity 2*, pp. 248–9
[42]   Penrose, *The Emperor's New Mind*, p. 344;
quoted in Meyer, "Evidence for Design in Physics and Biology", p. 61
[43]   Strobel, *The Case for a Creator*, p. 135
[44]   Strobel, *The Case for a Creator*, p. 135

zeros of this number on an *imaginary* piece of paper—never mind the astronomical amount of paper needed.

Early Universe's Expansion Rate

In a 1974 book entitled *Confrontation of Cosmological Theories with Observational Data*, consisting of a collection of scholarly articles of the *International Astronomical Union*, a most famous theoretical physicist and cosmologist Stephen Hawking made a very revealing estimate regarding the initial expansion rate of the universe. He stated that if the universe's expansion speed were reduced by just one part in a trillion (10^{12}) when the temperature was about ten billion degrees Kelvin (10^{10}K), the universe would have stopped expanding and instead started collapsing when the temperature was still about ten thousand degrees Kelvin (10^4K). [45] [46] Needless to say, this temperature is far too hot for *any* form of life, as we know it (carbon based), to be able to start or survive: we live in an environment of only about three hundred Kelvin (300K). Moreover, in contracting the universe would become even hotter than ten thousand degrees Kelvin (10^4K). One might recall the fact, here, that when one pumps a bicycle tire, the far end of the pump—that portion of the pump barrel where air happens to be compressed—gets hot: this is the normal result of reducing (contracting) an original volume (quantity) of air. Expansion of a given volume (quantity) of air produces the opposite effect: that is, cooling. The universe has cooled down because it expanded.

In 1978, nuclear physicist, astrophysicist, and cosmologist Robert Dicke made a similar calculation. He stated that if the universe's expansion speed at one second after the Big Bang decreased by just one part in a million (10^6), it would have caused the universe to stop expanding and instead start collapsing before its temperature had dropped to ten thousand degrees Kelvin (10^4K). If, on the other hand,

---

[45]   Hawking, "The Anisotropy of the Universe at Larger Times", p 285.

[46]   Leslie, "The Prerequisites of Life in Our Universe", p. 232.

a similar increase in the universe's expansion speed occurred at the same time, it would have dominated gravity altogether; consequently, there would not have been enough conglomeration of matter to form systems capable of producing stars. [47] [48] No stars, practically means no life.

Moreover, it is obvious that the fine-tuning had to be even more accurate than the above, because the more one pushes back the time at which these calculations are carried out, the more they become uncertain; so it is understandable that scientists refrain from committing themselves to earlier calculations.

Mass and Energy Density

One may recall that mass and energy are inter-convertible according to Einstein's equation $E=mc^2$. One should distinguish, here, between the *mass density*, which is the amount of *matter* per cubic meter, and the *mass and energy density*, which also includes the *dark energy* discussed in the first chapter on "Matter and Energy". While forming the universe, early cosmic mass and energy densities are closely related to its later expansion speeds because of their gravitational effects and other energy fields. The *critical density* is that quantity of matter and energy per cubic meter that places the universe precisely on the cusp (edge) between its continued expansion and collapse.

In a 1982 article published in the *Irish Astronomical Journal*, mathematician and astronomer Bernard Carr stated that the *initial* mass and energy density—that is, at the *Planck time*, 10^{-43} second after the Big Bang—must have been within one part in 10^{60} of the critical density. [49] [50] According to quantum physics, time cannot be

---

[47] Dicke & Peebles, *General Relativity*, p. 514.

[48] Leslie, "The Prerequisites of Life in Our Universe", p. 232.

[49] Carr, "On the Origin, Evolution and Purpose of the Physical Universe", p. 244.

[50] Leslie, "The Prerequisites of Life in Our Universe", p. 232.

divided any smaller than the Plank time, just as matter cannot be split smaller and smaller indefinitely.

According to theoretical physicist Albert Einstein's *relativity theory*, matter in space "warps" the space close by: like a bowling ball placed on a stretched, horizontal trampoline mat causes it to sag. If one rolls, with some speed, a small marble on a trampoline mat sagged by the weight of a bowling ball, in the vicinity of the ball, the former will not travel in a straight line as in the case of a perfectly flat trampoline. It will lean towards the bowling ball for a short distance before it continues to travel in a straight line. It is easy to understand the analogy of warping a two-dimensional (flat) surface in our three-dimensional space; however, it is not so easy to visualize how our three-dimensional space can be warped. We need a fourth (extra) dimension to appreciate it: again, as we have already seen, this can be achieved by using an area of mathematics known as matrix algebra.

In practice, warping—curvature or non-flatness—of space can be detected experimentally by light's travelling in a curved path. A *flat* space is one that is not curved: that is, one in which light travels in a straight line—far away from massive entities, of course. The mass and energy density of the universe thus affects the flatness or curvature of space—or rather *spacetime*. Observation shows that the current mass and energy density of our universe is slightly less than the critical value: the difference is less than 1%: thus resulting in a slight continued expansion of space. Space in our universe is almost perfectly flat; in other words, light practically travels in straight lines in free space.

The present mass and energy density, extrapolated backward to the beginning of the universe, is what allowed stars to form in our universe: it did not end up in a crunch or devoid of complex structures—such as galaxies, stars, and planets. Any minor departure from this initial value at the Planck time would have been magnified during the billions of years of our universe's expansion to create a current mass and energy density of the universe very far from the critical value.

How difficult would it have been to achieve this back then? According to Carr, the odds are one against 10^{60}. Recall how impossible it seemed to select a single white marble from a pool of a trillion (10^{12}) marbles? This is, by far, a much larger number, and 10^{48} times more difficult to beat. Alternatively, reverting to the example of the fair coin-flipping scenario described above, it would take about 200 consecutive flips, all the time obtaining heads: no tails in between; if a tails comes up, one has to restart the 200 consecutive flips again. Our universe has only existed for about 14 billion (1.4×10^{10}) years—that is, about 4.4×10^{17} seconds. I doubt very much if this is ever achievable, in practice, at one flip per second in the total time our universe has existed: mathematically, one needs the time of about 10^{42} similar universes to get close to a 50% chance of obtaining 200 identical coin flips. We shall have a closer look at the total probabilistic resources of the universe in chapter six on the "Origin of Life". Assuming, for the time being, that the universe had only one opportunity of getting it right, this estimate of the accuracy of the initial mass and energy density of the universe is further evidence of fine-tuning of the universe.

Cosmological Constant

We have already encountered the cosmological constant in chapter two on the "Big Bang Theory". In that chapter, for simplicity, we assumed the cosmological constant to be zero. As promised, we shall now have a closer look at this constant—or fixed number.

Nowadays, according to Wikipedia, the cosmological constant is defined as the equivalent of the *energy density* of free space, what scientists normally call the *vacuum* of space. We have already seen that theoretical physicist Albert Einstein conceptualized this constant (fixed number) in 1917: adding it as a fudge factor to his general relativity equations to hold back gravity, "preventing" the universe from collapsing. Thereby, he conformed to the then current consensus of scientists that the universe was static, that is, basically unchanging. Einstein later thought that it was his "greatest blunder" when in 1929 Hubble discovered that galaxies are receding from one another: which means that the universe's space is expanding as a whole. So from 1929 to 1990 most cosmologists deleted this number, or equivalently, set it to zero. [51]

In chapter one on "Matter and Energy" we also saw that about seventy-three percent (73%) of the matter and energy in our universe is *dark energy*. One may recall that the concept of dark energy is an *ad hoc* explanation for the accelerating expansion of the universe: we don't really understand what dark energy is all about. Wikipedia continues to explain that after these discoveries in the 1990s, the cosmological constant had to be re-introduced, at least in its simplest form, as a fixed number over all of space and time:

> While dark energy is poorly understood at a fundamental level, the main required properties of dark energy are that it dilutes much more slowly than matter as the universe expands, and that it clusters much more weakly than matter, or perhaps not at all. [52]

So Einstein turned out not to have made a blunder after all: but it was just luck, of course, not insight.

The cosmological constant could have had any value: positive or negative, large or small. From first principles it appears that this number should be very large because *dark energy* constitutes the greater part of our universe—about 73%. Moreover, most quantum field theories predict its having a huge value. [53] [54] In his book *A Designer Universe?*, physicist and avowed atheist Steven Weinberg expresses surprise at how it is set to favor our existence (life); he writes about this number:

---

[51]   https://en.wikipedia.org/wiki/Cosmological_constant (7 August 2015)

[52]   https://en.wikipedia.org/wiki/Cosmological_constant (7 August 2015)

[53]   Weinberg, "A Designer Universe?"

[54]   Strobel, *The Case for a Creator*, p. 133

If large and positive, the cosmological constant [number] would act as a repulsive force that increases with distance, a force that would prevent matter from clumping together in the early universe, the process that was the first step in forming galaxies and stars and planets and people [life]. If large and negative, the cosmological constant would act as an attractive force increasing with distance, a force that would almost immediately reverse the expansion of the universe and cause it to recollapse. [55] [56]

Either way, life is in jeopardy; but neither of these scenarios is what actually happened. Why? In actual fact, astronomical observations show that the cosmological constant is very small, extremely smaller than would have been anticipated from first-principles. [57] [58]

According to physicist Robin Collins, referenced in Strobel's book *The Case for a Creator*:

[T]he unexpected, counterintuitive and stunningly precise setting of the cosmological constant "is widely regarded as the single greatest problem facing physics and cosmology today. ... The fine-tuning has conservatively been estimated to be at least one part in a hundred million billion billion billion billion billion [10^{53}]. [59]

According to Collins, these odds are comparable to successfully hitting a single atom on earth from way out in space. [60]

[55]   Weinberg, "A Designer Universe?"
[56]   Strobel, *The Case for a Creator*, p. 133
[57]   Weinberg, "A Designer Universe?"
[58]   Strobel, *The Case for a Creator*, p. 133
[59]   Strobel, *The Case for a Creator*, p. 133
[60]   Strobel, *The Case for a Creator*, p. 134

General

There are more than thirty separate physical and cosmological parameters that require precise calibration in order to result in a life-sustaining universe, [61] [62] but it is not my intention to bore you with numbers over numbers. This chapter is not intended to be an exhaustive treatise on the fine-tuning of the universe: the purpose of this chapter is only to give a general idea of the evidence supporting the apparent fine-tuning of our universe. Incidentally, no modern scientist denies this.

In general, trying to make a slight change in one parameter treads on many other parameters and enters into the realm of limitation of various other vital relationships: it is a really impressive case of fine-tuning itself.

Are there combinations where *all* the parameters can be varied at once? This is what particle physicist and self-declared atheist Victor Stenger suggests in his book *God and the Folly of Faith*?

> It is also to be noted that in almost all the literature that advocates fine-tuning, the authors make a serious analytical mistake by varying only one parameter at a time and holding all the others constant. [63]

Admittedly, this may be the case; however, to me, it sounds more like wishful thinking or an atheist's escape hatch. From my experience in engineering, and I'm sure most engineers will agree with me, it is usually much easier to change one parameter at a time and try to accommodate for it, rather than to change a whole slew of parameters all at once; unless one wants to create a radically new design—like creating a new form of life, say. Another form of life, for example, could possibly have been based on the next higher tetravalent element, silicon ($_{14}Si$), of which we have an abundance on

---

[61]   Meyer, "Evidence for Design in Physics and Biology", p. 60

[62]   Strobel, *The Case for a Creator*, p. 132

[63]   Stenger, *God and the Folly of Faith*, p. 185.

earth (rocks); however, as far as I know, nobody has tried to come up with a coherent set of numbers that could have supported silicon-based life, say, in our universe. *That* would have been supporting evidence for the general statement Stenger makes above!

Anyway, to get a feel for the actual situation, following is a quote by philosopher John Leslie paraphrasing the opinion of a couple of experts in this field, namely, astronomer Virginia Trimble and physicist Iosif Rozental:

> To guard against one disaster by tinkering with these or those factors would be likely only to introduce some new disaster because each factor enters into so many other vital relationships. [64] [65] And even if disaster could in theory be avoided, actually avoiding it—compensating for variations by making appropriate changes elsewhere—could itself be a very impressive instance of fine-tuning! [66]

As we shall see in the next chapter entitled "Mainstream Scientific Hypotheses", scientists try to explain these observations naturally. I do agree that they should do so: as far as is possible! We shouldn't conclude that an omnipotent Creator is the cause every time we encounter an unexplained natural phenomenon; if we do so, the result would not be science at all. We have to exclude all possible natural explanations before we can conclude, beyond any reasonable doubt, that God is the *direct* cause. We shall discuss this concept further, later on in the book, and see where the dividing line happens to be.

---

[64]   Trimble, "Cosmology: Man's Place in the Universe", p. 85; and Rozental, "Physical Laws and the Numerical Values of Fundamental Constants", p. 303.

[65]   Leslie, "The Prerequisites of Life in Our Universe", p. 237.

[66]   Leslie, "The Prerequisites of Life in Our Universe", p. 237.

Physical Forces in Nature

Modern physics agrees that there are four main forces in our universe: *Gravity, Electromagnetism*, the *Strong Nuclear Force*, and the *Weak Nuclear Force*.

We are all familiar with gravity, and we have probably experienced the force of a magnet. We probably also heard that two unlike, one positive and one negative, charges attract each other while two like, both positive or both negative, charges repel each other; although we may not have experienced it first-hand. This type of force between charged particles is known as an *electrostatic* force. *Electromagnetism* is the force associated with both electric charges and magnetic fields: they are thought of as being a single force because we know that they are related to each other, and we also know how they are interconnected with each other.

The forces of gravity and electromagnetism both extend to infinity obeying the *inverse square* law. We have already explained this law briefly but we shall now explain it in a little more detail because it is so common and important in nature. The inverse square law means that at a distance of two meters the strength of either of the above forces reduces to one-quarter ($1/4=1/2^2$) of its strength at a distance of one meter ($1=1/1^2$); at a distance of three meters it reduces to one-ninth ($1/9=1/3^2$); at a distance of four meters it reduces to one-sixteenth ($1/16=1/4^2$); and so on, up to an infinite distance. Notice that they never reduce to zero. Needless to say, however, although there is still some influential effect due to these forces at a large distance away from the source, their strength at that particular location will be negligible, because they fall off quite rapidly.

We do know why the inverse square law is so common in nature: actually, it is the natural way radiation intensity and force field strength should decrease with distance from a source. Imagine a point source of light, say, in free space; this light is radiated equally in all directions in space. So, as it spreads out, its intensity will be distributed evenly over a sphere of radius "r", where "r" is the distance from the light source. The surface area "A" of such a sphere is given by the formula $A=4\pi r^2$, where "π" is about 3.142 (the number of times the diameter of a circle can be "fitted" onto its circumference) and, of course, "r" is the radius of the sphere. Since the area of this sphere increases proportional to the square of its radius (r^2) we would expect the intensity of the light source at any point on the surface of this sphere to be "diluted", or to fall off, accordingly: that is, as the inverse square of the distance ($1/r^2$) from the point source.

It feels good when we can explain something in science; but unfortunately there is so much we cannot explain. For example we cannot explain why like charges repel and unlike charges attract each other. We cannot even explain the existence of gravity in our universe; why all matter is attracted towards any other matter; or, if you want to be technical about it, why the presence of matter warps spacetime. We can explain why a bowling ball placed on a horizontal stretched trampoline causes it to sag: because there is gravity underneath "pulling" the ball towards the ground. But this is only a model to enable us to understand the warpage of space; we don't really know why the presence of matter in free space warps the space in its surroundings: that is, light in its vicinity does not propagate in a straight line any more. The strength (intensity distribution) of all natural forces in space *should* all obey the inverse square law as illustrated above; but, as we shall see presently, the strong and the weak nuclear forces don't—far from it. Why? We don't know. Why do they have the magnitudes they have and not others? We don't know. We don't even know why they exist. Unfortunately, what we scientists know about reality pales in comparison to what we still don't know!

The strong nuclear force is the force that holds protons and neutrons (refer to "Appendix II" on "Chemistry") in the nucleus of an atom together and also the quarks (refer to "Appendix III" on "Physics") inside the protons and neutrons themselves. It is a very powerful attractive force at a distance of 10^{-15}m, but at slightly shorter distances it becomes repulsive (and zero in between); this is what determines the size of the nuclei of atoms. The weak nuclear force is the force involved in nuclear reactions such as fusion and radioactivity; its range of influence is confined to about 10^{-18}m. [67] Why are their intensities so different from the naturally expected inverse square law? We simply don't know.

We know there are four forces in nature, and we know their relative strengths and the distances over which they have influence; but we don't know why they are only four, or how they were "selected" that way out of all possible force strengths and influential distance combinations possible. The wide range of possibilities, coupled with the narrow limitations for the possibility of life in our universe, suggest fine-tuning. It stands to reason that knowing how something works is a far cry from knowing how something came about to be that way.

Imagine you are in an automotive class. The instructor explains how a car engine works; however, this does not answer the begging question of how it came about to be that way, or who designed the intricate details with the tight tolerances necessary to make it functional. What would happen to a car if some part was slightly larger or smaller? It probably wouldn't work without making any changes; but, chances are, we would be able to accommodate for it by tweaking the design a little. But if big changes are made to one of its parts, then all bets are off; it will probably have to be redesigned from scratch in order to keep it within its physical constraints. In any design, there is no way of avoiding conforming to tight tolerances; most of the time it would not work otherwise.

What would have happened if the above four fundamental forces were any different? Following are some considerations, by experts in the field, of the catastrophic effects resulting from such changes in every one of the four forces, respectively.

(1) Gravity

Physicist Robin Collins, in Strobel's book *The Case for a Creator*, asks us to imagine a very long ruler, marked in one inch increments,

---

[67]   https://en.wikipedia.org/wiki/Fundamental_interaction (10 August 2015)

that goes all the way across our universe, and that the above four forces were to be selected from this scale (ruler) of force magnitudes. The strong nuclear force—the strongest force in nature—is placed at the high end of the scale, and gravity—the weakest force in nature—is proportionately placed at the low end. Now imagine one wants to move up (increase) the force of gravity by just one inch from where it currently is on this very long scale. This small adjustment would increase gravity a billion-fold! Animals anywhere near the size of human beings would be crushed. [68] Imagine your weight increased ten times; you would probably have a hard time moving around, even by a few inches. You probably would not move at all if your weight increased "only" a hundred times. We cannot even imagine our weight being a billion (10^9) times what it is now.

Even if the gravitational force was just about an order of a thousand (10^3) times what it is now, there could be no viable life-sustaining planet; and the lifetime of stars would be reduced to less than a tenth of that of our sun. In the same book Strobel quotes Collins:

> "[A] planet with a gravitational pull of a thousand times that of the Earth would have a diameter of only forty feet, which wouldn't be enough to sustain an ecosystem. ... [S]tars with lifetimes of more than a billion years—compared to ten billion years for our sun—couldn't exist if you increase gravity by just three thousand times." [69]

(2) Electromagnetism

Philosopher John Leslie, in his article "The Prerequisites of Life in Our Universe", explains that the magnitude of force of electromagnetism has to be confined within narrow limits for life as we know it (carbon-based) to thrive.

---

[68]   Strobel, *The Case for a Creator*, p. 132
[69]   Strobel, *The Case for a Creator*, p. 132

In stellar fusions (fuel-burning), electromagnetism and gravity act against each other. The former is a repulsive force: protons are positively charged, and like charges repel; while the latter is an attractive force: all matter is attracted to any other matter. Electromagnetic repulsion, therefore, interferes with protons' colliding against one another and thus prevents most proton-proton (diproton) fusions from happening.

Diproton fusions eventually give rise to helium ($_2$He) burning stars rather than hydrogen ($_1$H) burning stars. As we shall see in the next subsection on the "Weak Nuclear Force", the former burn a billion billion (10^{18}) times faster than the latter. The relative strengths of electromagnetism and gravity, therefore, affect the rate of stellar reactions. Economic supply of energy is crucial for the sustenance of life; in fact, it so happens that the sun is thousands of times more economical than our bodies. [70]

Besides economy, Leslie continues to explain that a star's surface temperature is crucial for the sustenance of the life of organisms that happen to exist on a nearby planet. Recall that a photon is an indivisible energy package, better known as a *quantum* of radiation. A photon obeys the equation E=hf, where "E" is the "energy" of the photon—or quantum of radiation—of frequency "f", and "h" is a very small fixed (or constant) number (6.626×10^{-34}) known as the *Planck constant*. It follows that the higher is the frequency of the light (violet, say) the higher is its photon energy; the lower is the frequency of the light (red, say) the lower is its photon energy.

The hotter is a star, the more capable it is of supplying a planet with higher frequencies of radiation: for example, light for plant photosynthesis rather than just heat (infra-red). Heat alone does not enable photosynthesis in plants; just like red light does not affect a photographic film in a dark room. It is no use placing the planet closer to the star (or sun); the quanta (various packages) of energy supplied by the star remain unchanged at any distance from the star. On the other hand, if a star is too hot, it produces excessive ultra-violet radiation;

---

[70] Leslie, "The Prerequisites of Life in Our Universe", p. 236.

which is normally harmful to living organisms. [71] Hence there is a very delicate balance between electromagnetism and gravity involved in a star's formation and its resulting surface temperature. [72] [73]

Leslie quotes from theoretical physicist, cosmologist, and astrobiologist Paul Davies' book *Superforce* to show the exquisite balance between electromagnetism and gravity in star formation. In this book Davies affirms that theoretical physicist, astrophysicist, and cosmologist Brandon Carter has shown that:

> [C]hanges either in electromagnetism or in gravity [74] "by only one part in 10^{40} would spell catastrophe for stars like the sun." [75]

In a journal article entitled "Gravitation without a Principle of Equivalence", physicist Robert Dicke wrote that if electromagnetism were appreciably stronger, all the stars in the universe would be so cold as to preclude the existence of man. [76] [77]

In a scholarly article entitled "Large Number Coincidences and the Anthropic Principle", theoretical physicist Brandon Carter points out that had electromagnetism been very slightly stronger, the light emitted from our sun would decrease immensely and the sun would turn dull. The surface temperature of the sun happens to be close to that at which electron-extraction from atoms (ionization) occurs; below this temperature imperviousness to light (opacity) increases significantly. If, on the other hand, electromagnetism were a little bit stronger than this, stars would spend most of their lifecycle as red stars: losing energy (heat) chiefly by convection, rather than radiation, and consequently, they would be too cold for the proper

---

[71] Leslie, "The Prerequisites of Life in Our Universe", p. 237.
[72] Press, Lightman, Peierls, & Gold, "Dependence of Macrophysical Phenomena on the Values of the Fundamental Constants"
[73] Leslie, "The Prerequisites of Life in Our Universe", p. 237.
[74] Leslie, "The Prerequisites of Life in Our Universe", p. 237.
[75] Davies, *Superforce*, p. 242.
[76] Dicke, "Gravitation without a Principle of Equivalence", pp. 375–6.
[77] Leslie, "The Prerequisites of Life in Our Universe", p. 237.

sustenance of life.[78] If a planet happens to be close enough to such a star, so as to be sufficiently warm for life to survive, it would experience tidal forces of such magnitude which would gradually reduce its rotation about its axis until it stopped spinning. Thus, the same hemisphere (half) of the planet would always face the star; then, all its liquids and gases would eventually collect and freeze on its dark—and hence unheated—side.[79] [80]

Leslie then considers what would happen if electromagnetism were very slightly weaker. In that case, all stars would look blue, rather than white, during the greater part of their much shorter lifecycle. They would be very hot and radiate a plethora of harmful ultra-violet rays onto their planets.[81]

Even at the current magnitude of the electromagnetic force, stars that are more than twenty percent greater than the mass of our sun (>1.2 times) would probably burn too fast; they would exist for too short a time for any kind of intelligent life form to be able to evolve on their planets—assuming life evolved from non-living chemicals, microorganisms, etc.[82] Blue giant stars exist for only a few million ($\sim 10^7$) years—compared to a few billion ($\sim 10^{10}$) years (like our sun).[83]

Now, if electromagnetism and gravity, somehow, stood in a different relationship to each other, so that the above delicacy of balance was still maintained, there are still other factors involved that would have to be considered and fine-tuned. The relative masses and sizes of the proton to the electron and the size of the smallest (indivisible) unit of electric charge—the quantity of charge carried by an electron or a proton—are all determining factors in the stability of atoms.[84] The interdependence of electromagnetism and gravity, in

---

[78]   Carter, "Large Number Coincidences and the Anthropic Principle", pp. 296–8.

[79]   Gale, "The Anthropic Principle", p. 155.

[80]   Leslie, "The Prerequisites of Life in Our Universe", p. 237.

[81]   Leslie, "The Prerequisites of Life in Our Universe", p. 237.

[82]   Rood & Trefil, *Are We Alone?*, p. 21

[83]   Leslie, "The Prerequisites of Life in Our Universe", p. 237.

[84]   Leslie, "The Prerequisites of Life in Our Universe", p. 237.

both the universe and the atom, is observed to be delicately balanced across the board.

(3) Weak Nuclear Force

Recall that the weak nuclear force is the force involved in nuclear reactions such as fusion and radioactivity; as opposed to chemical reactions in which the nuclei of the elements (atoms) involved do not change. The most striking feature of the weak nuclear force is its extreme weakness compared to the strong nuclear force. It is only a billion-billionth (10^{-18}) of the strong nuclear force.

> The weak force controls proton-proton fusion, a reaction 10^{18} times slower than [i.e., $1/10^{18}=10^{-18}$ times] one based on the other nuclear force, the strong force. [85]

Luckily, this slowed down and limited the conversion of hydrogen into helium in the early universe: the first five minutes after the Big Bang.

Leslie quotes theoretical physicist and mathematician Freeman Dyson, who in his book *Energy in the Universe* points out that if this were not the case:

> "[E]ssentially all matter in the universe would have been burned to helium before the first galaxies started to condense." [86]

It was very important to leave an abundance of hydrogen, rather than helium, in the early universe; without hydrogen there would have been no water, which partly consists of hydrogen (H_2O), and therefore no life. As we have already seen, the hydrogen fusion reaction is much slower than that of helium fusion; moreover, hydrogen fusion releases much more energy than helium fusion; and

---

[85] Leslie, "The Prerequisites of Life in Our Universe", p. 235.
[86] Dyson, "Energy in the Universe", p. 56.

finally, hydrogen fusion requires only half the matter: $4(^1H) \rightarrow 2(^2H)$ $\rightarrow (^4He)$ compared to $2(^4He) \rightarrow (^8Be)$. Therefore, hydrogen fusion is more economical on three levels and is thus more conducive to life. The relative weakness of the weak nuclear force, compared to the strong nuclear force, allows stars to last much longer; thus, it enables our sun to provide life-giving energy economically by allowing it to [87] "burn its hydrogen gently for billions of years instead of blowing up like a bomb." [88]

If the weak nuclear force were significantly stronger, the first twenty-five elements would have been "burned" into higher elements during the universe's *nucleosynthesis* period: that is, in the first twenty-odd minutes after the Big Bang. In particular, there would be no carbon ($_6C$)—atomic number six—which constitutes the basic element for any life form on earth. To this effect Leslie writes:

> Had the weak force been appreciably stronger then the Big Bang's nuclear burning would have proceeded past helium and all the way to iron [atomic number 26]. Fusion-powered [hydrogen and even helium fusion] stars would then be impossible. [89]

On the other hand, the weak nuclear force could not be too weak either. Leslie refers to cosmologist and astrobiologist Paul Davies' book *Other Worlds* to tell us the reason why. Since the mass of a neutron is approximately equal to the mass of a proton, their production would have been almost equally probable in the early universe: the surrounding environmental temperature was so hot that their very small mass difference would have been relatively unimportant. However, had the number of neutrons ended up roughly equal to the number of protons, helium (^4He)—with two protons and two neutrons—would eventually form, rather than hydrogen (^1H)—with only one proton; the result would have been a universe

---

[87] Leslie, "The Prerequisites of Life in Our Universe", p. 235.

[88] Dyson, "Energy in the Universe", p. 56.

[89] Leslie, "The Prerequisites of Life in Our Universe", p. 235.

that consisted of practically nothing but helium. It so happens, that Helium (^{4}He) has a lower binding energy than deuterium (^{2}H)—with one proton and one neutron; so the former would form more readily than the latter. In fact, there is only a small percentage, about 0.02%, of deuterium in a stable (general) population of hydrogen atoms. Since the weak nuclear force is the force mediating nuclear reactions, it is the force responsible for the neutrons' decaying or not decaying into protons. The magnitude of this force, therefore, had to be "chosen" carefully in order to produce an overabundance of hydrogen—about seventy percent (70%) by mass—in the early universe: thus requiring a tight setting for both the upper and the lower magnitude of the weak nuclear force. [90] [91]

A weaker weak nuclear force would also have unbalanced the ratio of heavier elements such as carbon, nitrogen, and oxygen; which are essential for all life forms. These elements are formed during supernovae explosions of stars at the end of their life cycle: we are, literally, stardust! Here, Leslie quotes cosmologists and astrophysicists Jacques Demaret and Christian Barbier:

> [W]eakening the weak force would ruin the proton-proton and carbon-nitrogen-oxygen cycles which make stars into sources of the heat, the light and the heavy elements which life appears to need. [92] [93]

Leslie then references an article written by cosmologist and astrophysicist Martin Rees, and astronomer and mathematician William McCrea:

> While the calculations are hard, it seems a safe bet that weakening the weak force by a factor of ten would have

---

[90] Davies, *Other Worlds*, pp. 176–7.
[91] Leslie, "The Prerequisites of Life in Our Universe", p. 235.
[92] Demaret & Barbier, "Le Principe Anthropique en Cosmologie", p. 500
[93] Leslie, "The Prerequisites of Life in Our Universe, p. 235.

led to a universe consisting mainly of helium and in which supernovae could not occur. [94] [95]

Free neutrons (that is, outside the nucleus of an atom) are unstable: decaying to protons (and other particles) in about fifteen minutes. Since the weak nuclear force mediates this decay, the binding energies of the weak nuclear force in the neutron and the proton affect the mass difference between a neutron and a proton—recall that energy and mass are interchangeable. In Strobel's book *The Case for a Creator*, physicist Robin Collins explains the consequences of increasing the weak nuclear force:

> "There are other examples of fine-tuning ... there's the difference in mass between neutrons and protons. Increase the mass of the neutron by one part in seven hundred and nuclear fusion in stars would stop. There would be no energy source for life." [96]

Increasing the mass of the neutron implies an increase in the weak nuclear force. An increase in this force resulting in the neutron's mass (weight) increasing by one seven hundredth (1/700) of its current mass would inhibit nuclear fusion in stars completely.

(4) Strong Nuclear Force

Recall that the strong nuclear force is the force that binds protons and neutrons together inside the nucleus of an atom; it also binds the quarks, or internal building blocks, of these protons and neutrons.

Leslie references articles written by cosmologist, theoretical physicist, and mathematician John Barrow, astronomer and physicist Joseph Silk, and theoretical physicist and mathematician Freeman

---

[94]   Rees & McCrea, "Large Numbers and Ratios in Astrophysics and Cosmology", p. 317.

[95]   Leslie, "The Prerequisites of Life in Our Universe, p. 236.

[96]   Strobel, *The Case for a Creator*, p. 134

Dyson showing that the magnitude of the strong nuclear force is limited on both sides; that is, it can neither be too strong nor too weak for stars to operate life-encouragingly. [97] Had this force been just two percent (2%) stronger, protons would be unable to form from quarks: only neutrons would be able to form. Consequently, *none* of the elements in our universe would have existed: not even hydrogen. [98] Alternatively, the same increase (2%) in the strength of this force would bind all protons into diprotons (proton pairs) in the first few minutes after the Big Bang: thus turning all hydrogen in the universe into helium. [99] As we have already seen, helium stars burn much faster because they make use of the strong nuclear force, rather than the weak nuclear force, in fusing two helium atoms together; [100] as a consequence, the former reaction takes place a billion billion (10^{18}) times faster than the latter. [101]

Even an increase of about one percent (1%) in the strength of the strong nuclear force, would preclude the formation of carbon—the basic element of all life forms. According to astronomer Fred Hoyle and astrophysicist Edwin Salpeter, most of the carbon that happens to form would be converted into oxygen:

> A yet tinier increase, perhaps of 1%, would so change nuclear resonance levels that almost all carbon would be burned to oxygen. [102] [103]

An increase in the magnitude of this force slightly more than 10% would cause nuclei to grow to incalculable sizes: even to the

---

[97]   Leslie, "The Prerequisites of Life in Our Universe", p. 236.

[98]   Barrow & Silk, "The Structure of the Early Universe", pp. 127–8.

[99]   Davies, "The Anthropic Principle" p. 8; and

Rozental, *Elementary Particles and the Structure of the Universe*, p. 85.

[100]   Dyson, "Energy in the Universe", p. 56.

[101]   Leslie, "The Prerequisites of Life in Our Universe", p. 236.

[102]   Hoyle, "On Nuclear Reactions Occurring in Very Hot Stars", p. 121; and

Salpeter, "Nuclear Reactions in Stars", p. 516

[103]   Leslie, "The Prerequisites of Life in Our Universe", p. 236.

size of small clumps of matter. According to theoretical physicist, cosmologist, and astrophysicist Brandon Carter:

> A somewhat greater increase … [o]ne a trifle greater than this [10%] would lead to "nuclei of almost unlimited size" [104] even small bodies becoming "mini neutron stars." [105] [106]

Needless to mention, perhaps, had this force a long range influence, it is so strong that the universe would have collapsed to an indistinguishable blob of matter, according to chemist Peter Atkins. [107] [108]

Decreasing the magnitude of the strong nuclear force is just as detrimental. As it stands, it just manages to bind the deuteron (^{2}H)— which is a combination of a proton and a neutron. A decrease of about five percent (5%) in its strength would allow the deuteron to fall apart. This isotope of hydrogen (deuterium) is crucial as a stepping stone in stellar fusion. Incidentally, it is also the one actually used in the hydrogen bomb. According to physicist Paul Davies:

> The deuteron, a combination of a neutron and a proton, which is essential to stellar nucleosynthesis, is only just bound; weakening the strong force by "about 5%" would unbind it, [109] leading to a universe of hydrogen only. [110]

Even if this force's strength is decreased by just one percent (1%), it would preclude the important nuclear formation of carbon (^{12}C)— the basic element of all life forms—from helium (^{4}He) and beryllium

---

[104]   Carr & Rees, "The Anthropic Principle and the Structure of the Physical World", p. 611.

[105]   Carter, "Understanding the Fundamental Constants", p. 652.

[106]   Leslie, "The Prerequisites of Life in Our Universe", p. 236.

[107]   Atkins, *The Creation*, p. 13

[108]   Leslie, "The Prerequisites of Life in Our Universe", p. 236.

[109]   Davies, "The Anthropic Principle" p. 7

[110]   Leslie, "The Prerequisites of Life in Our Universe", p. 236.

(^{8}Be). According to cosmologists Martin Rees, John Barrow and Frank Tipler:

> [E]ven a weakening of 1% could destroy [111] "a particular resonance in the carbon nucleus which allows carbon to form from ^{4}He [helium] and ^{8}Be [beryllium] despite the instability of ^{8}Be" [112]

As it is, ^{8}Be is stable for a long enough time to allow the above nuclear synthesis to occur; again suggesting another form of fine tuning. [113]

A 50% decrease in the magnitude of the strong nuclear force would, of course, be devastating to the formation of carbon, and hence to all life forms. According to Barrow and Tipler:

> "A 50% decrease would adversely affect the stability of all elements essential to living organisms": [114] any carbon, for example, which somehow managed to form, would fast disintegrate. [115]

According to physicist Iosif Rozental, a change in the magnitude of this force of about twenty percent either way (+/-20%), would prevent the formation of deuterons (^{2}H) or of any other element heavier than helium (^{4}He):

> Rozental estimates that the strong force had to be within 0.8 and 1.2 [i.e., within plus or minus 20%] of its actual strength for there to be deuterons [bonded neutron and

---

[111] Rees, "Our Universe and Others", p. 122.

[112] Barrow & Tipler, *The Anthropic Cosmological Principle*, (1986) pp. 252–3

[113] Leslie, "The Prerequisites of Life in Our Universe", p. 236.

[114] Barrow & Tipler, *The Anthropic Cosmological Principle*, (1986) p. 327

[115] Leslie, "The Prerequisites of Life in Our Universe", p. 236.

proton] and all elements of atomic weight greater than four [helium]. [116] [117]

Since the strong force is an attractive force while electromagnetism is a repulsive force, it stands to reason that their relative strengths determine the formation or non-formation of atoms; hence the importance of matching them together. Leslie concludes:

> [I]t is the strong force's strength by comparison with electromagnetism (it is some hundreds of [about 137] times stronger) which is the real topic of the above remarks about carbon synthesis and about the deuteron's [bonded neutron and proton] being luckily just bound while the diproton [bonded proton pair] being equally luckily just unbound. [118]

These tight tolerances and constraints show an uncanny resemblance to engineers' designs of machinery; [119] there is usually very little one can change without affecting the device's intended performance. One can, of course, change the whole design, as particle physicist Victor Stenger suggests: [120] like changing, say, from a steam engine, to an automobile, to an airplane. The whole design will have to be changed; but it will still have to be *designed*! Think of it, a steam engine needs rails, coal, and water; while a car needs gasoline;

---

[116] Rozental, *On Numerical Values of Fundamental Constants*, p. 9; for atomic weights above four he cites:
Salpeter, "Accretion of Interstellar Matter by Massive Objects", p. 796
[117] Leslie, "The Prerequisites of Life in Our Universe", p. 236.
[118] Leslie, "The Prerequisites of Life in Our Universe", p. 236.
[119] I, personally, do have some experience in and knowledge of machinery design: I possess a Canadian (CA2115700) and a United States (US5400463) patent relating to a quiet (noise dampened) canister vacuum cleaner. I also designed a vacuum cleaner containing a self-cleaning filter, for which a patent application was about to be officially submitted; however, it did not materialize because the company I was working for was sold, and the patent application process was immediately frozen.
[120] Stenger, *God and the Folly of Faith*, p. 185.

and an airplane needs wings. Something like this, in relation to our universe and life, will probably imply a completely different universe and, most probably, also a completely different form of life—reproductive entities.

As a conclusion to this chapter, it suffices here to say that no modern scientist denies the many-faceted fine-tuning of our universe; they only disagree about the source, or reason, for this observable phenomenon. So, scientists try to explain these remarkable coincidences as a result of chance and perhaps natural selection. The next chapter gives a summary of these attempts.

CHAPTER 5

MAINSTREAM SCIENTIFIC HYPOTHESES

Multiverse

All the fine-tunings of our universe, described in the previous chapter, are hard to explain away—unless one postulates that the number of universes is *infinite*. Philosophically, it is impossible to have an infinite number of objects, so scientists have suggested a very large number of universes: 10^{500} say; some of them containing life, most of them don't. This is far too small a number to beat all the various odds described in the previous chapter; this large number, however, seems to come from *string theory*—or rather *string hypothesis*. But let us assume, or the sake of argument, the existence of an extremely large number of universes: enough of them to make the hypothesis plausible.

In his article "Cosmos and Creator", science and theology analytic philosopher William Craig explains that, according to this hypothesis, our observable universe is a part of a wider *universe-as-a-whole*: one of many different mini-universes. These mini-universes are like soap bubbles in air, which inflated from the original larger universe-as-a-whole; these then grow into mini-universes within the whole. We only observe our own mini-universe expanding, but we know nothing about other similar mini-universes. [121] (I shall be using concepts and ideas from Craig's article extensively in this chapter.)

---

[121]   Craig, "Cosmos and Creator", p. 21 & p. 27.

This hypothesis assumes that the mother universe, the universe-as-a-whole, is not expanding and that it is eternal. As these mini-universes expand within a non-expanding mother universe, they may collide and even coalesce. Now, since the mother universe is assumed to be eternal—that is, its past is infinitely long—an infinite number of mini-universes would have spawned, and all the individual universes would have merged by now. Of course, this is not supported by any kind of observation. If, on the other hand, we additionally assume that the background space, the mother universe, is also expanding, then we must posit an origin for the wider universe and its expansion; thus, we are right back where we started: that is, in a position similar to explaining what initiated our Big Bang. [122] In his article "The Prerequisites of Life in Our Universe", science philosopher John Leslie, writes:

> [Mainstream scientists] typically propose that there exist countless "universes" (that is largely separate or entirely separate systems, perhaps of immense size ...) and that force strengths, particle masses, expansion speeds, and so on, vary from universe to universe. Sooner or later, somewhere, conditions will permit life to evolve. [123]

Historically, science has *always* assumed that our observations are *typical* of the whole universe, and not that distant galaxies obey different laws of physics from ours. This is a "new" kind of science. Unless there is evidence to the contrary, scientists should not try to have things both ways! Why should we suppose that the laws of physics we observe in our universe are different in the other mini-universes? And why should we suppose that only our part of the universe is expanding rather than the whole mother universe? [124]

Moreover, what guarantee is there that *all* possible universes are going to happen in the wider universe: including ours? Most

---

[122] Craig, "Cosmos and Creator", pp. 21–2.

[123] Leslie, "The Prerequisites of Life in Our Universe", p. 231.

[124] Craig, "Cosmos and Creator", p. 22.

probably, there will also be repetitions. So, mathematically, one needs an *infinite* number of randomly typing monkeys in order to eventually produce Shakespeare's complete works: not just a very very large number. But then again, philosophically, one cannot have an infinite number of things. [125]

In the words of theoretical physicist and cosmologist Alan Guth, the original proponent of the multiverse hypothesis, in these mini-universes "anything that can happen will happen—and it will happen infinitely many times." Will we be able to observe or prove otherwise the existence of any of these mini-universes as true science is supposed to do? Some scientists have suggested the idea of observable gravity effects from these other universes, which interested me immensely as a scientist; but no more details were given. In actual fact, there is no such evidence of gravity from other universes affecting our universe. [126]

Some scientists described this hypothesis as *questionable* science, as it is not falsifiable; one might say comparable to an unseen Creator, a catch-all phrase for what cannot be explained, or a "theory of anything". Apparently, by their *own* admission, these universes are completely inaccessible to us by *any* physical means. [127] If science assumes that these multiple universes are so far away as to be inaccessible to us, *by definition*, then, in my opinion, this concept has no place in our science: if a concept does not lend itself to being proved true or falsified, then it is not classifiable as science; it is *pseudoscience*. In 2014, Cosmologist Paul Steinhardt criticized this concept radically:

> A pervasive idea in fundamental physics and cosmology that should be retired [is] the notion that we live in a multiverse .... Different regions of space too distant to ever be observed have different laws and properties [randomly set by chance] .... No observation or combination of

---

[125]   Craig, "Cosmos and Creator", pp. 27–8.
[126]   https://en.wikipedia.org/wiki/Multiverse (13 February 2016)
[127]   https://en.wikipedia.org/wiki/Multiverse (13 February 2016)

observations can disprove it. ... The rest of the scientific community should be up in arms since an unfalsifiable idea lies beyond the bounds of normal science. Yet, except for a few voices, there has been surprising complacency .... The scientific journals are full of papers treating [it] seriously. What is going on? [128]

Believe it or not, this is the current "scientific" explanation for the fine-tuning of our universe; to me, it seems like clutching at straws. There is not a shred of evidence of a multiple universe; and, on top of that, there seems to be hardly any possibility of our obtaining any evidence for or against it: doesn't that sound like *faith* in religion? Are we supposed to have faith in (trust) our scientists and believe things they tell us on their gut feeling? But they have been wrong before!

In summary, any scientific hypothesis is supposed to be possibly provable as true or false experimentally or through other forms of evidence; otherwise, it is just fanciful imagination or pseudoscience. Will it ever be possible to infer phenomena outside our universe from evidence within our universe? We were able to achieve this in the case of stars; even though, we initially thought they were completely inaccessible to us; I guess we shall have to wait and see. I don't know whether the reader will buy this hypothesis; I'll leave it at that.

In this book, on the other hand, I contend that the existence or non-existence of God is a scientific question about reality, and that it lends itself to be statistically provable or falsifiable through the evidence in our universe and on our earth.

Anthropic Principle

Of course, science does not feel comfortable positing a *designer* for the fine-tuning of the universe; and, believe me, as a scientist I sympathize with them. However, rather than leave it at that: as a gap that science cannot yet explain—of which there are many, some

---

[128] Steinhardt, "Theories of Anything." https://www.edge.org/response-detail/25405 (13 August 2016)

scientists resorted to smoke-screening and confusing simple folk, in my opinion; and, I think, Craig agrees with me when in his article "Cosmos and Creator" he writes:

> Loath to admit design, however, some thinkers appeal to the so-called ... Weak Anthropic Principle (WAP). According to this principle, the astonishingly improbable properties of the universe are not surprising once we realize that we could not have observed anything else. [129]

This kind of thinking is not a scientific way of thinking, and it is only intended to create confusion among those who disagree with atheistic views. While it is true that we would be unable to observe a universe that was not conducive to intelligent life—because we would not be around to observe it; at the same time, marveling at the immense improbability of our actual existence is still in order: given the small odds of a universe being conducive to any kind of life, let alone intelligent life. [130]

Consider the following parable conceptualized by John Leslie. Suppose you are taken away to be executed by fifty sharp shooters. They blindfold you, and you hear the rifles firing. But you feel nothing, and you notice that you are still alive and unscathed. Every single one of the fifty sharpshooters, somehow, missed hitting you, completely! [131] Craig continues:

> Taking off the blindfold you do not observe that you are dead. There is no surprise there: you *could not* observe that you are dead. Nonetheless, you should be astonished to observe that you are alive. The entire firing squad missed you altogether! Surprise at the extremely improbable fact is wholly justified—and the fact calls for an explanation.

---

[129] Craig, "Cosmos and Creator", p. 26.
[130] Craig, "Cosmos and Creator", p. 26.
[131] Leslie, "Anthropic Principle, World Ensemble, Design." p. 150.

> You would immediately suspect that they missed you on
> purpose, by design. [132]

Now, the odds against a universe being conducive to life are much
much larger than the odds of one's surviving in the above parable.

However, most scientists believe that there exist ever so many
different universes, all with different physical properties; consequently,
it is possible for such a rarity to have happened. According to Craig:

> Proponents of the weak anthropic principle will contend
> that ... [o]ur surprise is justified ... only if the basic
> features of the observable universe are co-extensive
> with [identical to] the basic features of the Universe-
> as-a-Whole. Yet our observable universe ... is but one
> member of a collection of diverse universes ... including
> our observable universe ... within this wider Universe. [133]

It is worth repeating here, once again, science has traditionally always
assumed that we are not special in the universe: that the physical laws
we observe here on earth are the same all over the universe. So, why
not over the whole mother universe? Should science be allowed to
break its own rules and have it both ways? Would it remain science?
Or should it be called metaphysics or, even worse, pseudoscience?

Inflation

To circumvent the necessity of the combined fine-tuning of
the initial mass density and expansion rate resulting in the current
flatness of the universe, scientists proposed the idea of an extremely
rapid inflation in the early universe. (Recall that light travels in a
straight line in a *flat* space.) Craig explains:

---

[132]   Craig, "Cosmos and Creator", p. 27 (emphasis in original)
[133]   Craig, "Cosmos and Creator", p. 27.

Contemporary cosmologists have attempted to eliminate some of this fine-tuning via inflationary models of the early universe. On these models, around 10^{-35} second after the Big Bang, the universe inflated exponentially, pushing certain poorly-tuned features of the universe out beyond our event horizon. [134]

This expansion, supposedly, happened at a rate greater than the speed of light. In other words, since the expansion rate during this inflationary period exceeded the speed of light, a light ray from these locations can never reach us: so we can never tell what has happened over there—that's comfy—how can we prove it true or false?.

One might make a subtle distinction here: the general relativity theory precludes anything travelling at a speed greater than the speed of light *in* space; however, it does not preclude space *itself* expanding at a speed greater than the speed of light. Leslie explains:

> Inflation would have hurled them asunder faster than the speed of light. General relativity permits such velocities: when expansion of space produces them. [135]

According to this hypothesis, inflation of a single, initially separated region hurled its constituents apart much faster than the speed of light; after which, it settled down to its present rate of expansion. Analogous to a large inflated balloon, a small portion of such a balloon's surface looks flat; hence, this rapid expansion of the universe made space flat: in other words, light travels in a straight line in free space—that is, at significant distances away from matter. [136] Leslie continues to explain:

> Massive inflation could give an extremely flat space: a greatly inflated balloon has a very flat surface. [137]

---

[134] Craig, "Cosmos and Creator", p. 25.
[135] Leslie, "The Prerequisites of Life in Our Universe", p. 233.
[136] Leslie, "The Prerequisites of Life in Our Universe", p. 233.
[137] Leslie, "The Prerequisites of Life in Our Universe", p. 233.

However, this is all conjecture; there is no positive evidence for this rapid expansion having ever happened. On the contrary, evidence contradicts what inflationary models expect the universe to be like; the actual mass density of the universe is about a tenth of that predicted by this hypothesis. According to Craig:

> No positive observational evidence establishes that the universe underwent an inflationary phase. In fact, inflationary models predict a universe possessing critical mass density, whereas [actual] observation supports a value ten times lower than [a tenth of] that. [138]

This hypothesis, anyway, begs a lot of new questions. What causes this inflation to suddenly start happening? What makes it end without creating turbulence—that is, gracefully? How does it produce irregularities that are neither too small nor too large so as to allow galaxies, stars, and planets—rather than black holes—to grow later on in time? The answers to these questions reek of fine-tuning the inflationary period *itself*: the very problem inflation is supposed to render unnecessary! [139] Craig's conclusion is:

> [N]o inflationary model has yet succeeded in starting and stopping inflation so as to allow for galaxy formation. Most importantly, inflationary models require the same fine-tuning which some theorists had hoped to eliminate via such models. [140]

Quantum Vacuum

Quantum theory, also known as q*uantum physics* or *quantum mechanics*, is one of the fundamental branches of physics, which deals

---

138   Craig, "Cosmos and Creator", p. 25.
139   Leslie, "The Prerequisites of Life in Our Universe", p. 233.
140   Craig, "Cosmos and Creator", p. 26.

with physical phenomena at nanoscopic scales: that is, between one and one hundred nanometers (10^{-9} meter). It departs from *classical* mechanics primarily at the realm of atomic and subatomic length scales. [141]

For example, collisions of atoms and nuclei are *reversible*: in other words, particles are not damaged or deformed, like every-day objects, when they collide with one another. Otherwise, atoms would have got old and weary with use over the time of the universe's existence. So there is a dividing line, in the reality of our universe, where things change considerably from our everyday observances. Also, strangely enough, atomic size particles sometimes exhibit wave-like properties: that is, they behave like waves rather than like particles; and vice-versa waves occasionally behave like particles.

> Quantum physics also provides a mathematical description
> of much of the dual particle-like and wave-like behavior
> and interactions of matter and energy. [142]

Therefore, in considering what happened at the very early moments of the universe—at the Planck time: that is, 10^{-43} of a second after the Big Bang; it stands to reason, that quantum theory, rather than classical theories, needs to be invoked. Hypotheses that rely on basic principles of quantum theory normally invoke the concept of the *quantum vacuum fluctuations* of space. They also incorporate the *inflation* hypothesis from a *multiverse* scenario—both of these concepts have already been discussed above.

In his article "Cosmos and Creator", Craig gives us the following background information relating to such models of the universe. These models propose that our whole universe emerged spontaneously from a *quantum vacuum*; just as sub-atomic particles are *actually* known to emerge from such a vacuum: this phenomenon is technically known as a *fluctuation* of this vacuum. According to inflationary models, the Big Bang was immediately followed by a rapid expansion of

---

[141]   https://en.wikipedia.org/wiki/Quantum_mechanics (22 August 2015)

[142]   http://www.crystalinks.com/quantumechanics.html (22 August 2015)

the universe's space, at a rate greater than the speed of light. Prior to that, the mother universe was just a quantum vacuum that is *not* expanding; randomly, throughout this mother universe, particles materialize spontaneously through "sub-atomic energy fluctuations" as described above. These material particles, somehow, eventually grow into mini-universes like ours; our universe started when matter emerged from the energy in this quantum vacuum. The universe we inhabit is, supposedly, only a small portion of a much larger mother universe; it is only our observable universe and all these other mini-universes that are expanding, not the mother universe.

The catch, here, is that a quantum vacuum is not exactly what we, ordinary folk, understand by empty space: it is not just void space, but it contains a myriad of fleeting electromagnetic waves. From the above, it is obvious that a quantum vacuum is definitely *not* nothingness; if our universe actually started as a quantum fluctuation of space in the mother universe, one simply cannot really say that our universe came out of nothingness spontaneously: it would amount to a play on words. [143] As we have seen, in the first chapter, both matter and energy cannot be created out of nothingness.

So, according to the above hypotheses, mainstream science seems to be telling us that all the matter in the universe came to exist, spontaneously, from nothingness: from a vacuum. However, take a look at the definition of a quantum mechanical vacuum, given by Craig in the same article:

> A quantum mechanical vacuum spawning material particles is far from our ordinary idea of a "vacuum" (meaning nothing). Rather, a quantum vacuum is a sea of continually forming and dissolving particles, which borrow energy from the vacuum for their brief existence. This is not "nothing," and hence, material particles do not come into being out of nothing. Popular presentations of these models often do not explain that they require

---

[143]   Craig, "Cosmos and Creator", p. 21.

a specially fine-tuned, background space-time on the analogy of a quantum mechanical vacuum. [144]

According to quantum physics, the *vacuum state* is not truly empty: that is, space void of anything inside; it contains fleeting electromagnetic waves and particles that pop into and out of existence. In other words, the language used is *misleading*. The energy in the vacuum is not nothingness; it is a physical entity, which, as we have seen in the first chapter, cannot be created (from nothing) or destroyed (into nothingness): it can only change form (quality).

Here's an analogy. Imagine a heavy object "floating" in air on earth and a similar object "floating" in free space: both of them are unrestrained and free to move as they will. In the first case, the object would fall to the ground because of the gravitational force-field (attraction) close to the earth's surface; in the second case, it would not move at all (or maybe drift just a little) because there is no gravitational force-field (or hardly any) around it. In the first case, the object can gain kinetic (motion) energy at the expense of some of its potential (positional) energy. Conceivably, if the object falling under gravity accelerates enough, its kinetic (speed) energy could then be converted into another (much smaller) object: according to Einstein's equation $E=mc^2$, where "E" is the kinetic energy gained by the accelerated object and "m" is the mass of the newly created object. It *seems*, therefore, that we were able to create matter out of nothingness, in this case.

The space close to any planet only *looks* empty: there is, in fact, a gravitational field around it; if you place an object in this seemingly empty space, it will accelerate to the ground of that planet. This background gravitational field is not nothingness; the mass of the planet is in the background, and it is a necessary condition for such an energy field to exist: creating a whole planet is no small feat. Both the above scenarios can be described as a "free floating object", but they are not physically identical conditions. Similarly, the use of the

---

[144] Craig, "Cosmos and Creator", p. 21.

same word "vacuum" by scientists is confusing and unfair to use with ordinary people.

The "self-reproducing inflationary universe" model owes its origin to theoretical physicist Andre Linde; this model is also built on the quantum vacuum fluctuation phenomenon, which, as we have seen, is based on advanced principles of quantum physics. In his book *The Case for a Creator*, journalist Lee Strobel explains it as follows:

> Linde postulates a preexisting superspace that is rapidly expanding. A small part of this superspace is blown up by a theoretical *inflation* field, sort of like soap bubbles forming in an infinite ocean full of dish detergent. Each bubble becomes a new universe. In what's known as "chaotic inflation theory," a huge number of such universes are randomly birthed, thanks to quantum fluctuations, along various points of superspace. Thus each universe has a beginning and is finite in size, while the much larger superspace is infinite in size and endures forever. [145]

The major difference from other inflationary models is that it postulates an expanding mother universe.

Even if this were the case, mathematician, physicist, and philosopher Robin Collins thinks that it would only push the idea of fine-tuning to a higher level. It would necessitate something like a "many universes generator" that gives rise to these universes; one that possesses the following properties: (1) The hypothesized inflation field acts as a reservoir of unlimited energy creating the bubble universes. (2) Einstein's equation of general relativity acts as a mechanism to form these bubbles; it has to work harmoniously with the inflation field. [146] According to Collins:

> [W]ithout Einstein's equation and the inflation field working together harmoniously, it wouldn't work. If the

---

[145] Strobel, *The Case for a Creator*, p. 141 (emphasis in original)
[146] Strobel, *The Case for a Creator*, p. 142

universe obeyed Newton's theory of gravity instead of Einstein's, it wouldn't work. [147]

(3) A mechanism is required to convert the inflation field to the matter and energy we are familiar with in our universe. (4) Most importantly, another mechanism is required to produce enough variation of the physical constants—a random generator—so that eventually a life-sustaining (fine-tuned) universe occurs by chance. (5) The right background laws of physics also need to be in place. [148] According to Collins:

> For instance, without the so-called principle of quantization, all of the electrons in an atom would be sucked into the nuclei. That would make atoms impossible. Further, an eminent ... physicist Freeman Dyson has noted, without the Pauli-exclusion principle [explained presently] electrons would occupy the lowest orbit around the nucleus, and that would make complex atoms impossible. Finally, without a universally attractive force between masses—such as gravity—stars and planets couldn't form." [149]

Pauli's Exclusion Principle states that no two electrons in an atom can occupy completely identical *quantum states*; at least one of the four quantum states an electron in an atom can have must be different. The four possible quantum states are: *principal* (orbit size), *azimuthal* (orbital shape), *magnetic* (orbital orientation) and *spin*. *Orbitals* are similar to orbits but have a more complex shape: something like joined three-dimensional elongated leafs. This *quantization* of all electrons in an atom keeps them orderly—sort of like planets in a solar system—rather than their being all sucked into the nucleus by electrostatic attraction. Recall that the nuclei of atoms contain protons, which are positive, while electrons are negative; and also that unlike charges attract each other.

---

[147]   Strobel, *The Case for a Creator*, p. 143

[148]   Strobel, *The Case for a Creator*, p. 143

[149]   Strobel, *The Case for a Creator*, pp. 143–4

Mirror Universe

In his book *God and the Folly of Faith*, particle physicist and self-declared atheist Victor Stenger tries to explain the origin of our universe. Stenger congenially admits:

> Then where did the universe come from if it was not divinely created? The truth is that we do not know. [150]

He proposes the existence of an earlier mirror universe, identical to ours, at time zero in an eternal multiverse. The arrow of time, and hence the direction of increasing entropy, in this mirror universe flows in the opposite direction. Believe it or not, such a universe is not forbidden by any known principles of physics; moreover, it has the added feature that, supposedly, it has the same cosmological equations that we use for our universe. According to these equations, this earlier universe would have contracted down until it becomes as small as it can possibly get according to quantum physics: a sphere of the Planck length, that is, 1.62×10^{-35} meter. Through a process known as q*uantum tunneling*, which is routinely observed in nuclear fusion, radioactive decay, and solid state physics, our universe managed to overcome a barrier that is normally insurmountable in classical physics. That is, our universe could have started as a spec doomed to collapse to a singularity (*infinite* density) but instead it "tunneled" through the energy barrier to a larger radius, thus initiating inflation. The probability of this happening may have been infinitesimally small; but it is mathematically possible for it to have happened—as it actually does happen in the physical phenomena mentioned above in this paragraph. [151]

No doubt, this is some serious scientific thinking by a particle physicist who is honestly trying to find the truth about the origin of our universe. I must admit that his hypothesis is, at least, based on good physics, and there is also some evidence of its possibility of

---

[150]   Stenger, *God and the Folly of Faith*, p. 186.

[151]   Stenger, *God and the Folly of Faith*, pp. 188–9.

happening. What I definitely don't buy is the multiverse portion with its zillion varieties of physical laws and constants. Moreover, I'm not too keen on accepting the concept of matter or energy existing eternally. There is a shred of evidence supporting this hypothesis, but not enough for me to start thinking about it seriously.

CHAPTER 6

ORIGIN OF LIFE

Let us now change gear and consider the obvious evidence of an abundance of life on earth. There is no question that life on earth started quite some time after the universe was launched with the Big Bang; after it had cooled down quite a bit, of course, and the earth was formed. An obvious scientific question is therefore: how did life on earth start?

Chance

In real life, there are certain things that do happen as a result of chance alone, and, admittedly, it is sometimes difficult to determine the demarcation line where one can reasonably expect chance to produce an observed effect. Many people actually think, for example, that *anything* can happen by chance in a very long time like the age of the universe—that is, about 14 billion years; but science does not think so! Let me elaborate.

In a lottery, for example, one needs to know the odds of winning the jackpot; so that someone may actually win it, occasionally, but not allowing too many people to win it simultaneously: resulting in it's being a small and unattractive jackpot. There is a section of mathematics, called *probability*, which deals with the odds of something happening by chance alone and roughly determine a demarcation line. We have already had a primer on this subject; we shall now delve a little deeper into it.

Let me try to illustrate things by giving a numerical example. Imagine a combination lock, the type used to prevent bicycles from being stolen, each having three registers of ten numbers: zero to nine (0 to 9). The number of possible combinations of the lock is determined as follows. There are ten possibilities of selecting a number in the first register. Having chosen any number in the first register one is completely free to select any of the ten numbers in the second register. So, the number of ways of selecting the two numbers in the first two registers is one hundred ($10 \times 10 = 100 = 10^2$). Having chosen the two numbers in the first two registers, one is still completely free to select any number in the third (last) register, making the total number of possibilities one thousand ($100 \times 10 = 1000 = 10^3$). In fact this can be easily visualized by imaging the three registers being set in succession to any of the numbers from: 000, 001, 002, 003, and so on, and ending in 997, 998, 999, that is, one thousand ($1000 = 10^3$) combinations.

Now, suppose a thief sets the three registers at random. Is it *possible* that he can hit the right combination the first time? Yes. Is it *probable*? No. Is it *reasonable* to suppose that such a coincidence should happen by chance? No. Science follows the attitude described in the last set of question and answer: if the answer is "no" it has to look elsewhere for an explanation of the phenomenon in question. That is *good* science!

Now, suppose the thief tries more and more different combinations, at *random* (allowing repetitions); of course, he has a better chance now that he has more opportunities. When does science say that it is *probable* that he will hit the right combination? It is around the fifty percent (50%) mark of the total number (1000) of combinations: that is, when he has tried five hundred (500), or so, random combinations. If he tries more than 500 combinations, the odds are in his favor; if he tries less than 500 combinations, the odds are against him. I think everyone will agree that this criterion is *reasonable*.

This does not mean, however, that if he tries four hundred (40%) combinations instead, or even one hundred (10%) combinations, he cannot hit the right combination. But when it comes down to just

10 combinations, that is, one percent (1%) of the total number of possibilities, it starts to become unreasonable to expect chance to produce the observed effect; in such a case, therefore, one should look *elsewhere* for an explanation of the facts. To keep claiming that it happened by a fluke reveals scientific ignorance: comparable to claiming a miracle. Anything can be explained away this way; but it is not a satisfactory scientific explanation.

Now, a little bit more on calculating probabilities. If the above lock had 4 registers, instead of 3, then the number of possible combinations would be ten thousand $(10,000=10^4)$; with 10 registers the number of possible combinations would be ten billion (10^{10}); obviously, the more registers the harder it would be to break the combination. If, on the other hand, the 3 registers consisted of the 26 letters of the alphabet—instead of the 10 numbers: 0 to 9—the number of possible combinations would be 26^3 (or 17,576), with 4 such registers 26^4 (or 456,976) and with 10 such registers 26^{10} (about 1.4×10^{14}).

Proteins

Nowadays, almost everybody is aware of the importance of proteins in living organisms. Proteins are responsible for nearly every task of a living cell. They manufacture building blocks for structural elements such as are required to maintain cell shape and connective tissue like muscle, cartilage, and bone. They are also responsible for inner organization, extra cellular monitoring and communication, waste disposal and routine maintenance. Other proteins act as enzymes catalyzing (explained presently) biochemical reactions, speeding particular reactions up perhaps a trillion-fold. [152] In his book *Signature in the Cell*, science philosopher and geophysicist Stephen Meyer describes the main functions of proteins in the living cell. (I shall be using concepts and ideas from Meyer's book *Signature in the Cell* extensively in this chapter.)

---

[152]   http://www.nature.com/scitable/topicpage/protein-function-14123348 (01 January 2016)

> [P]rotein molecules perform most of the critical functions in the [living] cell. Proteins build cellular machines and structures, they carry and deliver cellular materials, and they catalyze [see below] chemical reactions that the cell needs to stay alive. Proteins also process genetic information. To accomplish this critical work, a typical cell uses thousands of different kinds of proteins [20,000 to 25,000 in humans]. And each protein has a distinctive shape related to its function, just as the different tools in a carpenter's toolbox have different shapes related to their functions. [153]

A *catalyst* is a substance that starts or speeds up a chemical reaction while undergoing no permanent change itself: that is, it does not, actually, take part in the chemical reaction. It only, sort of, helps bring or hold the chemicals together so that the reaction in question can take place through its assistance. An *enzyme* is a protein that is a biological catalyst. The basic function of an enzyme is to increase the rate of a reaction. Most cellular reactions occur a million times faster than they would in the absence of an enzyme. Most enzymes act specifically with only one type of starting molecule to produce a different molecule.

Later on in this chapter, we shall see how proteins are generated by the genetic code (*DNA genes*) in a living cell. For the time being, let us just examine some of the properties of proteins.

Proteins consist of long chains of *amino acids*; the minimum number of amino acids in a protein is about one hundred (100), and the average size of a protein is about three hundred (300) amino acids, but there are proteins that consist of thousands of amino acids.

There are about five hundred (500) known amino acids; however, only twenty (20) different amino acids occur most commonly in nature. For simplicity we shall only consider these 20 amino acids. Amino acids are biologically important organic compounds composed of amine (NH_2) and carboxylic acid (CO_2H) groups. In

---

[153] Meyer, *Signature in the Cell*, p. 92.

forming proteins the amine (NH_2) side of one amino acid molecule reacts with the carboxylic acid (CO_2H) side of the adjacent amino acid molecule forming what is called a *peptide bond* according to the chemical reaction: $NH_2 + CO_2H = CONH + H_2O$ (water).

The function of proteins depends on their three dimensional configuration—their folds. The various folds are produced by an *exact sequence* of amino acids: similar to the combination lock illustrated above. In some cases changing a single amino acid destroys, or compromises, the function of a protein; in a lot of cases, a few substitutions are tolerated.

Miller-Urey Experiment

In December of 1952, assuming the early earth's atmosphere consisted of a gaseous mixture of methane (CH_4), ammonia (NH_3), water (H_2O), and hydrogen (H_2), chemist Stanley Miller conducted the following experiment to produce amino acids, the basic building blocks of proteins, from these simpler gases. He introduced some water into a glass vessel and filled the rest with a mixture of these gases. He then boiled the water; the evaporating steam created a circulating flow of these gases. The vessel also contained a discharge chamber, which consisted of a pair of tungsten electrodes (wires) fused into the glass. Miller then applied a high enough voltage across the two electrodes so that it discharged (sparked) through the gases inside the vessel; [154] its purpose was to simulate the effect of lightning amidst the atmospheric gases he presumed present on the early earth. A couple of days later he collected several types of amino acids from a U-shaped water-trap at the bottom of the vessel. [155]

---

[154] Miller, "A Production of Amino Acids under Possible Primitive Earth Conditions".

[155] Meyer, *Signature in the Cell*, pp. 56–7.

Probability of Chance-Produced Functional Protein

What is usually grossly unappreciated is the extreme difficulty of finding even a single *functional* protein: the number of amino acid sequences possible for blind chance to choose from is enormous. (The following information is mostly taken from Meyer's book *Signature in the Cell*.)

Although the average size of a protein is about three hundred (300) amino acids, a modest protein of only one hundred and fifty (150) amino acids is considered in the following analysis. Why? There is a significant amount of research information available for this size of protein.

While forming a protein chain, an amino acid joining with another amino acid must do so *exclusively* through a peptide bond (see the above chemical reaction). It so happens that amino acids can also form many other types of chemical bonds with other amino acids. When normal amino acid mixtures are allowed to react freely in a test tube, they form both peptide and non-peptide bonds roughly half the time: that is, the probability of forming a peptide bond is roughly the same as forming a non-peptide bond. Thus every time an amino acid is attached to a protein chain, the probability of its forming another peptide bond is roughly fifty percent (50%), or one-half ($\frac{1}{2}$). It follows that the probability of two amino acids being joined by a peptide bond is roughly $\frac{1}{2}=(\frac{1}{2})^1$; the probability of three amino acids being joined exclusively by peptide bonds is about $\frac{1}{2}x\frac{1}{2}=(\frac{1}{2})^2$; four amino acids: $\frac{1}{2}x\frac{1}{2}x\frac{1}{2}=(\frac{1}{2})^3$; five amino acids: $(\frac{1}{2})^4$; and so on. Thus, from a general population of amino acids, the probability of randomly building a chain of 150 amino acids in which all the bonds are peptide bonds will be about $(\frac{1}{2})^{149}$: in other words, the odds are roughly 1 chance against $7.1x10^{44}$ possibilities. [156]

Now, nineteen of the twenty amino acids occurring naturally in proteins have a mirror-image of themselves: there is a left-handed version, or L-form, and a right-handed version, or D-form.

---

[156] Meyer, *Signature in the Cell*, p. 206.

(Incidentally, mirror-image versions are called *optical isomers*.) Biologically functional proteins can tolerate only left-handed (L-form) isomers; however, in a general amino acid population both the D-form and L-form isomers are produced in roughly equal quantities. This exclusive L-form requirement makes it even harder to build a biologically functional protein by chance alone. A similar calculation to the above yields the probability of attaining, randomly, only L-form amino acids in a chain 150 amino acids long; it is $(\frac{1}{2})^{150}$: or roughly 1 chance in about 1.4×10^{45}. Starting from a normal mixture of both D-form and L-form amino acids, the probability of building a 150 amino acid chain, randomly, in which all bonds are peptide bonds and all amino acids are L-form is, therefore, roughly 1 chance in 10^{90} ($=7.1 \times 10^{44} \times 1.4 \times 10^{45}$). [157]

Biologically functional proteins have a third independent requirement, which is definitely the most important. Their amino acid sequences, like the letters in the words of a meaningful paragraph, must link up in functionally specified sequential arrangements. In some cases, even changing one amino acid at a given site results in the loss of protein function. Since there are 20 naturally occurring amino acids, the probability of getting a specific amino acid at a given site is very small: only 1/20. (Actually the probability is even lower than this because there are many more naturally occurring non-protein-forming amino acids: about 500 of them, but 20 of them are the most commonly found). Assuming, for the time being, that each site in a protein chain requires a particular amino acid, the probability of constructing a particular protein 150 amino acids long would be $(1/20)^{150}$: that is, roughly 1 chance in about 1.4×10^{195}. [158]

However, it so happens that most sites along the protein chain can tolerate several of the twenty different amino acids found in proteins without destroying or compromising the function of the protein. [159]

In 1992, molecular biologist Douglas Axe realized that the most important feature in the structure of proteins is the intricate folds

---

[157] Meyer, *Signature in the Cell*, pp. 206–7.

[158] Meyer, *Signature in the Cell*, p. 207.

[159] Meyer, *Signature in the Cell*, p. 207.

they need in order to perform their function. He realized that what he needed to investigate was how rare are the amino acid sequences that produce stable folds enabling proteins to perform their biological functions. Without producing stable folds, chances are a protein cannot perform any useful biological function. [160]

In 2004, Axe performed experiments with the purpose of estimating the frequency of producing sequences of a 150 amino acid protein with stable folds. On the basis of his experimental results, he calculated the ratio (proportion) of the number of 150 amino acid sequences capable of folding into stable structures to the total number of possible amino acid sequences of the same length. He estimated that ratio to be about 1 to 10^{74}. [161] [162]

To be fair, therefore, instead of using the above figure of 1.4×10^{195}, we must use Axe's figure of 10^{74}. So the probability of a functional protein occurring at random is roughly 1 chance in 10^{164} ($=7.1 \times 10^{44} \times 1.4 \times 10^{45} \times 10^{74} = 10^{1+44+45+74}$). This is an extremely low probability indeed. [163] However, because the universe has existed for about 14 billion years, ordinary folk may think that anything can happen in such a length of time; at least once. But this is actually not the case. Let us look a little deeper into this and see how *scientifically* probable it really is.

Universe's Total Probabilistic Resources

Let us retrace our steps a little and reconsider the scenario of a thief trying to steal a bicycle by guessing the combination of a three-register lock. As we have seen the odds of his guessing it the first time are one to one thousand ($1000=10^3$): alternatively we can say that the probability is one-thousandth ($1/1000=10^{-3}$); which

[160] Meyer, *Signature in the Cell*, p. 209.
[161] Axe, "Estimating the Prevalence of Protein Sequences Adopting Functional Folds".
[162] Meyer, *Signature in the Cell*, p. 210.
[163] Meyer, *Signature in the Cell*, p. 212.

is a very small probability, so he will probably be unsuccessful. (Incidentally, probability is always less than or equal to one (1). When the probability of something happening is equal to one (1), it means that it will definitely happen—for sure. If, on the other hand, the probability is zero (0), it means that it will never happen.) However, if the thief has time and opportunity, he can try other combinations, over and over again: thus increasing the probability of his breaking the combination. As we have seen, science accepts as reasonable a probability of something happening by chance when it exceeds the half way mark: that is, ½ or 50%. (Probabilities can also be expressed as a percentage—which is always less than 100%). This does not mean that something cannot happen by a fluke; but if we assume this, it is unscientific because then we can explain everything this way: we wouldn't need any science. Science does not accept very low probabilities as a reasonable cause because it indicates *ignorance* in explaining the phenomenon observed; it tries to find tangible evidence and give probable answers to questions.

There is one additional minor point I have to make here. Chance is not intelligent like a thief might be: in fact we call it *dumb* luck. If a thief tries to break a combination lock by trial and error, he will probably try not to repeat combinations he has already tried. But it would be too time-consuming to write down the combinations he already tried, and refer to the list every time he tries a new combination. So what he probably ends up doing, assuming he has time, is to try combinations successively: 000, 001, 002, 003, etc. This is not dumb luck. We must imagine the thief to be ignorant and forgetful in order to properly simulate things happening by chance in nature.

Let us now try to calculate the *probabilistic resources* of the whole universe. There are about 10^{80} elementary particles—protons, neutrons, and electrons—in the observable universe; all of these have the opportunity of interacting with one another. Notice that I qualified the universe by the word "observable". For any of these particles to be able to interact with any other, the two particles cannot be too far away from each other because nothing travels faster than the speed of light in a vacuum ($3x10^8$m/s) in our universe. Since the universe has only existed for about 14 billion years, anything further away from the earth than light can travel in 14 billion years could never have interacted with the particles on earth, so they could not have been involved in the origin of life on earth. This maximum distance can be calculated by multiplying the speed of light by 14 billion years. Fourteen billion years turns out to be about $4.4x10^{17}$ seconds (which I shall round off to $5x10^{17}$ seconds to simplify our calculations). This is the absolute time limit for anything to have happened in the universe: including earth itself.

Now, according to quantum physics, space—just like matter—cannot be split into smaller and smaller divisions (lengths) indefinitely: the smallest distance space can be divided into is $1.6x10^{-35}$ meters, which is known as the *Planck length*. It follows that interactions between particles cannot take place in less time than it takes light to travel this distance: that is, this distance divided by the speed of light. So, elementary particles must take at least $5.4x10^{-44}$ ($=1.6x10^{-35}/3x10^8$) of a second to be able to interact with one another—this is known as the *Planck time*. Consequently, elementary particles can only interact about $1.9x10^{43}$ ($=1/5.4x10^{-44}$) times per second (which I shall round off to $2x10^{43}$ times per second to simplify our calculations).

Thus, as mathematician William Dembski shows in his book *The Design Inference*,[164] the total probabilistic resources of the observable universe, since the beginning of the universe, add up to 10^{141} ($=10^{80}x5x10^{17}x2x10^{43}$) opportunities. [165]

As we saw above, the odds of producing a functional protein consisting of only one hundred and fifty (150) amino acids by chance alone is about1 in 10^{164}. If used for the sole purpose of producing just one functional protein, the total probabilistic resources of the observable universe are only a very small fraction of this. Thus the probability of chance alone producing a protein of one hundred and fifty (150) amino acids is 10^{-23} ($=10^{141}/10^{164}$); that is, it has only

---

[164]   Dembski, *The Design Inference*, chap. 6.
[165]   Meyer, *Signature in the Cell*, pp. 216–7.

about 1 chance in 10^{23}—one chance in about one hundred billion trillion. This is an infinitesimally small probability compared to the scientific 50% requirement; even when compared to 10%, 1%, or 0.1%—1 chance in a thousand—discussed at the beginning of this chapter. This is like being confident of picking a single white ball out of a sea of one hundred billion trillion (10^{23}) black balls in a *single* opportunity. If the reader is confident that this is possible, then the reader can continue believing that anything can happen in fourteen billion years. In conclusion, I think the reader will agree, by now, that chance alone cannot explain the emergence of just *one* modest 150-amino-acid protein; so science requires that we look elsewhere for an explanation for the existence of functional proteins on earth.

Incidentally, since the absolute limit of the whole universe, from its very beginning, for making anything happen on earth is about 10^{141} opportunities, this translates to about 470 bits ("0" or "1" in binary or computer language: $2^{470}=3.0 \times 10^{141}$) of information as an upper limit of what the universe can or could have possibly produced by chance alone. That's 470 flips of a fair coin always getting heads, say, never getting tails in between; or, equivalently, a sequence of 470 previously specified flips. [166] It is ludicrous to shoot an arrow, at random, onto the trees in a forest and draw a circle around the arrow on the tree it happens to hit, pretending that it was the target: one has to specify the target *prior* to shooting the arrow

I would like to clarify concepts here. Keep in mind that in our present calculation only the Planck time (5.4×10^{-44}s) is taken for every flip of the coin, and that 10^{80}—the number of elementary particles— flips are occurring simultaneously in this time; this is equivalent to about 1.9×10^{123} flips per second. This result should, therefore, not be confused with someone's flipping a coin at *one* flip per second (as we were comparing to in previous chapters).

---

[166] Meyer, *Signature in the Cell*, p. 294.

Miller-Urey Experiment Revisited

Let us now have a second look at the famous Miller-Urey experiment in which chemist Stanley Miller was able to produce a variety of amino acids: molecules that constitute the building blocks of all life forms. As we have seen, he assumed that the earth's early atmosphere was composed of *reducing* gases, that is, gases rich in in hydrogen, namely, hydrogen (H_2), methane (CH_4), and ammonia (NH_3). He also assumed that hardly any free oxygen was present. [167] [168]

As it turned out, geochemical evidence, discovered some time after his experiment, proved him completely wrong. This new evidence strongly suggested that neutral (non-reducing) gases—such as, carbon dioxide (CO_2), nitrogen (N_2), and water vapor (H_2O)— dominated the early earth's atmosphere. [169] Moreover, several geochemical studies concluded that a significant amount of free oxygen was present even before plant life started—probably the bi-product of water-vapor (H_2O) being broken down by photons. [170] [171]

Now, amino acids form easily in a mixture of reducing (hydrogen-rich) gases. On the other hand, similar reactions do not occur so easily among mixtures of neutral (non-reducing) gases; if they do occur, their amino acid yields are very poor. [172] Even minute quantities of free oxygen among these gases will immediately stop the production of amino acids; not only that, but any amino acids already present will degrade in no time at all. [173]

In the opinion of two leading geochemists, James Brooks and Gordon Shaw, assuming that one, or more, of the oceans was rich in amino acids (which are found in proteins) and nucleic acids (which

---

[167] Miller, "A Production of Amino Acids under Possible Primitive Earth Conditions".

[168] Meyer, *Signature in the Cell*, p. 224.

[169] Abelson, "Chemical Events on the Primitive Earth"

[170] Towe, "Environmental Oxygen Conditions During the Origin and Early Evolution of Life".

[171] Meyer, *Signature in the Cell*, p. 224.

[172] Holland, *The Chemical Evolution of the Atmosphere and Oceans*, pp. 99–100

[173] Thaxton, Bradley, & Olsen, *The Mystery of Life's Origin*, pp. 69–98

are found in DNA) it would have left large deposits of nitrogen-rich solid fuels—called cokes, like coal—in sedimentary rocks. Cokes are *metamorphosed* rocks: they are formed by material transported, deposited and solidified by heat, pressure, and chemical activity without melting in the process. Since no such deposits exist anywhere on earth, Brooks concludes that [174] "… if such a soup ever existed it was for only a brief period of time." [175] Of course, this is very detrimental to Darwin's theory of evolution; long periods of elapsed time are of its very essence. Consequently, recent geological and geochemical evidence points to a prebiotic atmosphere that was hostile, rather than friendly, to the production of amino acids and other building blocks of life; besides, more than likely, there was never a prebiotic soup that was conducive to the formation of organic molecules on earth. [176]

Moreover, most researchers of the origin of life agree that even assuming that there existed a prebiotic soup, there were also many other destructive chemicals present at the same time: making their continued thriving for the eventual emergence of life extremely questionable. [177]

Without exception, chemical experiments similar to that originally performed by Miller and Urey to synthesize amino acids, have produced both biological building blocks—such as amino acids—and non-biological molecules mixed together. If scientists do not intervene at all, both types of molecules normally react with each other to form a chemically irreversible and meaningless sludge. Consequently, experimenters invariably filter (remove) those chemicals [178] that degrade amino acids and transform them into slimy goo. Furthermore, instead of using indiscriminate ultra-violet "light" as is naturally and realistically found in lightning—that is, containing

---

[174] Meyer, *Signature in the Cell*, p. 225.
[175] Brooks, *The Origins of Life*, p. 118
[176] Meyer, *Signature in the Cell*, p. 225.
[177] Schwartz, "Intractable Mixtures and the Origin of Life", p. 656.
[178] Kenyon, Forward in Thaxton, Bradley, & Olsen, *The Mystery of Life's Origin*, p. vi.

both short and long wavelengths of the ultra-violet region—they only use short-wavelength ultra-violet. Guess why! Because long-wavelength ultra-violet degrades amino acids rapidly. [179] [180]

So why do high school biology text books keep misleading students by depicting the Miller-Urey experiment as a highly probable origin of a prebiotic soup, leading to an origin-of-life scenario? It was all right, maybe, for the mid-twentieth-century when we did not know any better, and the living cell was thought to be a slimy blob of chemicals; but not for today when we know of the complexity of the genetic code and of the living cell. In his class biology textbook entitled *Biology*, biologist and teacher Matthew Distefano concludes a section headed "How did life start?" as follows:

> Today, it is believed that the chemicals used by Miller and Urey were probably not the ones present in the Earth's early atmosphere; but similar experiments with new chemicals also produced organic compounds. [181]

I wonder whether the reader thinks this is a fair statement after reading this section.

Probability of Chance-Produced Living Cell

Several theoretical and experimental studies have determined that the minimal set of proteins that are necessary and sufficient to sustain a functional cell under ideal conditions is about two hundred and fifty (250). Ideal conditions mean in the presence of unlimited amounts of essential nutrients and in the absence of any adverse factors: like predators and harmful chemicals, including competition.

Assuming the minimum size of a cell to be two hundred and fifty (250) proteins and every protein is of modest length—one

---

[179]   Thaxton, & Bradley, "Information and the Origin of Life", p. 184; and Thaxton, Bradley, & Olsen, *The Mystery of Life's Origin*, pp. 100–1

[180]   Meyer, *Signature in the Cell*, p. 226.

[181]   Distefano, *Biology*, p. 150.

hundred and fifty (150) amino acids long, say— then the odds against building a minimally complex functionally independent cell by chance (alone) is determined as follows. Recall that the odds against chance producing any one of these proteins is 1 to 10^{164}; so the odds against chance producing a minimally complex cell is equal to this number multiplied by itself two hundred and fifty (250) times: that is, $(10^{164})^{250}=10^{164\times250}=10^{41000}$. [182]

This is an unimaginably small probability, practically zero, which means it will never happen: not in the time span of very many universes like ours, anyway. The reader may recall how impossible it is to select, blindfolded, a single white marble from a large pool of a trillion black marbles. A trillion has only 12 zeros; this number has forty-one thousand (41,000) zeros. Recall, also, that every time a zero is added to the odds, the difficulty increases ten times what it was before. It is impossible to fully appreciate the infinitesimally small odds of the emergence of the simplest living cell by chance; yet, here we are!

If we apply all the probabilistic resources of the observable universe since the Big Bang to the sole purpose of producing a minimally complex living cell of only 250 proteins each only 150 amino acids long, the odds of actually ending up with such a cell are still infinitesimally small. Dividing the probability of building a minimally sized cell (10^{41000}) by the universe's total probabilistic resources (10^{141})—calculated above—would yield 1 chance in about 10^{40859} (= $10^{41000-141}$).

These astronomically high adverse odds are too daunting and undeniable even to atheistic scientists. Some scientists try to increase the probabilistic resources by proposing a multiverse and the anthropic principle (see the above corresponding paragraphs); as I already hinted there, notice that not even 10^{500} universes can cut it: far from it, we would require about 10^{40359} (=$10^{40859-500}$) such multiverses to produce just one simple living cell by chance alone.

---

[182] Meyer, *Signature in the Cell*, p. 213.

Other scientists admit that chance alone could not have achieved such a feat. One of the most famous evolutionary biologists and self-declared atheist Richard Dawkins graciously admits in his book *The God Delusion*:

> No indeed, chance is not the likely designer. That is one thing on which we can all agree. The statistical improbability of phenomena ... is the central problem that any theory of life must solve. ... But the candidate solutions to the riddle of improbability are not, as is falsely implied, design and chance. They are design and natural selection. Chance is not a solution, given the high levels of improbability we see in living organisms, and no sane biologist ever suggested that it was. [183]

I have to disagree with Dawkins' very last clause, but anyway, let's take only the first part of the quote, for the time being: what we agree on. So Dawkins basically says that we can *all* agree on the improbability of life arising by chance alone: even allowing it 14 billion years—the approximate time our universe has existed—to happen.

One can throw in the multiverse concept with that too, if one likes: it won't make much, if any, difference in the final probability—mathematically speaking. It is practically zero anyway, meaning it will never happen; so science must look elsewhere. Dawkins suggests *natural selection*. Besides chance and intelligent design there is Charles Darwin's concept of natural selection, which, as we shall see, can *mimic* design.

Apparent Design

What Dawkins is saying above means that what we normally call design is only apparent: it is not what is really happening; here's an example to show the concept.

---

[183] Dawkins, *The God Delusion*, p. 145

Imagine someone who owns a number of horses desiring to take part in the local horseracing. Realizing that his horses are not fast enough, he starts to breed horses for racing purposes. He does this by selecting the fastest horse and mare and allowing them to mate. He allows their fast offspring to mate only with other fast offspring, and he does not allow the slower horses to mate at all. He carries this on for a few horse generations; because of inheritance of traits, all his horses suddenly become fast horses in a relatively short time.

Now, on the other hand, imagine a herd of horses in the wild where there is a relatively high proportion of predators. Only the fast horses manage to escape from the predators: all the slower horses are killed. The result is that eventually the herd will similarly consist of only fast horses in a relatively short time.

Notice that the end result of the natural selection that occurred in the wild, without any prior planning, was the same as the designed plan of the breeder. Thus there are circumstances where natural selection mimics design. So, we have to be very careful not to jump to conclusions too fast when we see evidence of design: it may only appear designed.

Random variation and *natural selection* are thought to be the mechanisms of the *Theory of Evolution*; as a result of these two processes working together, organisms can improve themselves and become more adapted to their environment.

Evolution Question

Could life have *started* by evolution?

Before we proceed any further, let us first see what the theory of evolution says. In the final chapter of his most famous book *The Origin of Species*, naturalist and geologist Charles Darwin summarized the laws governing the theory of evolution as follows:

> Growth with Reproduction, Inheritance which is almost
> implied by reproduction, Variability from the indirect and
> direct action of the conditions of life, and from use and

> disuse; a Ratio of Increase so high as to lead to a Struggle for Life, and as a consequence to Natural Selection, entailing Divergence of Character and the Extinction of less improved forms. [184]

What this means is that all living organisms, by their very nature, consistently vary slightly. These slight variations are preserved, by inheritance, if they are conducive to an advantage in the surrounding conditions of life. Because of profuse birth rates—which is normally the case in every species—overpopulation of a species ordinarily ensues. A struggle for life follows among the varieties of a species; the fittest varieties survive while the weak become extinct. (To avoid confusion, the phrase *natural selection* is synonymous to *survival of the fittest*.)

Darwin concludes *The Origin of Species* as follows:

> There is grandeur in this view of life, with its several powers, having been originally breathed by the Creator into a few forms or into one; … from so simple a beginning endless forms most beautiful and most wonderful have been, and are being evolved. [185]

Notice that Darwin does not exclude the possibility of a Creator "breathing" first life into one (the first) form, or a few forms. However, just because Darwin did not exclude the possibility of a Creator, it does not follow that God must have initiated life on earth. Darwin could have been mistaken like the rest of us.

The reader will easily realize that evolution—random variation and natural selection—can only take place *after* life has started. It is therefore easy to see that evolution, really, has nothing to say about the beginning of life. In fact Darwin himself in the same book declares:

---

[184] Darwin, *The Origin of Species*, p. 648
[185] Darwin, *The Origin of Species*, p. 649

I have nothing to do with the origin of mental powers [the mind], any more than I have with that of life itself. [186]

Interpolation and Extrapolation

When one obtains scientific data experimentally and manages to find, say, a linear (straight line) relationship linking the various parameters involved, it is good science, or engineering, to *interpolate* (assume) data between the actual measurements; perhaps even to *extrapolate* (extend) the straight line in either direction for relatively short distances. However, it is bad science, or engineering, to extrapolate results as far as the straight line can be extended. Why? Because conditions may change far away from the measured data: the linear relationship may not necessarily be valid in the vicinity of those new extreme conditions. Extensive extrapolation is usually wishful thinking; however, I do admit, it may lead one to hypothesize a new concept, but one is still required to prove that the same relationship holds in those extreme conditions.

In fact, Darwin himself had misgivings about his own hypothesis of a universal ancestor for all living phyla (body plans); in the same book he writes:

> Analogy would lead me one step farther, namely, to the belief that all animals and plants are descended from some one prototype. But analogy may be a deceitful guide. [187]

Notice that Darwin basically admits that he did not have enough evidence to jump to the conclusion of a common ancestor for all species, even though he would have liked to.

On the other hand, modern scientists have become very comfortable assuming that evolution from a common ancestor is unassailable: nowadays, it is, sort of, cast in stone. They seem to have no problem with such an extensive extrapolation: it is assumed a

---

[186]   Darwin, *The Origin of Species*, p. 317
[187]   Darwin, *The Origin of Species*, p. 642

priori; but do they have enough evidence to support this assumption? We shall deal with this question in the next chapter on "Evolution".

In this chapter, we shall only investigate whether chemical evolution—that is, explaining the origin of life—is only atheistic wishful thinking, or whether there is enough evidence for it. Since evolution gives the impression of design, it follows quite naturally for one—who honestly does not believe in a supreme creator—to extrapolate the concept of apparent design far backward to chemical evolution, in order to explain the origin of life. However, one can easily appreciate the fact that what living organisms are capable of doing cannot be automatically extrapolated to what inanimate things can do; the latter follow completely different rules: somewhat more deterministic in nature and governed mainly by the laws of physics and chemistry. Living organisms, on the other hand, try to adapt to changing conditions around them.

Notice that, in the above quotes, although Darwin proposes a universal ancestor, he never even suggests anything akin to abiotic (non-living) evolution. In fact, he purposely steers away from it, as he steers away from the origin of the mind. Did the concept of chemical evolution occur to Darwin? Of course it did; but as a good scientist he realized that he did not have enough evidence and shied away from even publishing the idea: he only went as far as to propose the concept of a common ancestor. In fact, in a letter to botanist and explorer Joseph Dalton Hooker dated 1 February 1871, Darwin wrote the following: [188]

> It is often said that all the conditions for the first production of a living organism are now present, which could ever have been present. But if (and oh! what a big if!) we could conceive in some warm little pond, with all sorts of ammonia and phosphoric salts, light, heat, electricity, &c., present, that a proteine compound was chemically formed ready to undergo still more complex changes, at the present day such matter would be instantly devoured

---

[188] https://en.wikipedia.org/wiki/Abiogenesis (22 March 2016)

or absorbed, which would not have been the case before
living creatures were formed.

So Darwin believed in chemical evolution, but he did not publish the
concept because he realized that it might just be wishful thinking
that supports his theory—his baby—better. He simply did not have
enough evidence to support it and to go that one step further as to
propose it publicly. Do we now have enough evidence? We shall
address this question in this chapter.

Incidentally, before we proceed further, one should also be very
cautious in extrapolating evolutionary theory the other way: that is,
the evolution of societies, politics, philosophy, and religions; even
though, it might be tempting to do so. Darwin's theory applies strictly
to already living organisms: not to chemicals, molecules, beliefs, or
societies.

But anyway, let us go along, for the time being, with Dawkins'
hypothesis of chemical evolution triggering the origin of life and
examine its plausibility scientifically.

Abiogenesis (Chemical Evolution)

In this section, therefore, we shall look more closely at the
feasibility of Dawkins' hypothesis that there was some form of
natural selection prior to the emergence of life. *Abiogenesis* is the
hypothesis of life arising, spontaneously, from non-living matter
through a natural process: that is, one involving only the laws of
physics and chemistry in combination with some form of natural
preference or exclusion.

In 1938, biochemist Alexander Oparin proposed an evolutionary
abiogenesis hypothesis. In this book, I shall not call abiogenesis a
theory, as one may find it referred to in many books; I shall call it
a hypothesis. Some time ago it was very popular among scientists;
so much so, that it acquired the esteemed title of theory. Nowadays,
it apparently lost its former popularity: scientists do not have much

faith in it any longer; although Dawkins, apparently, still does—but unfortunately he does not have much else to lean on. (The following information is mostly taken from Stephen Meyer's book *Signature in the Cell*.)

Oparin divided his hypothesis in two phases. First, he describes how the basic building blocks of life were (chemically) synthesized from the much simpler chemicals present in the earth's atmosphere, ground, and oceans. Subsequently, he describes how the first living organisms emerged from these building blocks. [189]

Oparin believed that when the earth was initially formed, its core consisted mainly of heavy metals. [190] As it cooled down and contracted, its core would have exposed cracks in the earth's surface. The heavy metals in the core then combined with carbon forming iron carbides (Fe_3C). Eventually, these carbides were "squeezed" up to the earth's surface. [191]

Oparin, for some reason, believed that the early earth's atmosphere had no free oxygen. (As we have seen, later geochemical evidence suggests that this assumption is probably wrong.) Instead he believed the early earth atmosphere to be a noxious mixture of gases such as: ammonia (NH_3), di-carbon (C_2), cyanogen (C_2N_2), steam (H_2O), and simple hydrocarbons (C_mH_n)—like methane (CH) and methylene (CH_2). These simple hydrocarbons then reacted with the iron carbides that were squeezed to the earth's surface, resulting in the formation of heavy energy-rich hydrocarbons: the first organic molecules. [192] [193]

The compounds produced this way eventually reacted with the ammonia (NH_3) in the atmosphere, forming nitrogen-rich compounds. [194] These hydrocarbon derivatives reacted with one

---

[189] Meyer, *Signature in the Cell*, pp. 50, 52.
[190] Graham, *Science and Philosophy in the Soviet Union*, pp. 217–8, 221
[191] Meyer, *Signature in the Cell*, p. 52.
[192] Oparin, *The Origin of Life*, pp. 98, 107–8; and
Graham, *Science and Philosophy in the Soviet Union*, pp. 218–23.
[193] Meyer, *Signature in the Cell*, p. 52.
[194] Oparin, *The Origin of Life*, p. 108

another and with other chemicals in the oceans to produce amino acids, which then linked together to form proteins. [195]

The second phase of Oparin's abiogenesis hypothesis invoked Darwin's concepts of evolution: that is, random variation coupled with natural selection; thus explaining the emergence of living organisms from these organic molecules. Oparin proposed that competition for existence arose between small entities enclosing protein and carbohydrate $[C_m(H_2O)_n]$ molecules, and that this competition eventually produced primitive cells capable of complex chemical reactions inside them. [196]

As an enclosing membrane, or wall, he suggested an inanimate structure called a *coacervate*. A coacervate is a collection of fatty molecules that gather together forming a spherical structure because of the way they interact with water. These fatty molecules are elongated molecules that have a *hydrophilic* (water-attracting) side and a *hydrophobic* (water-repelling) side. They therefore form a spherical structure that both repels water on the outside and encloses water on the inside: somewhat similar, but opposite, to what soap molecules do to oil droplets in water. In effect these structures constitute a semi-permeable membrane: meaning that they allow molecules to pass in and out between them—similar to modern cell walls. Carbohydrates and proteins could have thus become enclosed in such spherical formations in the oceans; as these molecules reacted with one another inside these structures they developed a kind of primitive metabolism. [197]

Some coacervate structures grew larger than others and, conceivably, developed more efficient means for assimilating new substances from the environment. The result was that they grew bigger and developed even more efficient means of assimilating more and more substances, which effectively became their nutrients. When the supply of nutrients in the environment changed and became relatively scarce, those coacervate structures that failed to compete

---

[195]   Meyer, *Signature in the Cell*, p. 52.
[196]   Meyer, *Signature in the Cell*, pp. 52–3.
[197]   Meyer, *Signature in the Cell*, p. 53.

grew feeble and exhausted their internal supplies; eventually they decomposed and ceased to exist. Those coacervates that had, by chance, developed crude forms of metabolism continued to develop further; a Darwinian-style competition ensued which eventually resulted in the first living cell. [198] [199]

One has to appreciate the details Oparin went into when he proposed his hypothesis; no wonder many scientists adopted it as a theory for a while. As we shall presently see, this is probably one of the cases where science jumped to conclusions too soon and had to reconsider. But this, unlike dogmas in religion, is the normal way science works: it simply searches for the truth. Prior classical examples where science retracted concurrence were: the flat-earth and the geocentric (earth-centered) universe hypotheses.

Analysis of Abiogenesis (Chemical Evolution)

What undermined the plausibility of Oparin's proposals was the later discovery that any kind of protein is not so easy to build by natural chemical processes and random chance alone. The extreme improbability of amino acids forming strictly peptide bonds, using only left-handed isomers, and spelling an exact sequence to form stable-folding proteins, made it implausible to expect them to be formed naturally: that is, strictly by random, unmediated chemical processes. Unfortunately, Oparin's scenario relied too heavily on chance alone to explain the initial formation of the proteins involved in the primitive cellular metabolism; consequently, it fell out of favor with biochemists. [200]

In actual fact, in 1968, when most scientists came to know about the complexity of DNA and proteins, Oparin himself revised his hypothesis; in it he postulated natural selection kicking in earlier in the process: that is, while the molecules were still much simpler. His

---

[198]   Oparin, *The Origin of Life*, pp. 229–31

[199]   Meyer, *Signature in the Cell*, p. 54.

[200]   Meyer, *Signature in the Cell*, p. 273.

new version of the hypothesis proposed that natural selection acted on random chains of amino acids as they reacted within coacervate enclosures. [201] He, therefore, proposed that natural selection acted on both unspecified polypeptides (random amino acid chains) and unspecified polynucleotides (random DNA molecule chains) that arose by chance and somehow ended up inside coacervate enclosures. They then gradually increased their specificity (exact sequence) and improved their functionality; eventually, they achieved some form of primitive metabolism. Obviously, natural selection favored the coacervates that developed functional proteins and DNA sections over those that did not. [202] As the information of the DNA and protein molecules inside these proto-cells increased, they cooperated with each other; eventually they produced metabolic processes, information storage, and replication (reproduction) systems. [203]

Is this scenario feasible? It is only feasible to the uninformed; here is what Meyer has to say about it in his book *Signature in the Cell*:

> Oparin's concept of *prebiotic* natural selection begged the question. Natural selection occurs only in organisms capable of reproducing or replicating themselves. Yet in all extant cells, self-replication depends on functional and sequence-specific [i.e., exact sequence or information-rich] DNA and protein molecules. [204]

Before reproduction takes place, natural selection simply cannot start acting at all. In the above scenario, reproduction is the last stage that happens; it must be present in the first place for the hypothesis to hold water.

Meyer gives us an illustration to clarify the situation. Imagine a man falling into a deep pit; he realizes that he absolutely needs a

---

[201] Oparin, *Genesis and Evolutionary Development of Life*, pp. 146–7
[202] Kamminga, "Studies in the History of Ideas on the Origin of Life", p. 326; and Oparin, *Genesis and Evolutionary Development of Life*, pp. 146–7.
[203] Meyer, *Signature in the Cell*, pp. 273–4.
[204] Meyer, *Signature in the Cell*, p. 274 (emphasis in original).

ladder to be able to climb out of it. He therefore walks home and picks up a ladder; he then goes back to the pit and climbs back down into it; finally, he sets up his ladder and climbs out of it. But this parable fails to answer a very obvious question: how did he manage to get out of the pit in the first place in order to be able to walk home? [205]

When a system of molecules copies itself, it *invariably* makes minor errors: it *mutates*; this has the advantage of giving rise to diversity in a population of living organisms. A phenomenon known as an *error catastrophe* will occur when small errors (deviations) are amplified in successive replications of functionally necessary sequences. [206] This means that if the mutations that occur are serious enough, the organism may lose some or all of its former function and possibly even its ability to reproduce altogether. Unless the organism's system consists of molecules specified enough to ensure an error-free transmission (copying) of information, an error catastrophe is more than likely bound to happen eventually. Equivalently, if the transmitting system happens to be a crude one, major mutations rather than minor ones will occur; consequently, such a system of molecules would also be susceptible to an error catastrophe, sooner or later. Evidence from molecular biology shows that unspecified (random) polymers will not replicate genetic information at all; consequently, Oparin's proposal, of starting from unspecified polymers, would have been extremely vulnerable to a very early occurrence of an error catastrophe. [207]

There was no way out left for Oparin's chemical evolution hypothesis; it encountered a "catch-22" situation. If it invoked natural selection early in the process: that is, prior to the amino acid or nucleotide strings' attaining functional specificity, accurate replication would not have been possible. Consequently, improved reproduction cannot occur and the concept of natural selection becomes meaningless. On the other hand, if the hypothesis invoked

---

[205]   Meyer, *Signature in the Cell*, p. 275.

[206]   Joyce & Orgel, "Prospects for Understanding the Origin of the RNA World", pp. 8–13.

[207]   Meyer, *Signature in the Cell*, p. 275.

natural selection late in the scenario, it would have to rely on chance alone to produce the sequence-specific (information-rich) molecules required for faithful self-replication; as we have already seen, this is no trivial task. [208]

Darwinian natural selection only favors functionally advantageous varieties; thus, blind, untargeted, random variations have to occur *first*. Then, and only then, can it make a selection; furthermore, it can only select what random variations have already produced. However, as we have seen above, it is extremely unlikely that random molecular interactions would produce the sequence-specific information present in even a single functioning protein or DNA molecule of modest length. This initial threshold that random mutation needs to overstep is too high; in a Darwinian natural selection process, the random mutations are normally very small because the organism cannot lose functionality at any step in the meantime. [209]

In his book *The Selfish Gene*, evolutionary biologist Richard Dawkins writes:

> At some point a particularly remarkable molecule was formed by accident. We will call it the *Replicator.* [210]

Notice the phrase "by accident". So Dawkins seems to admit that the first replicators had to happen by chance alone. Is this a feasible proposition? Can a simple crude replicator of the type described by Dawkins in his book avoid the development of an error catastrophe? From our considerations above, the answer seems to be negative.

Moreover, given the results of the simulation experiments of the Stanley Miller type, it seems highly improbable for a replicator to arise from a prebiotic soup by chance alone; biologically significant molecules tend to react spontaneously with biologically insignificant molecules, resulting in an insoluble sludge.

---

[208]   Meyer, *Signature in the Cell*, pp. 275–6.
[209]   Meyer, *Signature in the Cell*, p. 276.
[210]   Dawkins, *The Selfish Gene*, p. 15 (emphasis in original).

To resolve this disagreement in opinion between experts in the field, I think we need first to have an idea as to what is actually involved in producing a faithful replicator, such as DNA or RNA. After reading the following few sections, the reader can decide for oneself.

There is one last point I would like to make before I move on to the next section. Silicon (sand and rock) is much more abundant than carbon on our earth (including the oceans and its atmosphere). The number of silicon atoms on earth is about one hundred and thirty-eight (138) times the number of carbon atoms. [211] Isn't it strange that, in spite of silicon's superior relative abundance, no life form has ever evolved from this element, even though its chemical properties are very similar to that of carbon? Its greater abundance should have lent itself to better probabilities of evolving into life; as carbon is claimed to have done. Yet silicon compounds are dead as a doornail. Why did the *lower* probability prevail? In my opinion, the laws of physics and chemistry should have produced some other silicon-based life form if abiogenesis were a feasible hypothesis.

DNA (Deoxyribonucleic Acid)

We saw above how extremely difficult it is to produce a single functional protein by chance alone. Consequently, given the importance of proteins, organisms need a method of reproducing functional proteins accurately and at will.

This happens through _deoxyribonucleic acid_ (or DNA), as evolutionary biologist Richard Dawkins explains in his book *The Greatest Show on Earth*. Proteins are outstandingly bad at replicating themselves; the best replicating molecule known is DNA. Proteins and DNA complement each other like a happily married couple. DNA takes care of the information storage and eventual transmission during replication and reproduction; proteins actually do the dirty

---

[211]   http://www.chem.wisc.edu/deptfiles/genchem/sstutorial/Text3/Tx33/tx33.html (13 February 2016)

work. From the information it stores, DNA produces proteins which serve as enzymes (biological catalysts) for the innumerable cell functions. Proteins fold into complex three-dimensional forms; hence the information on the inside of these folds cannot be easily accessed. These folds are the reason why proteins are such efficient catalysts; they hold reacting molecules in their crevices and electro-chemically facilitate a reaction to take place; proteins, in turn, assist in the formation of DNA molecules. DNA molecules are the exact opposite of proteins; they do not fold three-dimensionally, and therefore they perform terribly as enzymes. DNA molecules are linear molecules, lending themselves to be copied easily: the information is readily accessible and not hidden in three-dimensional folds. [212]

Double Helix

The physical structure of DNA resembles that of a ladder twisted about its length: resulting in something like a mild spiral staircase. The sides of the twisted ladder constitute the double helix, or double backbone. The two backbones consist of alternate deoxyribose sugar ($C_5H_{10}O_4$) molecules and phosphate (PO_4) groups. The rungs of the ladder consist of four nucleotide *bases* known as: adenine ($C_5H_5N_5$), cytosine ($C_4H_5N_3O$), guanine ($C_5H_5N_5O$), and thymine ($C_5H_6N_2O_2$) paired to another of the same four bases as follows. Adenine always pairs with thymine and vice-versa; while cytosine always pairs with guanine and vice-versa. The respective pairs are bound by a weak hydrogen bond that can easily unzip to enable replication (copying).

The four nucleotide bases adenine (A), cytosine (C), guanine (G) and thymine (T) constitute the four letters of the *genetic code*: like the alphabet in a language or like the binary numbers, zero and one (0 and 1), in computer programming. The *gene*, or protein-producing information, is determined by the specific (exact) sequence of these four nucleotide bases (letters) in DNA. In his book *Genome*, science writer Matt Ridley describes the human genome as follows:

---

[212]   Dawkins, *The Greatest Show on Earth*, p. 420

> Imagine the [human] genome is a book.
> There are twenty-three chapters, called *chromosomes*.
> Each chapter contains several thousand stories, called *genes*.
> Each story is made up of paragraphs, called *exons*, which
> are interrupted by advertisements called *introns*.
> Each paragraph is made up of words, called *codons*.
> Each word is written in letters called *bases*. [213]

We shall have something more to say about the "advertisements" mentioned above: the book *Genome* was written in 1999, and there have been some developments since then; more information about this subject is given in the section on "Junk DNA" below.

The next three sections are accounts of the work done by DNA, namely, reading, copying, and protein building; they are summarized from biologist and teacher Matthew Distefano's high school textbook entitled *Biology*.

Reading DNA

Reading the specific sequence of the nitrogen bases (letters) constituting a DNA molecule is not done one location at a time but in groups of three. Every *triplet*—three consecutive nitrogen bases taken together—forms an entity termed a *codon*. While the nitrogen bases can be thought of as the letters of the genetic code, codons can be thought of as the words of the genetic code. Since there are four possible letters for every one of the three positions of a triplet, it follows that there are sixty-four (64=4x4x4) different codons possible. The sixty-four (64) different triplets of nitrogen bases represent, in code form, all the twenty (20) essential amino acids found in living organisms, including "start" and "stop" instructions plus several repetitions.

When a particular codon is encountered, the corresponding amino acid is picked from inside the cell (we shall see exactly how

---

[213] Ridley, *Genome*, p. 7 (emphasis in original).

later) and placed in a specific sequence following the DNA script as template. The various amino acids are thus bound together in the required sequences later, producing the many proteins necessary for the various functions of living cells: a process known as *translation* (explained right after the next section). [214] Every DNA protein-building sequence, from start to finish, is commonly known as a *gene*. According to Matt Ridley, these make up the thousands of stories in the DNA book of life—the *genome*.

Transcription (Copying DNA)

Since the DNA molecule contains the information for both the present and the future life of the cell, it must be protected at all costs; consequently, it never wanders about the cell but stays in the safety of the cell's *nucleus*. However, the cell needs to construct proteins, in order to accomplish the numerous chemical processes life requires. Since the various *organelles* (internal cell structures) assist these chemical processes, they need to use the information in DNA. These organelles are located in the *cytoplasm*: the relatively wide space, far away from the nucleus, containing the cell's environmental fluid and nutrients.

So the DNA molecule satisfies this requirement by creating a new chemical: a messenger that relays the required information to these relatively remote locations spread across the cytoplasm; meanwhile, it holds the vital genetic information safely tucked in the nucleus. (Incidentally, notice the extreme care taken to avoid an error catastrophe; the replicator is not allowed to wander about freely among other chemicals.) This chemical is known as *messenger ribonucleic acid* (or mRNA); the mRNA contains of a copy of the DNA code which is safely delivered to the organelles requiring it: without compromising the integrity of the DNA molecule itself. While copying takes place, the DNA molecule unzips at the weak hydrogen bonds that hold the nitrogen base pairs (rungs) together.

---

[214]   Distefano, *Biology*, p. 129.

Each of the two sides of the DNA molecule (backbone pair) is called a *template*; since they are both used to create complementary copies of each strand.

When the DNA molecule unzips, it exposes the nitrogen bases attached to both of its backbones, making them available (like templates) for creating a complementary pair: something like a photographic negative, as explained below. This is accomplished through the agency of an *enzyme* (biological catalyst) called *DNA polymerase*. The two copies produced are known as mRNA strands; these strands consist of the complementary bases of the original DNA template pairs, except that the base uracil ($U=C_4H_4N_2O_2$) is used as complement to adenine ($A=C_5H_5N_5$), instead of thymine ($T=C_5H_6N_2O_2$) which is the normal chemical found in a DNA molecule—thus constituting a distinguishing feature between DNA and RNA. For example, if a DNA strand has the bases "A-C-G-T", then the strand produced by the mRNA will have the complementary bases "U-G-C-A". The biological process of making an mRNA copy of an original DNA strand is known as *transcription*. [215]

Translation (Making Proteins)

The chemical process of making proteins is termed *translation* because it converts (translates) information from the four-character alphabet of DNA and RNA into the twenty-character amino-acid alphabet. Translation creates the various proteins required by a living cell. [216]

When transcription is completed, inside the nucleus of the cell, the mRNA possesses a complementary base code of the original DNA strand. The mRNA then leaves the safety of the nucleus and makes its way out into the cytoplasm, the relatively remote space of the cell. The mRNA travels through the cytoplasm and towards the *ribosomes*; these are molecular work benches on which protein

---

[215] Distefano, *Biology*, pp. 129–30.

[216] Meyer, *Signature in the Cell*, p. 127.

synthesis takes place. (Obvious enough, this journey of delivering the DNA code to the ribosomes gives this particular RNA its name: messenger RNA.) The ribosomes act as a surface area where the bonding of amino acids into chains occurs, of course, following the template (code) that the mRNA brings along. This happens through the assistance of another type of RNA called *tRNA*—for *transfer RNA*—which leads these amino acids to the ribosomes and the mRNA, thus bringing them close together.

All tRNA molecules look like a (three-leafed) cloverleaf, and they are between seventy-three (73) and ninety-three (93) nucleotides long. There are several types of tRNA molecules; every type of tRNA molecule engages (latches on to) a particular amino acid. Now, every one of the tRNA molecules acts as a carrier that picks up a unique amino acid at the stem of the cloverleaf shape; it also has a corresponding key (code) at the other end: the center leaf of the cloverleaf shape. The key at the center leaf of every tRNA molecule consists of a sequence of three nitrogen bases, which reflects the type of amino acid attached to that tRNA molecule stem. This sequence of three nitrogen bases is termed an *anticodon*, and these three bases act complementary to the codon bases on the mRNA molecule. If, for example, an mRNA portion contains a codon with a base sequence "U-G-C", it will bind with a tRNA anticodon that has a base sequence "A-C-G".

The tRNA molecules enter the ribosome with all the various (twenty) amino acids attached at one end; their anticodons at the other end will then hook up to the mRNA, following the codon code on the mRNA: that is, in the right sequence. The tRNA molecules remain at the bonding site of the ribosomes for just a short while: only until the amino acids they carry bond together. Consequently, the amino acids are joined in the right order to build specific proteins. [217]

The reader should also realize here, that the above account of the translation process is an over-simplified version; not insignificantly, Distefano concludes his account of the translation process as follows:

---

[217] Distefano, *Biology*, pp. 130–1.

> The chemical process of translation, illustrated here, requires the action of many enzymes [protein-based biological catalysts] working in many locations throughout the surface of the ribosome. [218]

In his book *Signature in the Cell*, Meyer elaborates on how amino acids are attached uniquely to particular tRNA molecules. He notes that there is nothing that is chemically preferential to attach a particular amino acid rather than another to the stem of the cloverleaf shape of a tRNA molecule; yet only one of the twenty amino acids is carried by a particular tRNA molecule. The pairing (attachment) of specific amino acids to their tRNA molecules depends upon the assistance of twenty specific enzymes (biological catalysts), one for every amino acid. These adapter enzymes are termed *aminoacyl-tRNA synthetases*. These specific adapters fit both a particular amino acid and a particular tRNA molecule like a glove and thus engage them together.

These adapters are indispensable to latch on every single amino acid to a particular tRNA molecule with its corresponding anticodon (code); because, as mentioned above, there is nothing about the chemical properties of the bases in DNA, mRNA or tRNA that prefers to chemically bond with a particular amino acid rather than another. There is complete freedom of action; like placing magnetic letters onto a refrigerator door to leave a message. The letters have no preferential magnetic attraction; they have no special preference to be at the top, bottom, left, right, alone, or near others on the refrigerator door: their sequence follows only the originator's thoughts.

Moreover, in actual fact, the cloverleaf-shaped tRNA molecule attaches to the mRNA at one end and carries an amino acid on the other end; this physical distance ensures that amino acids do not react chemically with either the codons on the mRNA molecule or the anticodons on the tRNA molecule. [219]

---

[218] Distefano, *Biology*, p. 131.
[219] Meyer, *Signature in the Cell*, pp. 129–30.

Meyer, in describing the various functions of these enzymatic adapters, calls them "marvels of specificity":

> The synthetases have several active sites that enable them to: recognize a specific amino acid, recognize a specific corresponding tRNA (with a specific anticodon), react the amino acid with ATP to form an AMP derivative [see next section], and then, finally, link the specific tRNA molecule in question to its corresponding amino acid. ... In virtue of the specificity of the features they must recognize, individual synthetases have highly distinctive shapes ... [and] are themselves marvels of specificity. [220]

One might ask: why all this complexity for a molecule to copy only a portion of itself? Does the reader think that nature would complicate things with coordinated functions unnecessarily? Or does it imply that replicating molecules chemically is not such an easy task? One might add here, that we haven't arrived at a molecule replicating itself in its entirety yet (Dawkins' replicator); later on in this chapter, we shall examine what the minimum requirements are in such a case.

Cell Energy

A living cell needs energy for everything it does: it is scientifically intuitive that nothing comes from nothing. The energy of the cell comes from a molecule inside it called _adenosine triphosphate_ (ATP), which consists of an adenosine ($C_{10}H_{13}N_5O_4$) molecule with three phosphate (PO_4) groups attached to it. The chemical bond between every one of the three (tri-) phosphate groups and the adenosine molecule holds energy. That is, energy is stored and made available to the cell when such a bond is formed inside the cell, and when such a bond is broken the energy is released to the cell. When an ATP molecule loses a phosphate group, releasing energy to the cell,

---

[220] Meyer, *Signature in the Cell*, p. 130.

it becomes an *adenosine diphosphate* (ADP) molecule, since only two (di-) phosphate groups remain attached to it. Inside the cell, ATP may be likened to a fully charged battery, while ADP may be considered an uncharged battery. The reverse can take place: ADP can be recharged and once again have a third phosphate group joined to it, thus becoming an ATP molecule again and making energy readily available to the cell. So the whole process of energy production in the cell is very similar to charging and discharging a rechargeable battery. A second phosphate group may be removed from an ADP molecule, thus becoming an *adenosine monophosphate* (AMP) molecule. Although it can have this second phosphate group removed from it, ADP does not usually provide energy to the cell this way. [221]

ATP is produced in the cell from glucose ($C_6H_{12}O_6$) by a very complex process known as *glycolysis*. However, glycolysis consists of ten steps every one of which is catalyzed (assisted) by specific proteins. So we have a curious situation here: in order to do anything the cell needs energy, which is available through the ATP molecule; however, in order to have ATP available, it needs a complex system of proteins; but to produce these proteins in the first place, it needs ATP. Thus, one function depends on the other, and vice-versa—a catch-22 situation. This is like the old chicken and egg riddle: a chicken comes from an egg, but only a chicken can produce an egg; so, which came first, the chicken or the egg? [222]

RNA World Hypothesis

Similar chicken and egg situations are strewn across the whole landscape of living organisms. For example, the production of proteins through transcription and translation requires DNA information; however, the production of DNA, whenever cells divide to copy themselves, requires proteins; so which came first, DNA

---

[221]   Distefano, *Biology*, pp. 94–5.

[222]   Meyer, *Signature in the Cell*, p. 131–2.

or proteins? [223] This is a major catch-22 situation which Dawkins, admirably, tries to address in his book *The Greatest Show on Earth* when he writes:

> The 'Catch-22' of the origin of life is this. DNA can replicate, but it needs enzymes [protein-based biological catalysts] in order to catalyze the process. Proteins can catalyze DNA formation, but they need DNA to specify the correct sequence of amino acids. How could molecules of the early Earth break out of this bind and allow natural selection to get started? Enter RNA. [224]

The reader may recall here that Dawkins believes in natural selection starting from inanimate chemicals (abiogenesis). Here he proposes that natural selection acted on RNA molecules prior to the existence of DNA molecules. Very interestingly, following scientific principles, Dawkins adds tangible evidence in support of RNA starting the whole process. He mentions the fact that some viruses have no DNA, and that they have survived for a very long time with just RNA in their genetic code. Undoubtedly, this seems like strong supporting evidence. He concludes his hypothesis by presenting a mechanism through which this catch-22 situation was overcome in proto-molecules.

He admits that RNA may not be the perfect tool; however, it can perform, albeit crudely, both the job that proteins do and the job that DNA does. He adds that RNA enzymes that fold three-dimensionally and act as catalysts do actually exist; while the information storage and replicating properties of RNA are well known. He concludes by proposing that RNA might have done the job of both molecules; that is, until DNA had had a chance to evolve together with a more sophisticated protein suite. This basically constitutes the *RNA world hypothesis*—it is *not* a theory as Dawkins promotes it to in his book. [225]

---

[223] Meyer, *Signature in the Cell*, p. 132–4.
[224] Dawkins, *The Greatest Show on Earth*, p. 420
[225] Dawkins, *The Greatest Show on Earth*, p. 421

Before leaving the subject he, somewhat apologetically, makes the following observation, which I hope I am not misunderstanding:

> I must repeat the warning I have given in earlier books. We don't actually need a plausible theory of the origin of life ... The theory that we seek, of the origin of life on this planet, should therefore positively *not* be a plausible theory! If it were, life should be common in the galaxy. [226]

In my humble opinion, this last paragraph is *bad* science! A scientific hypothesis should have a high probability of its happening; even more so, if it is to be considered as a theory by the scientific community. It should be supported by a vast body of evidence in its favor; consider the amount of corroborating evidence there is for the theory of evolution, for example, or for the big bang theory. It should also exclude a chance hypothesis unless the odds are reasonably, preferably heavily, in its favor; as was explained in the section on "Chance" near the beginning of this chapter. If the odds are vastly against a certain hypothesis or theory, an alternative explanation should be sought. This is the normally accepted way of science; anything can be explained as the result of chance. But when we do explain things this way, it boils down to scientific *ignorance*.

Before we start our analysis of the RNA world hypothesis, here's a short description of the hypothesis summarized from Meyer's book *Signature in the Cell*. (1) RNA building blocks arose on earth. (2) They linked to form short nucleotide chains. (3) An RNA molecule capable of copying itself arose by chance. (4) Natural selection ensued: favoring more efficient methods of replication. (5) Ribozymes (RNA-based biological catalysts) form proteins from RNA templates. (6) A simple self-replicating membrane enclosed the initial ribozymes along with some amino acids. [227] (7) Ribozymes were replaced by more efficient proteins: ones possessing modern

---

[226] Dawkins, *The Greatest Show on Earth*, pp. 421–2 (emphasis in original)

[227] Szostak, Bartel, & Luisi, "Synthesizing Life", pp. 387–8.

enzymatic functions. [228] (8) Protein-based translation takes over RNA-based production of proteins. (9) Genetic information was transferred from RNA into DNA: RNA was relegated to its current intermediate role. [229] (10) The modern DNA-protein system arose within a primitive cell membrane. [230]

The next section involves somewhat complex chemical analyses: the observations of more than thirty-years of still ongoing research. It is rather involved; if one is only interested in reading the book leisurely, one may get a general idea of what it says by just reading the headings (in bold font) of every subsection. Alternately, one may possibly have to read it a couple of times to really savor the relevance and strength of every subsection. This section is included mainly for the benefit of the scientifically oriented reader who is interested in the chemical details of the RNA world hypothesis.

Analysis of RNA World Hypothesis

In his book *Signature in the Cell*, Meyer enlists the following objections to the RNA world hypothesis:

(1) RNA Building Blocks Do Not Synthesize Spontaneously but Decompose Easily

The first problem the RNA world hypothesis encounters is that the chemical and physical laws of nature are not conducive to constructing, spontaneously, the building blocks of RNA molecules, namely, nucleotide bases and ribose sugars: that is, without providing them with any engineering assistance—if left on their own. Chemists Robert Shapiro and Stanley Miller, of the Miller-Urey experiment

---

[228] Gilbert, "Origin of Life: The RNA World", p.618.

[229] Gilbert, "Origin of Life: The RNA World", p.618.

[230] Meyer, *Signature in the Cell*, pp. 298–300.

fame, have done extensive chemical analyses on the building blocks of RNA molecules confirming the above statement. [231]

Shapiro reports that, unlike amino acid production, there is no known instance of anybody producing nucleotide bases of *any* kind from spark-discharge experiments of the Miller-Urey experiment type. [232]

Miller and Shapiro also noted that the nucleotide bases of RNA are unstable at the high temperatures currently accepted by scientists as having been the case in origin of life scenarios. What happens at these temperatures is that they lose their all-important amine (NH_2) groups, a chemical process known as *deamination*. The *half-life* of a substance is the length of time it takes for half of the total amount of that substance to change or deteriorate into some other substance or substances. At one hundred degrees Celsius (100C), the half-life of uracil is about twelve years, the half-lives of adenine and guanine are about one year, and the half-life of cytosine is only about nineteen days. These half-lives are all far too short in comparison with evolutionary processes: like finding functional RNA catalysts blindly (by trial and error) through natural selection. [233] Consequently, in 1988, Miller came to the conclusion that in the case of RNA bases, [234] "a high temperature origin of life involving these compounds therefore is unlikely." [235]

Miller further concluded, and Shapiro agreed with him, that the fact that one of the four nucleotide bases, cytosine, has such an extremely short half-life, at both high and low temperatures, implies the CG pair was probably not present in RNA. [236]

> "[Perhaps] the GC pair may not have been used in the first genetic material. ... [The assumption that] the

[231]   Meyer, *Signature in the Cell*, p. 301.

[232]   Meyer, *Signature in the Cell*, p. 302.

[233]   Meyer, *Signature in the Cell*, p. 302.

[234]   Meyer, *Signature in the Cell*, p. 302.

[235]   Levy & Miller, "The Stability of the RNA Bases"

[236]   Meyer, *Signature in the Cell*, p. 302.

bases, adenine, cytosine, guanine and uracil were readily available on the early earth [is] not supported by existing knowledge of the basic chemistry of these substances." [237]

I wonder what would be left of the RNA world hypothesis if this were the case.

Moreover, producing ribose sugar under realistic natural conditions has proven even more questionable. Most prebiotic chemists proposed ribose sugar arose as the by-product of the *formose reaction*: which consists of a long series of chemical reactions that start with molecules of formaldehyde (CH_2O) in water (H_2O) reacting with one another. However, Shapiro pointed out that this reaction will not produce sugars in the presence of nitrogenous substances. [238] This would exclude the presence of amines (NH_2) and amino acids (NH_2CO_2H): a category of molecules that includes nucleotide bases; hardly a situation where both RNA building blocks could coexist and still form spontaneously.

Thus the RNA world hypothesis here faces a catch-22 situation—a situation it was supposed to avoid in the first place—namely, that the presence of nitrogen-rich chemicals, which are necessary for the production of nucleotide bases, prevents the production of ribose sugars. The bottom line is, the building blocks of RNA would have reacted readily with other molecules present in the prebiotic environment. These interfering cross-reactions would have destroyed the RNA building blocks and produced biologically irrelevant substances. Consequently, the assembly of RNA from its constituents would have been stumped, and, needless to say, there would have been no evolution of RNA molecules toward more complex molecules. [239]

Incidentally, one might add at this point, that Miller and Shapiro are not scientists that support the design hypothesis of the origin of life, so the above is definitely unbiased scientific evidence.

---

[237]   Shapiro, "Prebiotic Cytosine Synthesis".
[238]   Shapiro, "Prebiotic Ribose Synthesis".
[239]   Meyer, *Signature in the Cell*, pp. 302–3.

Finally, Meyer concludes that the hard reality is:

> [I]n many cases, reactions ... that produce desirable by-products ... also produce many undesirable chemical by-products. Unless chemists actively intervene, undesirable and desirable chemical by-products of the same reaction react with each other to alter the composition of the desired chemicals in ways that would inhibit the origin of life. [240]

(2) RNA Catalysts Are Inadequate Substitutes for Proteins

Scientists, including Dawkins, claim that RNA catalysts could have done the same job as proteins; but in reality the situation is comparable to that of a drowning person clutching at straws. [241] Sure, straws can float, in theory, and one can envisage that they can help a drowning person to float better: only to a very minor degree though. However, to propose that a handful of straws can save a human being from drowning is ludicrous—that is why language has come up with such a metaphor. Similarly, the RNA world hypothesis is probably an atheist's clutching at straws, and here is the reason why.

Ribozymes (RNA catalysts) can only perform a handful of the thousands of functions performed by modern proteins. Scientists claim that some RNA molecules can split other RNA molecules, and that others can link separate strands of RNA molecules. The *ribosomal* RNA (*rRNA*), which is the RNA found in ribosomes—the protein-making machinery—promotes peptide (amino acid) bonding within the ribosome. [242] It can also promote peptide bonding outside the ribosome, but only in association with an additional chemical catalyst. [243] However, RNA catalysts can perform these functions only if scientists purposely manipulate the chemical reactions: thus

---

[240]   Meyer, *Signature in the Cell*, pp. 303–4.

[241]   Meyer, *Signature in the Cell*, p. 304.

[242]   Noller, Hoffarth, & Zimniak, "Unusual Resistance of Peptidyl Transferase to Protein Extraction Procedures".

[243]   Zhang & Cech, "Peptide Bond Formation by *In Vitro* selected Ribozymes".

leading them in the desired direction; they seem to forget that natural selection is both blind and has no foresight.[244]

Meyer portrays this situation by giving us an analogy. To claim that, at the dawn of chemical evolution, a ribozyme could have replaced proteins is similar to claiming that a carpenter could build a whole house by using only a hammer; simply because it can perform two or three functions. RNA catalysts can only perform a handful of the thousands of different cellular functions proteins perform, even in the simplest single-celled organisms like bacteria. It does not follow, therefore, that the former could conceivably perform all the indispensable functions required for the survival of a living cell. [245]

(3) RNA-Rooted Translation and Coding Systems Are Implausible

The whole idea seems to be impractical, in the sense that it expects too much from a basically inefficient system—that is, RNA alone: it is expected to do triple the work of a modern efficient system—namely, DNA, RNA, and proteins working together. From an engineering standpoint, the system would be overloaded and give up altogether. Here are some details exposing the practical difficulties.

To evolve beyond the RNA world, the primitive RNA replicator would eventually need to come up with a suite of protein machinery capable of performing all the functions modern proteins perform during translation (protein construction). [246] Therefore, the evolving system of RNA molecules, besides doing its normal job of self-replication, also needs to accomplish three other basic functions meanwhile, practically simultaneously: it needs to develop coding, translation, and information storage systems, all based entirely on RNA molecules. In other words, these *primitive* RNA molecules would need to do the work of the fifty ribosomal proteins (the protein machinery involved in translation, i.e., building proteins in general)

---

[244] Meyer, *Signature in the Cell*, p. 304.
[245] Meyer, *Signature in the Cell*, p. 304.
[246] Meyer, *Signature in the Cell*, p. 305.

and twenty different tRNA synthetases (the specific adapters which assist in attaching a particular amino acid to a given tRNA); it would also need to produce the many mRNAs that carry information for building the initial protein suite and eventually the tRNAs themselves. This is quite a tall order for an evolving system because everything has to happen randomly, blindly, simultaneously, and without a hint of foresight.

For example, to build proteins a bacterial cell (which is one of the most basic living cells) requires a translation and coding system of one hundred and six (106) different and *functionally integrated* proteins (meaning, the existence of some proteins depends on the presence of several others—again a catch-22 situation); in conjunction with several mRNAs, rRNAs, and tRNAs; plus free floating amino acids and ATP (energy) molecules. [247] This is quite possibly the minimum requirement for a viable system: for a living cell to survive and reproduce. [248]

Now, let us compare what the scientists report to have achieved and what they have actually achieved. Observe their bold claims:

> [M]olecular biologists have ... engineered ribozymes that can catalyze "all three elementary reactions" [249] required in translation, including amino-acylation (the formation of a bond between an amino acid and an RNA), the peptidyl-transferase reaction (which forms the peptide bond between amino acids), and amino-acid activation (in which adenosine monophosphate [AMP] is attached to an amino acid [explained presently]). [250]

---

[247] Gil, Silva, Pereto, & Moyal, "Determination of the Core of a Minimal Bacterial Gene Set", Table 1, pp. 521–2.

[248] Meyer, *Signature in the Cell*, p. 305.

[249] Wolf & Koonin "On the Origin of the Translation System and the Genetic Code in the RNA World by Means of Natural Selection, Exaptation, and Subfunctionalization"

[250] Meyer, *Signature in the Cell*, p. 306.

This language, however, is misleading. Claiming to have achieved making a chemical reaction happen does not describe the engineering (intelligent intervention) that went into it to make it happen, or the extreme improbability of its happening under natural conditions; nor does it describe whether such a reaction happens across the board, or only with just a few of the twenty amino acids—the exception rather than the rule.

This is what I meant, in the last subsection, by clutching at straws. For example, ribozyme engineers were able to design catalysts that could bond only two of the twenty amino acids to an RNA molecule: which leaves a lot to be desired. There is not much one can do with just two letters from a twenty letter alphabet. How can RNA molecules, therefore, produce the specificity required to build proteins useful enough during translation? [251]

Meyer comments on this quote as follows:

> At first glance, these results seem to support the feasibility of an RNA-based translation system. Nevertheless ... the gap between "some" and "all" reactions of a given type remains significant. [252]

Protein-based enzymes (biological catalysts) that participate in the building of other proteins (translation) perform multiple functions, many of which are coordinated together: that is, one's success depends on the assistance of another's, and vice versa. On the other hand, ribozymes (RNA-based catalysts) can typically perform only a single *sub-function* of the several integrated functions a protein-based enzyme, biologists might compare it to, can perform. [253]

For example, the synthetase enzymes (specific adapters) involved in the binding of tRNA molecules to their associated amino acids (amino-acylation) during protein building (translation) must catalyze (assist) a double chemical reaction: one reaction is immediately

---

[251] Meyer, *Signature in the Cell*, p. 306, 308.
[252] Meyer, *Signature in the Cell*, p. 306.
[253] Meyer, *Signature in the Cell*, p. 308.

followed by another. First, a specific synthetase engages a particular amino acid to an ATP (adenosine triphosphate); the amino acid ends up attached to an AMP (adenosine monophosphate), thus giving it the stored energy it needs to eventually bond with a tRNA molecule in the second stage of the reaction.

The first reaction is as follows:

AA (amino acid) + ATP = AA–AMP (bonded) + PP (pyrophosphate = P_2O_7).

Then, the synthetase couples the AMP-charged amino acid to a specific tRNA.

The second reaction is as follows:

AA–AMP (bonded) +tRNA = AA–tRNA (bonded) + AMP.

These tRNA molecules are all different and, in addition, include specific anticodons that bond to the respective mRNA codon at the ribosome (protein-producing site).

Does the proposed ribozyme, the supposed precursor to the synthetases, approximate the above functionality? [254] Here's the bottom line as described by Meyer. First, the proposed ribozyme does not bond ATP to an amino acid; the engineer fiddles the chemistry by providing amino acids already linked to AMP. Second, it does not attach an amino acid to a specific tRNA with its corresponding anticodon; the ribozyme binds a particular amino acid to itself: a molecule that does not have a specific anticodon like a tRNA. [255] He emphasizes:

> [T]his RNA does not carry an anticodon binding site corresponding to a specific codon on a separate mRNA transcript. Thus, *it has no functional significance within a system of molecules for performing translation.* [256]

Notice how misleading the scientists' language is: the whole claim is just a farce! If one does not know, in detail, what is going on, one

---

[254] Meyer, *Signature in the Cell*, p. 308.

[255] Meyer, *Signature in the Cell*, p. 309.

[256] Meyer, *Signature in the Cell*, p. 309 (emphasis in original).

almost concludes that we are on the verge of producing a living cell; in fact some atheistic books claim that it is just around the corner.

Let us retrace our steps a little. Remember Dawkins claim regarding viruses having survived solely through RNA? Here's his exact quote:

> Some viruses have no DNA at all. RNA is their genetic molecule, solely responsible for carrying information from generation to generation. [257]

His statement omits to mention that viruses survive only by hijacking the replicating system of the host cell: the latter is obliged to dedicate itself to copying the virus and prevented from doing what it is supposed to do. A virus, by definition, cannot survive in isolation from its host; it is doubtful whether science considers viruses as living organisms because of this dependence on another living organism: they cannot reproduce themselves; they can only multiply with the help of other organisms. Cars multiply because human beings are infatuated by them; but one cannot claim that they are alive.

To claim that viruses have survived from generation to generation and propose it as an origin of life scenario is, to say the least, misleading as a scientific statement. Why? Because viruses need a fully functional cell to replicate, and this (as we have seen) is not an easy luxury to come by. I am sure Dawkins knows all this; so why doesn't he tell the whole story to the ordinary public?

As an aside, it makes one wonder whether viruses are ever advantageous. Some scientists believe that viruses kill many bacteria that may be very dangerous to living organisms; others believe that they sometimes help in the evolution process by introducing significant changes in an organism's DNA; still others believe they keep immune systems taxed and vibrant; finally, others believe they help in genetic research. Funny how something seemingly completely useless and a real pain in the butt can inadvertently be so useful to us.

---

[257] Dawkins, *The Greatest Show on Earth*, p. 421

Now let us have another quick look at the world of probabilities. Meyer notes that in an article entitled "Aminoacyl-RNA Synthesis Catalyzed by an RNA", published in the *Science* periodical in 1995, biologist Mali Illangasekare and colleagues reported that the first researchers who found a single (just one) RNA molecule capable of *self-amino-acylation* (attaching an amino-acid to itself) had to sift through a pre-engineered pool of $1.7x10^{14}$ RNA molecules. [258] The odds against collecting, in close proximity, twenty such ribozymes, one for every amino acid, so they may function as a combined system corresponding to the genetic code, would be no better than 1 chance in about $4.1x10^{284}$ $[=(1.7x10^{14})^{20}]$. I don't think it is necessary for me to re-iterate how prohibitively small a probability this is. [259] However, recall that according to mathematician William Dembski, the total probabilistic resources of the observable universe (since its beginning) add up to only 10^{141} opportunities; so we would need about $4.1x10^{143}$ universes, similar to ours, to end up with something like the above situation.

We shall now analyze the attributes of the synthesized free-standing ribosomal RNA (rRNA) molecules that have exhibited a *peptidyl transferase* quality (an ability to assist bonding amino acids linearly). Firstly, they can only catalyze peptide (linear) bonding in the presence of another catalyst: they cannot accomplish it on their own. Secondly, when isolated from the proteins they normally associate with (inside the ribosome) rRNAs do not force amino acids to link together exclusively into linear chains (which is essential to form properly functional proteins): many other side chains with multiple side branching also occur. Thirdly, there is absolutely no mention of any mechanism showing how the exact sequence specificity, which is required by all proteins, is achieved. [260]

It is a very common experience in life that cooperation produces magnified results, where the whole is obviously much greater than the

---

[258] See Illangasekare, Sanchez, Nickles, & Yarus, "Aminoacyl-RNA Synthesis Catalyzed by an RNA"

[259] Meyer, *Signature in the Cell*, p. 536, n. 23.

[260] Meyer, *Signature in the Cell*, p. 309.

sum of the parts: that is, what a whole team can achieve far outweighs what its individuals, added together, can achieve in isolation. What two people can accomplish together is unbelievably more than the sum of what the same two people can accomplish separately. This experience is evident in warfare, at work, and even in games like chess: for example, one cannot expect to infiltrate enemy territory using one piece at a time. Likewise, water never goes up a hill; however, if there is a lot of water, it gradually accumulates and rises up the banks of a river.

Enzymes (protein-based biological catalysts) can accomplish a complex two-stage reaction where ordinarily only one would occur. If the first stage of such a reaction releases energy, it will most probably happen spontaneously; however, if the second stage of the reaction requires energy, the entire reaction may not happen. Enzymes team up the two stages of a complex reaction together; they use the energy available from the first stage of the reaction to supply, immediately afterwards, the energy required by the second stage of the reaction. Provided the energy released by the first stage of the reaction is greater than the energy required by the second stage of the reaction, the complex two-stage reaction will probably take place. Enzymes are able to accomplish this feat as a result of their complex three-dimensional geometry; they are able to hold all the molecules involved in each step of a complex reaction in close proximity, coordinate their interactions, and combine the energy from one stage with the requirement of the other. A compound catalyst has an obvious advantage over two separate catalysts acting independently: the latter combination, most probably, cannot accomplish what the former can. [261]

In conclusion, the current research on the RNA world translation system, after more than thirty years of it, does not establish the plausibility of ribozyme-based (RNA-catalyst-based) protein building, never mind the eventual transition to enzyme-based

---

[261] Meyer, *Signature in the Cell*, p. 310.

(protein-catalyst-based) protein building that the RNA world scenario expects. [262]

(4) Origin of Genetic Information Remains Unexplained

The RNA world hypothesis proposes short chains of RNA molecules forming randomly in a prebiotic soup of chemicals on the earth. After a large population and variety of these molecules came into existence, some varieties would acquire the ability to copy themselves. This capacity to self-replicate would enable these molecules to make more and more copies of their own sequences, and consequently favor their survival over all other RNA molecules that could not self-replicate. In other words, natural selection would pick the specific sequences that the first self-replicating molecules happened to have. This sounds perfectly reasonable.

The problem with this scenario is that there is no known RNA self-replicating molecule; the current evidence does not support the theory: far from it! [263] No one has yet engineered a fully self-replicating RNA molecule. A molecule that can copy a portion of its own sequence has been engineered; however, only about ten percent (10%) of the molecule could be copied. But that was not the only problem. In order to be able to do so, it needed a complementary copy (similar to a photographic negative) of itself nearby; naturally, this was provided by human intervention. And still, that is not the whole story, either. In order to find this partial (10%) self-replicator, scientists had to sift through a pool of one thousand trillion (10^{15}) other RNA molecules specifically created for this purpose. Experimental evidence, therefore, seems to indicate that RNA molecules with the capacity to copy their own sequences, if they exist at all, are extremely rare among the possible RNA base sequences: that is, in a random (non-engineered) pool of RNA molecules. [264]

---

[262] Meyer, *Signature in the Cell*, p. 312.

[263] Meyer, *Signature in the Cell*, p. 313.

[264] Meyer, *Signature in the Cell*, pp. 313–4.

As if this was not enough of an obstacle, another even more serious problem the self-replicating molecule hypothesis faces is the fact that for an RNA molecule to be able to self-replicate, it must possess enzymatic (protein-based catalytic) properties; which means that it must be long enough to form a complex three-dimensional structure. Two prominent origin-of-life researchers, Gerald Joyce and Leslie Orgel, who studied the RNA world hypothesis in detail, believed that such a self-replicating RNA molecule could, theoretically, form if it consisted of a strand just fifty (50) bases long; however, it seems that they were rather skeptical that an RNA molecule of such length would do the job. [265]

A prominent ribozyme engineer, Jack Szostak, and his colleagues confirmed their skepticism experimentally: they found that it typically takes at least one hundred bases to form structures capable of catalyzing simple binding reactions. They estimated that for a *ligase* (a strand-binder) to be also capable of performing the other functions that synthesizers of long chains of nucleic acids must perform, that is, "proper template binding, fidelity and strand separation" may require a sequence of between two hundred (200) and three hundred (300) nucleotides long. [266] The ribozyme mentioned above, the one that could copy about ten percent (10%) of its sequence, was actually one hundred and eighty-nine (189) nucleotide bases long; so the actual evidence seems to agree. [267] [268]

The issue may be even more fundamental than just basic length. RNA molecules have only four building blocks (bases) to choose from; proteins have twenty building blocks (amino acids) to choose from. The four bases are all hydrophilic (water-attracting) molecules; on the other hand, some amino acids are hydrophilic, some are hydrophobic (water-repelling), some are acidic, and some

---

[265] Joyce & Orgel, "Progress toward Understanding the Origin of the RNA World", p. 33.

[266] Szostak, Bartel, & Luisi, "Synthesizing Life", p. 389.

[267] Johnston, Unrau, Lawrence, Glasner, & Bartel, "RNA-Catalyzed RNA Polymerization", p. 1321.

[268] Meyer, *Signature in the Cell*, p. 314.

are basic (anti-acidic). The greater diversity of choice proteins have, over nucleotide bases, gives them the ability to form more intricate three-dimensional geometries and create more variable electrostatic (charge) fields. With its limited choice of just four bases, it may just be impossible for RNA to achieve the physical properties necessary to perform binding and copying operations. [269]

Assuming, for the time being, that RNA-based chain synthesizers capable of copying their own sequence are possible, or existed in the remote past; from experimental evidence, we know that they would have to consist of a complicated exactly-defined sequence. [270] Chance has to act alone until the first self-replicating molecules have come into existence; natural selection simply cannot help reduce the odds against building a self-replicator since natural selection kicks in only after self-replication has first occurred. Thus, invoking natural selection in evolving or developing a self-replicating RNA base sequence doesn't really make sense: it's meaningless! [271] As we saw above, even self-declared atheist and evolutionary biologist Richard Dawkins seems to agree that the first "replicator" has to happen "by accident". Recall that in his book *The Selfish Gene*, Dawkins writes:

> At some point a particularly remarkable molecule was formed by accident. We will call it the *Replicator*. [272]

Since chance is the only game in town here, let us have another quick look at the world of probabilities. The odds against a two hundred (200) base RNA molecule arising randomly in a prebiotic soup of chemicals are 1 to about 2.6×10^{120} ($=4^{200}$). Furthermore, the odds against this happening are increased significantly by the fact that destructive cross-reactions are bound to occur between both

---

[269] Meyer, *Signature in the Cell*, p. 314.

[270] Meyer, *Signature in the Cell*, p. 314.

[271] Meyer, *Signature in the Cell*, p. 316.

[272] Dawkins, *The Selfish Gene*, p. 15 (emphasis in original).

biologically significant and biologically insignificant molecules, which normally coexist in any realistic prebiotic soup. [273]

Moreover, origin-of-life researchers Gerald Joyce and Leslie Orgel draw attention to the fact that a single-stranded RNA molecule cannot act as both copying machine (*replicase*) and copying material (*template*). For a single-stranded RNA catalyst to produce an RNA molecule identical to itself, there must be, nearby, an appropriate RNA molecule to function as template. However, this template cannot be an exact copy of itself. Joyce and Orgel also point out that this RNA template would have to be the exact complement copy (like a photographic negative) of the replicase. [274] Hence, we do not only need a self-replicating RNA molecule, but we also need its exact complement to happen to be nearby. Once this rare encounter of the two complementary strands happened by chance, the replicase could then make a copy of itself by making a complement copy of its complement (that is, by transcribing the template).

Theoretically, it is possible to copy a replicator in two steps if it encounters a universal RNA replicator: meaning, a molecule capable of copying any other RNA molecule. This universal replicator would not need to be the exact complement of an ordinary (template-dependent) replicator. This way, after the universal replicator had produced a transcript of the template-dependent replicator, the original (template-dependent) replicator could then transcribe the transcript. The end result would be an exact copy of the original (template-dependent) RNA replicator. Of course, the universal replicator would have a more complex sequence and structure than the template-dependent replicator because it possesses a more demanding feature.

Ribozyme engineers were, so far, unable to design an RNA molecule that can copy any RNA sequence; all known replicators to date are template-dependent: they can only copy complements of themselves with appreciable fidelity. This may mean that, in practice, it is more probable that an RNA replicase and its exact complement

---

[273]   Meyer, *Signature in the Cell*, p. 315.

[274]   Joyce & Orgel, "Progress toward Understanding the Origin of the RNA World", p. 33.

arise by accident in close proximity than for a universal replicator to arise alone: it is more chemically realistic though perhaps less feasible theoretically. [275]

The odds against a two hundred (200) base RNA molecule and its exact complement happening together in close proximity is 1 to about 6.7×10^{240} [$=(4^{200})^2$]. Recall that the probabilistic limit of the whole observable universe (since its beginning) is only about 10^{141} possibilities. [276] So we will need about 6.7×10^{99} universes like ours to encounter something like this.

Calculations by Orgel and Joyce show that for this to be a plausible scenario, we would need a collection of the order of 10^{48} RNA molecules. [277] The weight of this amount of RNA molecules would, by far, exceed the weight of the whole earth; which, of course, boils down to the fact that finding a primitive replicator by chance is highly improbable. [278] Out of the whole size of the earth the two complementary strands must happen to be in very close proximity: within the size of a living cell—a size so small that it is invisible to the human eye. How probable do you think this might be? Furthermore, the above calculation is rather optimistic because, in the above calculation, Joyce & Orgel assume that a fifty (50) base RNA molecule could function as a self-replicator. As we have seen above, more than likely a two hundred and fifty (250) base RNA molecule is required.

Reading between the lines, it is probably quite transparent to the reader that Joyce and Orgel tend to down-play the origin-of-life problem; so the information they provide qualifies as solid scientific evidence if not optimistically biased towards the RNA world hypothesis. [279]

---

[275] Meyer, *Signature in the Cell*, pp. 538–9.

[276] Meyer, *Signature in the Cell*, p. 315.

[277] Joyce & Orgel, "Progress Toward Understanding the Origin of the RNA World", p. 33

[278] Meyer, *Signature in the Cell*, p. 315.

[279] Meyer, *Signature in the Cell*, pp. 315–6.

From the above, we may conclude that neither chance nor chance in combination with natural selection is able to explain the origin of the information-rich sequencing that is necessary for a self-replicating RNA molecule. Can self-organization, like chemical or physical attraction, be the cause? Again this is a dead end. In actual fact, RNA bases (and also DNA bases) do not manifest any bonding preferences that can explain their specific arrangements. One can compare them to a message consisting of magnetic letters on a refrigerator door: the message depends entirely on the individual's thoughts. [280]

According to the RNA world hypothesis, the original RNA-based translation system would, at some point, have to transition to the current protein-based translation system. This would eventually require the production of more than one hundred (100) different proteins; every one of these proteins would require an information-rich RNA template for its construction. Therefore, it is not only one replicator pair that has to evolve, in close proximity to one another, but over a hundred (100) of them. [281]

Recall that, the RNA world hypothesis was originally proposed by origin-of-life scientists as an explanation for the chicken and egg functional interdependence problem between DNA and proteins. Dawkins' introduction of the RNA world hypothesis was:

> DNA can replicate, but it needs enzymes [protein-based catalysts] in order to catalyze the process. Proteins can catalyze DNA formation, but they need DNA to specify the correct sequence of amino acids. How could molecules of the early Earth break out of this bind and allow natural selection to get started? Enter RNA. [282]

The RNA world hypothesis assumes that natural selection would, somehow, gradually produce first simple and then more and more complex *replicators*. This may sound plausible at first blush;

---

[280] Meyer, *Signature in the Cell*, p. 316.

[281] Meyer, *Signature in the Cell*, pp. 316–7.

[282] Dawkins, *The Greatest Show on Earth*, p. 420

however, as we have seen, an RNA molecule will not copy itself (replicate) unless it is already a complex molecule: it has to be at least about two hundred (200) sequence specific nucleotide bases long; as we have also seen, even one molecule like that is practically impossible to encounter by accident (chance). So, in reality, the RNA world hypothesis does not even try to address the specific (rich) information problem: the most obviously demanding question is simply overlooked!

(5) Ribozyme Engineers Do Not Simulate Natural Chemical Conditions

RNA world theorists realized that complexity (length of molecules) is a key attribute in a self-replicating molecule. Hence, they believed that molecules which chemically bond (link) RNA bases together (*ligases*) were the first molecules that arose; they therefore proposed that ligases were the progenitors from which the first self-replicating RNA molecules evolved. This is probably a reasonable assumption since the RNA world hypothesis assumes that the molecules grow gradually in complexity, and growth is the direct result of ligation. Consequently, most ribozyme-engineering was performed on ligases: ribozymes that can link two short RNA chains (*oligomers*) together.

In the *directed-evolution* method, scientists try to simulate a prebiotic natural selection process among ligases. They use chemical traps to screen pools of RNA molecules and isolate those molecules that perform linking ability. They then randomly mutate (alter) some part of the sequence of these molecules and select the molecules with the best linking abilities from this lot. This process is then repeated over and over again until they hit on something interesting. [283]

Ribozyme engineers, therefore, intervene intelligently and with foresight to produce the desired results. This is unscientific and a form of cheating; the reason is that ribozyme engineers are actually

---

[283]   Meyer, *Signature in the Cell*, p. 318.

doing what natural selection itself is supposed to do, instead of allowing the laws of nature do it. Natural selection acts blindly and opportunistically. Natural selection wouldn't know that ligases are important for replication until replication actually happens. It will totally ignore the relevance of the bonding of molecules because it confers no functional advantage, by itself, until a self-replicating molecule arises; an irrational or unintelligent process obviously has no such foresight. Are scientists trying to do science or to, somehow, prove their belief? Has science turned to a religion now, a matter of faith: what mainstream scientists believe? If not, why do scientists overlook such meddling with the laws of nature? [284] Is this research? Initially, perhaps it is: until one finds a few things out; but not as a final conclusion. Meyer summarizes current ribozyme engineering methods as follows:

> Thus even if ribozyme experiments succeed in significantly enhancing the capacities of RNA catalysts, it does not follow they will have demonstrated the plausibility of an undirected process of chemical evolution. [285]

If anything, if you think about it, it is more an indication of design: the guidance of intelligent scientists. The designer of life may have been an extraterrestrial for all I care; or life could have been ejected or transported from another part of the universe. The point is that, as far as science can tell currently, life seems to have been initially designed.

One cannot extrapolate one's conclusions to a so called beginning that exists only in one's own imagination. For example, in his book *The God Delusion*, Richard Dawkins asks: "who designed the designer?" meaning God. [286] This kind of question is absolute scientific nonsense. Science should only follow the evidence in question, one step at a time, to its logical conclusion. We have a

---

[284] Meyer, *Signature in the Cell*, pp. 319–20.
[285] Meyer, *Signature in the Cell*, p. 321.
[286] Dawkins, *The God Delusion*, p. 147

saying where I originally come from that states: "Oil always floats on water"; meaning that, little by little, the truth will surface and eventually prevail. The bottom line is current scientific evidence strongly points to life's being designed.

Dawkins' labeling "Intelligent design" with "creationism in a cheap tuxedo" [287] does not refute their scientific arguments. *Intelligent Design* proponents disagree, in many ways, with mainstream scientists; but they also disagree, in just as many ways, with *Creationists*. Intelligent Design proponents have one very important belief in agreement with Creationists, namely, belief in a supernatural intelligent being who is the first cause of everything in the universe.

This paragraph is an informatory note on *Creationism*. Basically, Creationists believe in the infallibility of the bible; Creationism is, therefore, a derogatory term in scientific circles because Creationists' only, so called, scientific evidence is practically limited to quoting the bible. Creationism has also been called a pseudoscience. For example, they believe that the universe and the earth have only existed a few—less than ten—thousand years, because that is what it adds up to in the bible accounts of past human generations. They argue that most of the geological features of the earth's surface are the result of a great flood as described in the bible—estimated at roughly four thousand five hundred (4500) years ago. They also believe that dinosaurs and people lived together at some time, because the bible says that all animals and man were created within days of each other. They therefore conclude that the human fossil record must extend over the same period of time as the fossil record of all other animals, since they also assert a very limited variability of species—if any. [288]

Intelligent Design proponents do not agree with most of the above biblically inferred "science"; but through their scientific observations they conclude that the universe and life were designed by an intelligence of some sort.

---

[287] Dawkins, *The God Delusion*, p. 138 (footnote).
[288] Meyer, *Signature in the Cell*, p. 553, n. 34.

In conclusion of the above analyses, therefore, chemical (abiotic) natural selection in the origin-of-life question appears to be atheistic wishful thinking, and the RNA world hypothesis appears to be atheistic clutching at straws; they both seem plausible at first blush, but they are actually full of holes when examined carefully.

Information and Probability

Intelligent information is inversely related to probability: that is, the higher the probability of something happening, the less information one can acquire from it and vice versa. For example, imagine a country where it never rains in summer and it is the middle of summer; if someone predicts that in a couple of weeks it is going to be sunny there, no one is going to be surprised by that prediction. Why? Because the probability of such a thing happening is high. On the other hand, imagine a country where it never snowed in the past; if someone predicts that it is going to snow for the first time in a couple of weeks, that would be significant information—assuming it actually happens, of course. Why? Because the probability of such a thing happening is low, very low. This is also the case in other natural phenomena like crystal-forming or wind-vortices; there is an apparent order in nature that follows physical or chemical laws, but they do not lend themselves to deliver intelligent information.

For example, common salt (sodium chloride—NaCl) forms crystals in the shape of an equally-sided cube with every sodium (Na) atom surrounded by six chlorine (Cl) atoms and vice versa. Why? Because of the electrostatic attraction (recall: unlike charges attract) between the Sodium ion (Na^+), which loses its outermost (isolated) electron from its atom, and the Chlorine ion (Cl^-), whose atom gains an electron—thereby both acquiring a complete (stable) outer electron shell. This kind of order in nature does not lend itself to delivering intelligent information. Why? This is like writing a whole book with the word "love" as opposed to writing a whole romantic novel.

Another example of a natural force is a tornado: winds from various directions collide together thereby creating violently rotating winds; when combined with the force of gravity they produce a funnel-shaped vortex, which looks beautiful and organized, in a way. Atheist and astrophysicist Fred Hoyle asks whether anyone would even consider the possibility of a crude and brute natural force, like a hurricane or a tornado, ever putting together a complete "747" jumbo-jet from the spare parts in a junkyard; since all the parts required by a jumbo-jet are actually present in a junkyard—in some form or another. [289] He famously compared this scenario to life's arising from its building blocks by using strictly natural phenomena. The specificity found in life (DNA) cannot imaginably be achieved by one of nature's blunt and undifferentiated processes. [290]

What I am writing in this chapter, for example, does not depend on the paper on which it is written, or the ink used, or the language used, or the font used, or the storage device used, or the program used; in other words, the physical and chemical properties of the medium on which information is expressed is totally independent of the information conveyed. Why then do we expect intelligent information to follow the laws of nature? Its natural environment is only a means of expressing the information.

In view of the above considerations, to expect the natural laws to be conducive to creating intelligent (specified) information is a contradiction in terms, in my opinion. The probability of the physical and chemical laws happening is close to one hundred percent (100% or 1); as we have seen in the introductory paragraph of this section, we cannot expect something like this to give us any intelligent information—the non-obvious. In such cases, the laws of nature will probably interfere with the information one wants to convey, rather than help. To look for the origin of intelligent information, such as is required in living organisms, arising from the laws of physics and chemistry (abiogenesis) is, therefore, an exercise in

---

[289] Hoyle, *The Intelligent Universe*, p. 19
[290] Meyer, *Signature in the Cell*, p. 257.

futility. Because of this logical exclusion—the inverse relationship of intelligent information and probability—I do not think we will ever explain the origin of life through abiogenesis: we are literally looking in the wrong direction. How do we know that there is intelligent information in a living cell? We simply have to look at the high improbability of its happening by chance: similar to our looking at a written book, say. Does the reader honestly think that the laws of physics and chemistry will ever explain how and why Shakespeare wrote his works?

Natural selection may be a bit of an exception compared to the other two natural laws; but then it is not a law where the probability is close to 100%; there is an element of uncertainty: an element of choice in living organisms—and living organisms, especially intelligent species, are always unpredictable.

On theoretical grounds, therefore, future discoveries explained by natural causes will most probably not reveal the cause of specified information because of the inverse relationship described above. Necessity increases probability and decreases information carrying capacity. Natural laws lack complexity: they are associated with necessity, the probability of which is close to one hundred percent (100% or 1). Having expressed my opinion, as a scientist, I shall still keep an open mind: I shall always listen to solid scientific evidence or to someone's making a good subtle point; I've been wrong before!

Living Cell

The *cell* is the smallest building block of any living organism; no independent life can exist below this level. It can be compared to the most efficient factory imaginable that: (1) makes building materials (proteins) for the organism, (2) places them in the right location and in the right amount (operating system), (3) converts nutrients into energy, (4) copies itself for repair and reproduction purposes, (5) checks and corrects for errors in the genetic code, and (6) recycles worn out material. Following are some details on how it works.

The living cell factory is enclosed inside a wall, commonly known as a *membrane*, separating it from the outside world. Cell membranes are made from *amphiphilic* molecules, which are comfortable in both a watery and an oily environment: they are similar to soaps and detergents. As we have already seen, they are rather elongated molecules: one end of which prefers a water environment while the opposite end would rather be out of water. (This is the direct result of electrostatic attraction and repulsion.) If a significant number of amphiphilic molecules happen to come together in a watery environment, the *hydrophobic* (water-fearing) ends will all huddle together to exclude any water in between while the *hydrophilic* (water-loving) ends make intimate contact with the watery environment; thus, they form two sheets of oppositely oriented molecules, commonly known as a *lipid bilayer*. The two sheets will eventually close up like a (spherical) soap bubble. Since the middle section (between the two sheets) of the membrane is an oily environment, molecules that are soluble in water, such as salts and sugars, find it rather difficult to cross the membrane. [291]

About ninety percent (90%) of the cell's interior is filled with a jelly-like clear substance, commonly known as the *cytoplasm*, which contains dissolved materials such as salts, sugars, amino acids, and proteins floating freely around; still, about ninety percent (90%) of the cytoplasm itself is just water: thus, about eighty percent (80%) of the living cell's total content is simply water. The cytoplasm is the free space where most of the cell's chemical reactions take place. [292]

Eukaryotic cells contain a single *nucleus*; such cells are the building blocks of all multicellular organisms. These cells also have a number of subcellular entities, called *organelles*, which are separated by their own membranes. There are many organelles scattered around the cytoplasm.

In particular, the nucleus is such an organelle; it holds the cell's DNA. Because DNA contains the life code of both current and future

---

[291]   Behe, *Darwin's Black Box*, pp. 274–5.
[292]   Distefano, *Biology*, p. 75.

generations, it is overly protected by its membrane; no unwanted material enters the nucleus and, conversely, no important material leaves it and gets lost or compromised. The nucleus' membrane acts as the traffic gatekeeper allowing ingress to or exit from the nucleus only through a password. The nucleus' membrane is perforated by large eight-sided holes, called *nuclear pores*; these pores are not just passive holes but they are active gatekeepers. No large molecule, like proteins or RNA, gets past the nuclear pores without the correct password: these are special sequences (codes) that indicate that they belong in the nucleus. Thus large molecules that belong in the cytoplasm are kept out of the nucleus and vice versa. Notice the extra protection against an error catastrophe happening in the future: a special kind of genetic code protection that the RNA world hypothesis completely disregards. [293] Control through a password is definitely a sign of sophisticated intelligence at work: there is nothing like it in the natural laws.

There may be hundreds, even thousands, of *mitochondria* strewn inside a cell (depending on the cell type); they supply the energy required by the cell. They use the oxygen, introduced into the organism through respiration, to transform nutrients into the cell's chemical energy currency (ATP). Mitochondria have two membranes; the acidity of the space between the two membranes is controlled by the quantity of nutrients burnt (chemically combined with oxygen). This generates an acidity gradient between the space enclosed by the innermost membrane and that enclosed between the two membranes. Acid flow across the innermost membrane into the space between the two membranes produces energy: the same way water allowed to flow over a dam generates power; the amount of energy the cell requires is thus controlled by the magnitude of this acidity gradient. [294] As I already hinted, the cell is truly a self-sufficient and most complex factory engineered to the minutest detail; even membranes serve different forms of designed functions.

---

[293] Behe, *Darwin's Black Box*, p. 276.
[294] Behe, *Darwin's Black Box*, p. 276.

There may be several hundred *lysosomes* scattered inside a single animal cell (depending on the cell type); these are small, single-membraned organelles that specialize in waste disposal. The cell recycles its waste products economically: in a way that most modern factories have not even dreamed of. Lysosomes contain more than fifty (50) different enzymes (protein-based catalysts) which have the ability to break down molecules that have deteriorated. These enzymes are assisted by the acidity in the lysosomes, which is between one hundred (100) and one thousand (1000) times that in the cytoplasm. The higher acidity in the lysosomes first helps untangle the damaged molecules; after which, the degenerative enzymes are able to attack and decompose them back to more basic chemicals such as amino acids. Molecules that have outlived their usefulness are transported through the cytoplasm to the lysosomes in membranes, called *vesicles*. [295]

Ribosomes have no membrane; they are found free in the cytoplasm or bound to the *endoplasmic reticulum* (ER). Ribosomes are the protein builders of a cell; so several ribosomes may be attached to the same RNA strand. Ribosomes have only a temporary existence; when they have synthesized a polypeptide chain, the two sub-units separate and are re-used or broken up. In a mammalian cell there may be as many as ten million (10^7) ribosomes. [296]

There is only one *endoplasmic reticulum* (ER) in a cell; however, its membrane constitutes about half of the total membrane surface of an animal cell. The ER is organized in a netlike labyrinth of branching flattened tubules and sacs extending across the cell. The ER is divided into two components: the rough ER and the smooth ER. The rough ER and the smooth ER constitute two separate complex networks of membrane-enclosed flattened tubules; the rough ER gets its uneven appearance from the millions of ribosomes studded onto it. The rough ER, with its attached ribosomes (protein builders), is involved with the production, folding, quality control, and dispatch of

---

[295]   Behe, *Darwin's Black Box*, p. 276.

[296]   http://bscb.org/learning-resources/softcell-e-learning/ribosome/   (19 September 2015)

proteins. The smooth ER is largely associated with *lipid* (fat) synthesis and metabolism; in addition, it is involved in steroid and hormone production, and it also exhibits detoxification functions. [297] [298] The reader's attention is drawn here to the additional design features of the cell, namely, quality control and detoxification; not to mention the sophisticated immune (defense) system it possesses to combat pathogens (which I shall not address).

The *Golgi apparatus* consists of a stack of membrane-enclosed flattened tubules, located between the ER and the cell outer membrane, into which proteins made in the ER are sent for packaging and labeling; afterwards, they are expedited to their proper destination in different parts of the cell. The Golgi apparatus is thus a major collection and dispatch station (similar to a post office) of proteins received from the ER. [299]

The cell's shape is supported by a structural framework called the *cytoskeleton*; it consists of three major types of structural materials: *microtubules, microfilaments*, and *intermediate filaments*.

Microtubules have several functions. (1) First, microtubules form the *mitotic spindle*, which pushes one copy of every pair of chromosomes into each daughter cell during cell division. (Chromosomes are genetic material consisting of whole DNA strands: complete, from start to finish. There are twenty-three identical pairs, for a total of forty-six, such strands in the human genome.) (2) Second, microtubules constitute the backbone of hair-like structures projecting from the surface of a cell, called *cilia*, which are used like oars to propel the cell through its environment. (3) Finally, microtubules act as tracks (like railroad) over which molecular motors can carry cargo to distant parts of the cell.

Microfilaments are the thinnest filaments of the cytoskeleton; they keep the organelles in place inside the cell. Microfilaments consist of a twisted pair of linear polymers (strings) of the protein

---

[297] http://bscb.org/learning-resources/softcell-e-learning/endoplasmic-reticulum-rough-and-smooth/ (19 September 2015)

[298] Behe, *Darwin's Black Box*, p. 276.

[299] Behe, *Darwin's Black Box*, p. 276.

actin; they grab onto each other, twist and slide alongside each other, and thus they contract. Pinching or folding the cell's membrane at the right places gives the cell the desired shape. (The protein actin is also a major component of muscle.)

Intermediate filaments have a diameter between that of microfilaments and that of microtubules; while exhibiting considerable diversity in size, they act like steel girders: providing mechanical strength to cells. [300]

Plant cells contain additional organelles, namely, *Chloroplasts* and a *Vacuole*. Chloroplasts are organelles where *photosynthesis* takes place; a plant cell contains between ten (10) and one hundred (100) chloroplasts. Photosynthesis is the process which converts light energy into chemical energy and stores it in sugar molecules. This energy conversion takes place through the agency of the green pigment called *chlorophyll* which acts as an antenna to catch light. Chlorophyll absorbs red and blue light; so these colors are unavailable and are therefore invisible to us. It is the energy from the absorbed red and blue light that can be used for photosynthesis; green light is not absorbed by the plant and thus cannot be used for photosynthesis: it is reflected by the plant and is therefore visible to us, thus making chlorophyll look green. [301] Chloroplasts have a similar function to mitochondria: they both generate energy. They are also similar in other ways: they also have a double membrane, and the acidity of the space inside the double membrane controls the energy supply to the cell. Somewhat differently, however, the acidity inside the space between membranes is increased by passing the light energy through extremely complex machinery.

The *vacuole* is a membrane-enclosed reservoir for wastes, nutrients, and pigments. It occupies about ninety percent (90%) of the volume of some plant cells and is under high osmotic pressure: the tendency of a solution to take in water (see next paragraph). This

---

[300]   Behe, *Darwin's Black Box*, p. 277.
[301]   http://biology.clc.uc.edu/courses/bio104/photosyn.htm (19 September 2015)

pressure, pushing against the cell wall, stiffens the cell, thus also serving a structural role. [302]

This paragraph is a short explanation of osmosis and osmotic pressure. If a semi-permeable (porous) membrane (barrier) separates a solution from pure water, many molecules on both sides of the barrier will hit the membrane continuously; as a result, some of these molecules manage to cross the barrier. Since pure water molecules are usually smaller than the dissolved molecules (that is, solute molecules with water molecules clinging to them all around), more pure water molecules succeed in crossing the barrier than dissolved molecules, especially if the dissolved molecules are so large that they cannot pass through the pores of the barrier. This phenomenon is known as *osmosis*. So there is a tendency for the solution to take in pure water, thus diluting itself. This creates a measureable pressure, called *osmotic pressure*, from the water side to the solution side; it can be measured by applying a pressure to the solution side: just enough to stop the ingress of water. In general, if two solutions of different concentrations are separated by a porous barrier, the stronger solution will dilute over time, and the weaker solution will become more concentrated—until they both reach the same concentration.

Last Word

As far as we currently know, life on earth cannot be generated from inanimate matter: it can only come from other life. Scientists still cannot produce one single living cell yet, even if they start with all its constituent chemicals.

The complexity and intelligence in the genetic code is mind-boggling. The DNA of a living cell is very much like a computer program that not only runs the machine but also constructs the machine itself; it is much more complex than any computer program ever created by human beings. A single set of the 23 chromosomes (complete DNA strands) in the human genome consists of about three

---

[302]   Behe, *Darwin's Black Box*, p. 277.

billion (3x10⁹) nucleic acid base pairs (biological letters), which is equivalent to a billion (10^9) biological words. If written out on paper this code will take about a million (10^6) pages of a thousand (10^3) letters each, with no spaces; it is the equivalent of about a thousand (10^3) bibles. And all this (actually double this because chromosomes come in pairs) fits into the nucleus, an organelle so small that it's only visible under a microscope.

In living organisms, besides matter (atoms), there is stored another element that is not easy to define or describe: it is *information*. Much like a computer program or literary work, it is independent of the medium on which it is stored. Information needs a medium to be expressed on, and the format in which it is expressed has to match the particular medium used; however, that same information can theoretically be expressed on a totally different medium. For example, imagine a centuries-old stone inscription in Greek; it can be translated into English and typed on paper. Its format has changed completely, but the ideas conveyed are the same. The same thing with a computer program: when used on a basically different machine (computer), the machine language has to be changed, and possibly also the operating system, but ultimately it can achieve the same functions.

In this chapter we have practically ruled out all possibilities of the sequence-specific (intelligent) information, present in DNA, arising naturally: that is, by chance, physical and chemical laws, and natural selection—all inclusively. Since the only known source of sequence-specific, or intelligent, information (like computer programs and literature) is human intelligence, the best and most reasonable explanation for the source of the genetic code in DNA is some kind of intelligent designer: comparable, perhaps even superior, to human intelligence.

CHAPTER 7

EVOLUTION

Breeding

Charles Darwin's most famous book *The Origin of Species* consists of an integrated collection of factual evidence and phenomena from very diverse fields, namely, comparative anatomy, paleontology (fossil study), embryology, and biogeography. He starts his book by considering what is achievable through breeding animals.

Science philosopher Stephen Meyer gives us the following simple example in his book *Darwin's Doubt*. Imagine a sheepherder living in the north of Scotland. He wants his sheep to be woollier; perhaps he wants more wool for his own use, or maybe he also wants to sell it as an obvious commodity in a cold country. What he does is allow only the woolliest rams and ewes to breed, but not the others. This way, through heredity, he ends up with a predominantly wooly herd. Now suppose instead of the sheepherder's breeding woollier sheep, a number of successive extremely cold winters occur that result in the survival of only the woolliest sheep. The end result would be that natural selection has done the same job as that of an intelligent mind: the sheepherder's. [303] (I shall be using concepts and ideas from Meyer's book *Darwin's Doubt* extensively in this chapter.)

This actually means that what might seem like design or intelligence placing species in their right environment may only have been the consequence of a natural process of selection, or survival

---

[303] Meyer, *Darwin's Doubt*, pp. 5–6.

of the fittest in a particular environment. In other words, God does not necessarily have to be called into the picture as a requirement to explain how well animals are suited to their environment.

Is there any tangible experimental evidence for evolution? There is a lot. In particular, I think the following experiment with Escherichia coli (E. coli) bacteria—the bacteria that lives symbiotically (benefiting both host and parasite) in our intestines—is noteworthy.

Bacteria Experiment

In 1988, evolutionary biologist Richard Lenski started a laboratory experiment with E. coli bacteria that is still ongoing (in February, 2015). E. coli bacteria reproduce asexually: that is, by simple cell division. The experiment is described in detail by evolutionary biologist Richard Dawkins in his book *The Greatest Show on Earth*; I shall here summarize the highlights of his account.

Lenski took a genetically identical E. coli bacteria population and with it he infected twelve identical flasks containing the same nutrient broth: which included glucose ($C_6H_{12}O_6$) as the main food source. The bacteria populations in the original twelve flasks were called "tribes" (the twelve tribes of Israel come to mind). The twelve flasks were then placed in an incubator where they were kept warm and shaken to keep the bacteria well distributed throughout the liquid. A new set of twelve flasks, containing an identical nutrient broth as the original flasks, was prepared every day. The new flasks were then infected with liquid from the previous day: one new flask for every tribe; this was done by taking exactly one-hundredth of the volume of the old flask and adding it to the new flask. Since the new flasks contained a fresh supply of food, the population of bacteria in every new flask thrived and multiplied; however, the experiment was set up in such a way that the food always ran out and starvation set in well within the twenty-four hour period (the time for the next set of new flasks), hence the population in every flask always levelled off by the next day.

Thus, the experiment simulated the equivalent of (fast-forwarded) geological time; the bacteria went through repeated cycles of abundance followed by scarcity. Every now and then, at certain milestone populations, bacterial samples from every tribe were frozen; these samples were occasionally unfrozen and resuscitated: bacteria behave as if nothing happened to them when they are thus unfrozen. This unfreezing exercise was done to perform sideline comparative experiments with ancestors in the bacterial evolutionary history without stopping the main run of the experiment. [304]

In a bonanza environment both the strong and the weak individuals of a species will probably survive. However, when prolonged droughts and starvation set in, most probably, only the strong and the fittest of a species will survive. In Lenski's bacteria experiment, running out of glucose was the key factor in the experiment. There was obvious evidence that starvation was occurring every day; the population in every fresh flask was observed to start climbing, then it eventually stopped climbing and reached a plateau: the plateauing of the population was an indication of starvation setting in. (Hourly samples of a single drop of well mixed liquid could be weighed—before any water has had any time to evaporate—and the number of bacteria in the drop counted under a microscope: this could tell whether the number of bacteria per unit volume had increased in the meantime.)

Similar stress conditions are the ideal situation for Darwinian natural selection to kick in. Under these circumstances, one would expect that if a random mutation in an individual bacterium occurred that happened to help it exploit the use of glucose more efficiently, it would reproduce more profusely than the rest of the population in the flask when starvation set in. Thus, there would be a higher proportion of its kind in times of scarcity; eventually, in subsequent flasks, its offspring would monopolize the whole population of the tribe. Consistent with expectation and prediction, this is exactly what happened: in subsequent generations all twelve tribes got better than

---

[304]   Dawkins, *The Greatest Show on Earth*, pp. 117–9

their ancestors at exploiting glucose as a food source: the populations plateaued faster (in less time) after the flasks were renewed: their "fitness" improved. [305]

Incidentally, this is the way scientific experiments should work out: one predicts something and it actually happens, thus proving one right; one may also disprove the prediction, of course. There is no manipulation of conditions or data; and the odds ... well the odds must be significantly in one's favor for it to happen—if one is right, of course.

Also very interesting was the fact that although all twelve tribes improved their food exploitation abilities, or fitness, they did so in different ways: in other words, the twelve tribes did not achieve better fitness by the same genetic route, but all twelve tribes mutated differently. This was indicated by the fact that the daily population plateau times of all twelve tribes ended up all different. [306] This fascinating result is positive evidence in support of the randomness of mutation, which is a basic assumption in the process of Darwinian-style natural selection.

Now, how did Lenski make this fitness comparison? He took a sample of the latest population and put it in a virgin flask; then he put a same-sized sample of the unfrozen ancestral population in the same flask. But, wait a minute, how could he compare fitness when both bacterial samples were mixed in the same flask: how could he tell which is which? In order to compare fitness between modern bacteria and ancestral bacteria, they must have had some kind of distinguishing feature. How did Lenski achieve this? Everything was pre-planned at the start of the experiment; to appreciate how this was done we have to retrace our steps a little to the initial set-up of the experiment. [307]

The initial selection of the twelve tribes was from a genetically identical bacterial population, except for a single gene: "Ara": six of the tribes had the "Ara-" gene, while the other six tribes had

---

[305] Dawkins, *The Greatest Show on Earth*, pp. 120–1
[306] Dawkins, *The Greatest Show on Earth*, p. 121
[307] Dawkins, *The Greatest Show on Earth*, pp. 121–2

the "Ara+" gene. There is no known difference between these two bacterial strands: they behave identically, and the Ara genes have no known adverse or enhancing effect on fitness. The two strands exhibit no other known difference, except that when a certain indicator dye is applied to them the former turns red while the latter turns white. Competitive abilities were therefore determined by using Ara-modern bacteria, say, mixed with Ara+ ancestral bacteria (or vice versa) in the same flask. The ratio (proportion) of the two colored bacteria could then be determined by counting (under a microscope) the total number of each color in just a small drop of well-mixed infected liquid. Summarizing the results for all twelve tribes, Lenski's experiment showed that populations grew faster in later generations: that is, they plateaued (daily) in a shorter time after the flasks were renewed; and the average size of the bacteria also increased. In short, over time, they got better (more fit) at assimilating glucose. [308]

Although all twelve tribes grew fitter over the generations, they did not achieve this through the same genetic mutations. Currently, all twelve tribes seem to be approaching a growth (population and body size) plateau; however, all twelve growth curves (over time) had different shapes: some started growing slower but ended up bigger, while others started growing faster but ended up smaller. This observation indicates that the various tribes were following different genetic paths in reaching their maximum size. Again, this is direct evidence supporting random mutation while species evolve. [309]

Dawkins then makes a very interesting observation, which I shall quote to preserve its authenticity:

> But perhaps even more interesting is that sometimes a pair of tribes seem to have independently discovered the *same* way of getting bigger. Lenski … investigated this phenomenon by … looking at their DNA. The astonishing result … was that 59 genes had changed … in both tribes and *all 59 had changed in the same direction.* The odds

---

[308]   Dawkins, *The Greatest Show on Earth*, pp. 119–22
[309]   Dawkins, *The Greatest Show on Earth*, pp. 123–4

against its happening by chance are stupefyingly large. This is exactly the kind of thing creationists say cannot happen, because they think it is too improbable to have happened by chance. Yet it actually happened. And the explanation, of course, is it did *not* happen by chance, but because gradual, step-by-step, cumulative natural selection favored the same—literally the same—beneficial changes in both lines independently. [310]

I have to agree with what Dawkins says here: the odds are astronomically against this happening by chance alone. But it did happen: the evidence is there, and no one can deny it. Organisms are extremely well adapted to fine-tune themselves to any changing environment, and they are capable of great feats that chance alone cannot produce; provided, however, the organism can *first* accurately replicate (copy) itself.

So we do have a significant amount of conclusive evidence in favor of *microevolution*; but what about *macroevolution*? What about sudden, rather than just gradual, changes in a species? After all some form of sudden change is necessary if, according to Darwinian evolution, a *common ancestor* is supposed to give rise to all the different organisms on earth. We shall answer this question later on in this chapter; for now, I'll allow Dawkins to continue his account of Lenski's experiment.

Around generation thirty-three thousand (precisely 33,127) one of the tribes behaved very anomalously: the plateauing population of this tribe increased to about *six* times (wow!) what it had been, and also roughly what the other tribes were still plateauing to. This higher plateau was then reached by all its subsequent generations, but the other eleven tribes kept plateauing to their previous levels. From the very beginning of the experiment, besides glucose, there was a second nutrient in the broth: citrate ($C_6H_5O_7$). As one can guess from the sudden population increase, there was a lot of citrate in the broth; but normally, that is, in the presence of the oxygen in the air, E. coli

---

[310] Dawkins, *The Greatest Show on Earth*, pp. 124–5 (emphasis in original)

cannot use it. However, this tribe, and only this tribe, suddenly learnt to consume citrate besides glucose. [311]

The obvious question now arose: was this just a single mutation that eleven out of twelve tribes were not lucky enough to stumble on? Knowing, roughly, the mutation rate of genes in the bacteria's genome, Lenski figured that, after more than thirty-three thousand (33,000) generations, it was unlikely for this to be the case. His calculations showed that the odds were in favor of every single base pair having the opportunity of mutating at least once, in every one of the twelve tribes (see next paragraph). [312]

For those readers who may be interested, this paragraph estimates the approximate (1) daily bacteria populations, (2) gene mutation opportunities, and (3) base pair location mutation opportunities in Lenski's flasks. According to *Wikipedia*, "… each population … generated hundreds of millions of mutations over the first 20,000 generations". [313] So if we divide the total number of mutations ($Nx10^2x10^6=Nx10^8$, where N is a number from 2 to 9 inclusive, I guess) by the number of generations ($2x10^4$) of every tribe, we obtain an average number of mutations per day: so every tribe experienced $Nx10^8/2x10^4=Nx5.0x10^3$ mutations per day. Since one hundredth (1/100) of the total population was transferred to the fresh flasks every day, the daily initial population of the infected flasks increased a hundred times by the end of the day; hence, the daily number of generations in every flask would have been 6.46 (=2/log2). If one doubles a number 6.46 times ($=2^{6.46}$) one obtains a hundred (100) times that number. (One can easily see this: $2^6=64$ and $2^7=128$, now 100 is somewhere between 64 and 128 while 6.46 is somewhere between 6 and 7.) If we divide every tribe's number of mutations per day ($Nx5.0x10^3$) by the number of daily generations of every tribe (6.46) we obtain the number of mutations experienced by the infecting sample introduced into a fresh flask every day: ($Nx5.0x10^3/6.46=Nx7.5x10^2$),

---

[311]   Dawkins, *The Greatest Show on Earth*, pp. 126–7

[312]   Dawkins, *The Greatest Show on Earth*, p. 128

[313]   https://en.wikipedia.org/wiki/E._coli_long-term_evolution_experiment (23 September, 2015)

which works out to about Nx753. Again from *Wikipedia* we know that "… the mutation rate … in bacteria is roughly 0.003 mutations per genome per cell generation." [314] This means that we need about 333 (=1/0.003) bacteria at any one time to experience a single (just one) mutation in that group (notice that 1/333=0.003). So if we multiply this number required for one mutation (333) by the number of mutations the initial daily infecting sample experiences (Nx753), we obtain the infecting sample population, which turns out to be Nx753x333 or Nx2.5x10⁵. If we take N=4 (for rounding off purposes) we obtain about a million (4x2.5x10⁵=10⁶) bacteria starting the day, which will increase one hundred times to about a hundred million (10²x10⁶=10⁸) by the end of the day. (This is the result for item (1) above.) Now, let us see what would have happened in thirty-three thousand (33,000) generations. The total number of mutations in thirty-three thousand (33,000) generations is obtained by multiplying the number of daily mutations experienced by every tribe (Nx5.0x10³) by this number of generations (3.3x10⁴): Nx5.0x10³x3.3x10⁴, or a total of Nx1.7x10⁸ mutations were experienced by every one of the twelve tribes in thirty-three thousand (33,000) generations. From the "Table of Genome Sizes" we know that the E. coli genome has about five thousand (5,000) genes. [315] We can therefore determine the average number of opportunities any particular gene had to mutate if we divide the total number of mutations (Nx1.7x10⁸) by the number of genes in a bacterial genome (5.0x10³). It follows that every single one of the bacteria's genes had the opportunity to mutate Nx1.7x10⁸/5x10³, or about Nx3.3x10⁴ times. If we take N=3 (for rounding off purposes) we obtain 3x3.3x10⁴, or about 10⁵; that is, there were about one hundred thousand mutation opportunities per gene in thirty-three thousand (33,000) generations. (This is the result for item (2) above.) Again, the E. coli genome contains about five million (5,000,000) base pair

---

[314] https://en.wikipedia.org/wiki/Mutation_rate (23 September, 2015)

[315] http://users.rcn.com/jkimball.ma.ultranet/BiologyPages/G/GenomeSizes.html (23 September, 2015)

locations; [316] the number of opportunities of every base pair location in the bacterial genome can similarly be obtained if we divide the total number of mutations ($Nx1.7x10^8$) every tribe experienced by the number of base pair locations ($5.0x10^6$): so in thirty-three thousand (33,000) generations every single base pair location experienced $Nx1.7x10^8/5x10^6$, or about $Nx33$ mutations per base pair location. Again, if we take N=3 (for rounding off purposes), in thirty-three thousand (33,000) generations, every base-pair location should have mutated about 3x33, or about one hundred (100) times, in every flask. (This is the result for item (3) above.) So it is highly improbable for any base-pair location not to have mutated at least once. If you have a jar of one hundred (100) black marbles and drop one white marble among them, it would be very improbable to randomly pick the white marble. So one has to agree with Lenski's conclusion: namely, that all possible single mutations should have been experienced by every tribe.

There is another possibility however: that two (or three) mutations are required to take place together; in other words, without both mutations occurring together there will be no observed effect. For example, if, in order to assimilate citrate, the chemicals produced by a first mutation are used by a second mutation, then nothing would happen in the absence of either of them. If they were independent of each other, either one mutation would produce half of the effect, say, or some significant fraction of it; furthermore, the order in which they happened would not really matter if the observed phenomenon was simply an additive result of the two mutations. But that's not what happened here; the way things happened, so suddenly and convincingly increasing the population to about six times, suggests that the two mutations are *linked*. This also explains their rarity, namely, their occurring in only one of the twelve tribes. Consequently, Dawkins makes a very interesting conclusion, which I shall again quote to preserve authenticity. [317]

---

[316] http://users.rcn.com/jkimball.ma.ultranet/BiologyPages/G/GenomeSizes.html (23 September, 2015)

[317] Dawkins, *The Greatest Show on Earth*, pp. 128–9

> [T]he rarity of citrate metabolism suggests that we
> are looking for something more like the 'irreducible
> complexity' of creationist propaganda. This might be the
> biochemical pathway in which the product of one chemical
> reaction feeds into a second chemical reaction, and *neither
> can make any inroads at all without the other.* [318]

We shall discuss the concept of *irreducible complexity* later
on in this chapter; however, first I would like to give Dawkins'
opinion uninterrupted before proceeding any further. Basically
Dawkins believes that, given the fact that the age of the earth is such
a long time (about four and one-half (4.5) billion years), there was
enough time for anything classified as irreducible complexity to have
happened naturally: that is, through chance mutation in combination
with Darwinian-style natural selection. The time encompassed by
Lenski's experiment, twenty-five (25) odd years, is insignificant
compared to the age of the earth—and so it is!

Further evidence of there being some kind of link between a pair
of mutations is the fact that thawed-out ancestors of the anomalous
tribe after generation twenty thousand (20,000) showed increased
probability of subsequently evolving citrate absorption ability: no
ancestors prior to this generation did. One may, therefore, probably
conclude that after generation twenty thousand (20,000) the ancestors
of the anomalous tribe were sort of "primed" with a first mutation
(personally, I would guess at least two prior mutations) to be able
to take advantage of a second (or third) mutation whenever it
happened. [319]

Dawkins ends his account of Lenski's bacteria experiment with
the following conclusion:

> [N]ot only does it demonstrate the power of natural
> selection to put together combinations of genes that, by
> the naïve calculations so beloved of creationists, should be

---

[318]   Dawkins, *The Greatest Show on Earth*, pp. 128–9 (emphasis in original)
[319]   Dawkins, *The Greatest Show on Earth*, p. 130

tantamount to impossible; it also undermines their central
dogma of 'irreducible complexity'. [320]

One must admit that the above experiment is a beautiful scientific experiment in support of the theory of evolution; one might even call it "down-up" evolution: that is, from a less fit to a more fit organism. It also challenges the notion of irreducible complexity. Very importantly though, notice that currently, even after about 60,000 generations, E. coli bacteria still remained E.coli bacteria: it did not even change *species* (see the next section dealing with *taxonomy* or classification of organisms). Neither did it increase its number of genetic bases (locations) to become a "higher" organism, like a very tiny worm, say. Will it ever? I doubt it. We'll see if they continue the experiment.

I think that every scientist believes in microevolution: all species are obviously naturally equipped with survival and adaptive techniques. The big question is whether there is enough scientific evidence to justify belief in macroevolution: that is, evolution between bird, fish, insect, mammal, and reptile, say; on the flip side of the coin, it might just be the outcome of a vivid imagination, or atheistic wishful thinking—evidence is the determining factor.

The way I see it, evolution is basically a non-invasive and superficial process; it does not invade the basic structure of the animal's body plan: it is comparable to a band aid, repair surgery, or perhaps even constructive surgery. From an engineering point of view, a good design has to be in such a way as to possess a feedback loop capable of overcoming the natural tendency of things to deteriorate, or perhaps go wrong, in nature. According to *Murphy's Law*, if things can go wrong, they eventually will go wrong—at some time in the future; that is why organisms are equipped with micro-evolutionary capabilities. I think it is enough of a chore for an organism to combat the everyday problems and surprises in life; if it is a good design, its basic structure and function do not necessarily have to change; they only have to change drastically if it is a bad design.

---

[320] Dawkins, *The Greatest Show on Earth*, p. 131

Living organisms, as isolated systems, do not have to obey the second law of thermodynamics (the fact that everything runs down in the universe) all the time—like inanimate objects, which cannot help themselves: organisms can choose to utilize outside energy by ingesting food. In fact, while an organism grows to adulthood, it certainly does not appear to be obeying this law by deteriorating. However, in my opinion, the universe is made in such a way that the second law of thermodynamics, in the long run, trumps both growth and a down-up type of evolution: entropy may be decreased locally and temporally, but in the long run it will have to increase. The second law of thermodynamics seems stronger and more ingrained in the universe than evolution; no scientist in his right mind would dare challenge this law, while there is a significant number of scientists who challenge macroevolution. It seems that senility and macroevolution is where the second law of thermodynamics draws the line in living organisms. In the course of this chapter, I shall give a significant amount of evidence suggesting that macroevolution is implausible.

Taxonomy (Classification of Organisms)

The main taxonomical divisions are: *domain, kingdom, phylum, class, order, family, genus* and *species*: the latter representing the smallest differences. (1) Examples of domains are: eukaryotes, which have cells with nuclei, and prokaryotes, which have cells without nuclei. (2) Examples of kingdoms are: bacteria, fungi, plants and animals. [321] (3) Examples of animal phyla (the Latin plural of "phylum"), indicating different body plans, are: arthropods, which have an exoskeleton, segmented bodies, and jointed limbs; brachiopods, which have ciliated tentacles and a pedicle, attaching the organism to a solid substrate; chordates, which have a hollow dorsal nerve cord; cnidarians, which have stinging cells, like jelly-fish; echinoderms, which have spiny skins, like sea-urchins; mollusks,

---

[321]   http://waynesword.palomar.edu/trfeb98.htm (26 September 2015)

which are soft-bodied and have a mantle, like squids; and sponges, which have a perforated interior wall. [322] (4) Examples of classes, or animals with common traits, are: birds, fish, insects, mammals and reptiles. (5) Examples of orders are: carnivores, or meat-eaters; herbivores, or plant-eaters; insectivores, or insect-eaters; primates, which live on trees; and rodents, or gnawing animals. (6) Examples of families are: ape; bovid, which are cloven-hoofed, including cattle and sheep; and murid, which are small rodents like mice or rats. (7) Examples of genera (the Latin plural of "genus"), usually identified by the main species in Latin form, are: canine or dog-like; equine, or horse-like; and feline, or cat-like. [323] (8) Examples of species are: cat, cheetah, lion, puma and tiger in the feline genus; or donkey, horse and zebra in the equine genus; or coyote, dog, fox, jackal and wolf in the canine genus. It is probably obvious to the reader that there are overlaps between these classifications.

Let us take the tiger as a specific example and classify it completely to appreciate how taxonomy works—most of the names in taxonomy still use Latin words with an upper-key first letter; but since English is a Latin language, it is still possible to decipher their meaning. The tiger's species is, of course, "tigris" (tiger), of the genus "Panthera" (panther, or large cat), of the family "Felidae" (feline, or cat-like), of the order "Carnivora" (carnivore, or meat-eating), of the class "Mammalia" (mammal), of the sub-phylum "Vertebrata" (vertebrate, or having a backbone), of the phylum "Chordata" (chordate, or having a dorsal nerve cord), of the kingdom "Animalia" (animal), of the domain "Eukaria" (eukaryote).

Irreducible Complexity

Irreducible Complexity is a rather intuitive concept conceptualized and elucidated by biochemist Michael Behe in his book *Darwin's*

---

[322]   https://en.wikipedia.org/wiki/Phylum (08 February, 2015)
[323]   https://carm.org/dictionary-taxonomic-rank (08 February, 2015)

Black Box. In it, for simplicity and to convey the basic idea, he describes a mousetrap:

> The function of a mousetrap is to immobilize a mouse ...
> mousetraps ... consist of a number of parts: ... (1) a flat
> wooden platform to act as a base; (2) a metal hammer,
> which does the actual job of crushing the little mouse; (3)
> a spring with extended ends to press against the platform
> and the hammer when the trap is charged; (4) a sensitive
> catch that releases when slight pressure is applied, (5) a
> metal bar that connects to the catch and holds the hammer
> back when the trap is charged. (There are also assorted
> staples to hold the system together.) ...
> [D]etermining if a system is irreducibly complex is to ask
> if all the components are required for the function. In this
> example the answer is clearly yes. [324]

Not only are all the items of the mousetrap described above individually necessary for the operation of the system as a whole, but also the materials used and the dimensions of the individual parts are just as critical. If the metal bar were significantly shorter, for example, it would not be able to hold the hammer in the charged position; similarly, if it were made of a soft or thin metal. This does not mean that no use at all can be made of the parts in isolation; however, if one part is missing, or is changed significantly, the mousetrap fails to work as a system.

This makes things particularly difficult for the concept of Darwinian evolution, because any change experienced by the organism has not only to keep it alive, but it also has to give it an advantage over other current similar organisms for natural selection to be able to kick in. Now if one part out of a system of five parts, say, changes significantly so that the function of the five part system is compromised, the organism will end up with four or five useless or cumbersome parts. Obviously, this scenario is not conducive to creating a superior organism; hence, natural selection will not kick in!

---

[324] Behe, *Darwin's Black Box*, p. 42.

Alternatively, all five parts have to change together, simultaneously and in harmony with one another: for example, they can all become proportionately larger together. However, for this to happen, the function of the system in question has to be maintained. This has to be done ever so gradually; first one part increases very slightly in size so that it still fits and then another part, and so on for all five parts. And, remember, all this time natural selection has no foresight: it does not know that it has to grow bigger. But, admittedly, this does seem to happen as shown in the above experiment: where all twelve tribes grew bigger even though they might have needed more food; and also 59 genes changed together in the same direction. Even so, despite all these changes the function of the system did not change: the design did not change; it only adds up to an adaptation to circumstances—basically, a normal reaction to a feedback loop.

In the above experiment, one of the twelve tribes seems to have developed a system of two (or three) parts: namely, a system for using citrate as food. Notice, however, how long it took: about thirty-three thousand (33,000) generations in the exact same controlled environment, which is not usually the case in real life. There is usually an element of randomness in the environment; this might have changed the outcome of the above experiment in a real life situation. In any case, let us, for the sake of argument, assume that it can happen in real life since there is evidence in its favor.

Needless to say, it is even more difficult to imagine how an organism can ever develop a *new* functional system of five parts, say, rather than make minor, or even significant, changes to them simultaneously. It becomes much more of a challenge for an organism to develop a system of one hundred (100) *coordinated* genes, for example, the cell's translation system (for building proteins)—not just two, three, or even five parts working together.

In Ontario's "Lotto 6/49" lottery, six numbered balls are randomly drawn out from forty-nine differently numbered balls (1–49); every ticket played by the public consists of six numbers selected by the individual player. The jackpot consists of guessing all the six numbers drawn. It is not very difficult to guess two or three of the numbers drawn; hardly any money is paid for such a guess. The difficulty is getting all six numbers, as everybody knows—even a non-scientifically-oriented individual. The odds against guessing two numbers out of forty-nine are 1 to 1,176, while the odds of guessing six numbers out of forty-nine are 1 to 13,983,816. Imagine how difficult it would be to guess a hundred numbers out of a few billion, say. So for Dawkins to shoot down the concept of irreducible complexity with a phrase like "naïve calculations" is tantamount to inexcusable bias, in my opinion, because I don't think it's ignorance.

At the risk of being naïve myself, therefore, let us look at some numbers. In my opinion, mathematics is the strongest tool all the sciences have to determine whether or not they are on the right track in their hypotheses: barring experimental evidence, of course. Without numbers, science is sloppy: in fact, I define science as, "the formulation of observed phenomena in mathematical terms". Let's see what other experts in the field have to say about this question.

In his book *Darwin's Doubt*, commenting on coordinated mutations in proteins, Stephen Meyer writes the following.

> If coordinated mutations were necessary, then evolution at the genetic [protein-producing] level faced a catch-22 [chicken and egg question]: for the standard neo-Darwinian mechanism to generate just two coordinated mutations, it typically either needed unreasonably long waiting times, times that exceeded the duration of life on earth, or it needed unreasonably large populations, populations exceeding the number of multicellular organisms that have ever lived. [325]

Exactly, how improbable is this to happen? Meyer quotes probabilities from studies made by biochemist Michael Behe and physicist David Snoke.

> Behe and Snoke did, however, find one tiny "sweet spot" in which a gene requiring only two coordinated mutations

---

[325] Meyer, *Darwin's Doubt*, p. 245.

could arise. ... Such a gene could conceivably arise from 1
billion organisms in a "mere" 100 million generations. [326]

This means that $10^9 \times 10^8$, or 10^{17}, opportunities, in total, are required
for a combination of two coordinated mutations to take place in
constructing a new gene. In Lenski's experiment, I calculated (see
above) that the flasks started the day with about one million (10^6)
organisms and ended the day with about one hundred million (10^8)
organisms. So after thirty-three thousand (33,000) generations
evolution had ($3.3 \times 10^4 \times 10^8$), or 3.3×10^{12}, opportunities. Obviously,
Behe and Snoke seem to be far out, by a factor of about thirty
thousand ($3.0 \times 10^4 = 10^{17}/3.3 \times 10^{12}$) in their calculations; because
experimental evidence showed that it actually *did* happen in about
3.3×10^{12} opportunities. However, one must remember that Lenski's
experiment does not replicate a real life scenario: it is a controlled
experiment, and maybe this is why there is such a discrepancy in
numbers.

I don't want the reader to get me wrong, here. Lenski's experiment
is still a beautiful and very valid scientific, experiment, and I have no
intention of trying to knock it down or downplay its usefulness. In
any scientific experiment, the conditions under which an experiment
is carried out are probably more important than the actual results; that
is how anybody else can reproduce the results stated as discovered
experimentally. I am only trying to explain the discrepancy between
Behe's study and Lenski's experimental results. For example, in a
fire-conformance specification for a particular material, very rigid
conditions of flame quality are all defined, like: the ignited chemical
(fuel) used, its rate of flow, the rate of oxygen supply, orifice size,
and application method (horizontal, vertical, or tilted). But when a
fire occurs, nobody really knows what will actually happen with that
material; still, the specification gives us a rough idea, and a method
of comparing materials.

---

[326]   Meyer, *Darwin's Doubt*, p. 246.

I, personally, tend to lean towards Lenski's results because it is experimental evidence, rather than a theoretical calculation; however, Behe's and Snoke's, calculations will still hold water if we attribute the difference to the lack of randomness in Lenski's experiment; or if we (probably more appropriately) simply increase the number of coordinated mutations from two to three or four, say; which is no big deal. Probabilities decrease exponentially as the number of combinations is increased. Moreover, the difficulty of producing a few (two, three, or four) coordinated molecular mutations (using exclusively Darwinian mechanisms) pale when compared to producing a complex coordinated system consisting of, say, more than one hundred (100) genes: such as is required by a simple translation system. So Behe's and Snoke's point, in my opinion, still holds water —perhaps not so tightly.

Tree of Life

As we saw in the previous chapter on the "Origin of Life", Darwin imagined that all living organisms started from a single organism, nowadays called a *common ancestor*—or perhaps a few organisms:

> There is grandeur in this view of life, with its several powers, having been originally breathed by the Creator into a few forms or into one; ... from so simple a beginning endless forms most beautiful and most wonderful have been, and are being evolved. [327]

According to Darwin, from this simple common ancestor, other (slightly) different organisms arose by random variations. Through the process of natural selection (or survival of the fittest), other forms established themselves, probably in somewhat different environments.

---

[327] Darwin, *The Origin of Species*, p. 649

It stands to reason, and as I shall also show later in this chapter, major variations in form usually result in monstrosities that die out, probably without offspring; it follows that viable variations must be modest. Not all random modest variations will necessarily be more beneficial, only a few will be more successful than the then current general population; this means that permanent changes in form will be a slow, hit-and-miss process that takes a long time to happen. Over time, these forms diverge, more and more, giving rise to the various forms in the animal and plant kingdoms. So Darwin thought of all the various living creatures as being connected together by a tree-like structure, the *tree of life*: a down-up structure, that is, one in which the simplest organism gives rise to more and more complex and sophisticated organisms.

The nodes of the branches of this tree should show features common to both branches. Following are a few simple examples of the concept: the archaeopteryx has fingered wings, a combination of a wing and a hand; the penguin's flippers, look like something between a bird's wings and a fish's fins; the bat's wing has a five-finger bone structure, like that of a hand; the horse's hoof has four additional useless toes; humans have a useless tailbone; etc. Needless to say and this point is of extreme importance, according to Darwin's theory there should be very numerous intermediate organisms. Why? Because the random variations are always minimal to start with, and there is a blind element of hit-and-miss before natural selection settles on the final form. This is what Darwin's theory of evolution predicts; does the evidence we find in the fossil record confirm this?

Cambrian Explosion

The fossil record of between five hundred and thirty million (5.30×10^8) and five hundred and twenty-five million (5.25×10^8) years ago, in the *Cambrian period* ranging from five hundred and forty (5.40×10^8) to four hundred and ninety (4.90×10^8) million years ago, shows a period during which many new and anatomically sophisticated creatures appeared suddenly: that is, without evidence of any ancestral forms in the earlier (Precambrian) layers. Paleontologists (fossil experts) today call this event the *Cambrian explosion*. Interestingly, Darwin knew about this fossil record, and this is what he had to say about it:

> The abrupt manner in which whole groups of species suddenly appear in certain formations, has been urged by several paleontologists ... as a fatal objection to the belief in the transmutation [evolution] of species. If numerous species belonging to the same genera or families have really started into life at once, the fact would be fatal to the theory of evolution through natural selection. [328]

And Darwin's expectation was a *multitude* of animal fossils in earlier geological levels:

> Consequently, if the theory [of evolution] be true, it is indisputable that before the lowest Cambrian stratum was deposited long periods elapsed, as long as, or probably far longer than the whole interval from the Cambrian age to the present day; and that during these vast periods the world swarmed with living creatures. [329]

Regarding this piece of tangible evidence, the Cambrian fossils, Darwin had only one possible explanation that could still jive with his theory; namely, that the fossil record was incomplete. His only hope was that, over time, fossils supporting his theory would eventually be found; to this effect he writes:

> I look at the geological record as a history of the world imperfectly kept, and written in a changing dialect; of this history we possess the last volume alone, relating only to two or three countries. Of this volume, only here and there a short chapter has been preserved; and of each

---

[328] Darwin, *The Origin of Species*, p. 432
[329] Darwin, *The Origin of Species*, p. 438

page, only here and there a few lines. ... On this view, the difficulties above discussed are greatly diminished, or even disappear. [330]

Fossil Record

Well, guess what: none have been found yet, despite numerous searches for over one hundred and fifty (150) years! Although there should be innumerable generations of ancestors and transition members of the Cambrian animals, not one obvious fossil of such progenitors has yet been found. A few obvious fossils would not have proved much either—they should be innumerable.

The scientists' explanation of this fact is that probably the Precambrian organisms were too small or too soft to allow their fossilization. But this theory does not hold water at all because we do have many fossils of very small (bacteria) and very soft (embryos) organisms from Precambrian times. Moreover, the animals with hard parts, like shells or bones, could not have developed the hard parts overnight; their immediate ancestors must have had, albeit perhaps softer, parts that could easily fossilize.

The fossil record shows that different, not the same, phyla (top classifications with different body plans) were fossilized in the Precambrian period. For example, only three animal phyla were fossilized in the Precambrian period; compare this to twenty (20) new phyla fossilized in the Cambrian period. Darwin's theory of evolution postulates that when organisms evolve from one form to another, they do so very gradually through better chances of survival of the fitter organisms: that is, by constantly acquiring minor advantages, thus advancing from the less complex to the more complex structures—a down-up form of sophistication. [331] However, as Stephen Meyer notes in his book *Darwin's Doubt*:

---

[330] Darwin, *The Origin of Species*, p. 443
[331] Meyer, *Darwin's Doubt*, p. 40.

> The actual pattern in the fossil record, however, contradicts this expectation ... the fossil record shows representatives of separate phyla appearing first followed by lower-level diversification on those basic themes ... The Cambrian explosion appears to conform to ... the top-down effect. [332]

The number of branches of the tree of evolution should increase over time, and the fossil record does show an increase in complexity from the Precambrian to the Cambrian times; however, only the top branches of the tree of life are found: no nodes, or transition types.

It is hard to understand why the fossil record, in general, would be selective in this manner. Although, one must admit, there might be some difficulty in identifying a node, there does not seem to be any disagreement among paleontologists about the following fact: all through the ages, organisms seem to appear suddenly, thrive and evolve for a while, and then suddenly become extinct (like the dinosaurs); at the same time, or later, newcomers appear on the scene.

The Cambrian Explosion was only the most *outstanding* instance of this general pattern of discontinuity of organisms throughout the geological column. In fact, geologists have used this fossil discontinuity extensively as one of the standard methods, in conjunction with superposition (that is, that lower levels are usually older) and radiometric (radioactive decay) methods, of determining the age of rocks; so much so, that this discontinuity has given rise to the names of the major distinct periods of geological (earth) time.

There is evidence of organisms becoming more competitive in a particular function they already have, like assimilating food, as in Lenski's experiment described above. Selective breeding does manage to improve particular aspects (or functions) of an animal. However, one might note here, that, when all is said and done, after selective breeding dogs remain dogs, horses remain horses, pigeons remain pigeons, birds remain birds, and sheep remain sheep. Selective breeding and laboratory experiments never succeeded in changing

---

[332] Meyer, *Darwin's Doubt*, p. 41.

just the genus of an animal: from a cat obtaining a dog, or from a pig obtaining an ape, say.

Moreover, in general, there is hardly any evidence of an organism developing a new function or feature: like starting to fly or developing sight. According to the theory of evolution, it should be the *norm*, not the exception! On the other hand, we have a lot of evidence of organisms losing an unused feature. Such phenomena, again, represent a top-down effect—unlike Darwin's expectations of down-up effect.

The fact that, over the last one hundred and fifty (150) years or so, no Precambrian ancestral fossils for the Cambrian fossils have been found, has practically established what one might call a *statistical fact*: namely, that there isn't a myriad of such fossils. Conclusions drawn from statistical analyses are never one hundred percent (100%) certain, but they do give a level of certainty: scientists do accept their validity while keeping an open mind, of course.

As a practical illustration, let us suppose we have an enormous barrel full of marbles, but we do not know what color of marbles they are. However, every time we pick one out, either a black or a white marble comes up—no matter from where inside the barrel we reach out for the sample. If we do this a hundred times, say, and no colored marble comes up, we can almost certainly conclude that there are no colored marbles inside the barrel. If we take a thousand samples and still no colored marble comes up, then our certainty that there are no colored marbles inside the barrel increases; but we will never be one hundred (100%) certain—unless we check them all. But one can conclude that it will be improbable to pick out a colored marble from the barrel, and one can practically exclude the possibility of there being a multitude of colored marbles, for it is even more improbable. Such a situation is what one might call a statistical fact.

The theory of evolution cannot explain the actual fossil record, much less the Cambrian explosion; it seems, therefore, that a large number of new animal phyla were suddenly, somehow, placed on our earth—not only once but several times. Recall that geological times (ages) are named (distinguished) by the phyla found in them, before which they were not present and after which they became extinct. That is the *tangible* evidence; make of it what you like.

Morphology (Structural Affinity)

Morphology deals with the form of living organisms and with relationships between their structures; similarities in organisms play a prominent part in the theory of evolution. In his book *The Origin of Species*, Darwin writes:

> What can be more curious than that the hand of a man, formed for grasping, that of a mole for digging, the leg of a horse, the paddle of the porpoise [a dolphin relative], and the wing of a bat, should all be constructed on the same pattern, and should include similar bones, in the same relative positions? [333]

All the above fore-limbs have five digits (fingers or toes). In particular, the horse still has the remains of four tiny vestigial (functionless) toes on the bones above the hoofs: the hoof is an overgrown middle toe. The *pentadactyl* limb has five digits on the hand or foot; it also has a specific pattern of bones.

Here's an interesting question: are the above similarities in construction the result of evolution from a common ancestor or of common elements in design? The wheel is an example of a common element in design, which is used in so many mechanical devices; so is the use of an internal combustion engine in the motor-cycle, the lawn-mower, the snow-blower, etc.

The tailbone of a human being is a vestigial (useless) tail; we supposedly used it to help us hang from trees, as some monkeys do, while doing something else (like eating) with our hands. There was a time when I used to believe in macroevolution strongly; and to me, personally, the human tailbone was a very convincing piece of evidence in support of human evolution from apes. I still believe

---

[333]   Darwin, *The Origin of Species*, p. 579

this is the case: that our bodies evolved from that of apes; notice, however, that our body structure is practically the same as theirs. It will become clearer as one reads this chapter where I draw the line. In my opinion, a designer wouldn't design something as useless as a tailbone; it is more likely the result of adaptation, and perhaps sexual preference, that it gets lost. Nor would four useless toes be placed around a horse's hoof by an intelligent designer. So I do believe that also in this case evolution played a significant part. These instances, and others I shall mention in this chapter, look like strong evidence in favor of evolution to a significant degree. I'm open-minded to the possibility; I don't care where the evidence takes me: I'm only searching for the truth.

The loss of a function as a result of disuse or environmental conditions is very common in evolution, and it is well-documented. Animals consistently living in dark underground caves, for example, are known to become blind since they do not need sight any more. Beetles on windy small islands are known to have lost their wings because there are few predators there, and a (defective) wingless beetle is not so easily blown out into the sea by strong winds; so it tends to thrive (survive better) in that environment. This type of evolution is an up-down type, which is not difficult to understand: it is an adaptation. However, most modern biologists believe in a down-up type of evolution.

Evidence in favor of obtaining a new function through evolution is far less common, however; although in a down-up type of evolution from a common ancestor, as Darwin proposes, it should be the norm not the exception. Did the bat, for example, acquire the ability to fly; as in Lenski's experiment the E-coli bacteria acquired the ability to assimilate citrate as food? It's hard to say. The bat's wing, with its five-digit internal structure, could just be an element of common design, too. But let's say it did evolve from a hand to a wing (Darwinists say it did). There are not too many such down-up examples of evolution; the main point is that they should be very common. The fact that they are so rare makes the concepts of macroevolution and descent from a common ancestor very doubtful and implausible. It is not supported by a pillar of the theory of evolution: namely, that there should be a multitude of such instances—that it should be the norm rather than the exception.

Even so, notice that the basic body-plan of the bat's wing did not change significantly: every one of the digits grew longer independently, and flesh grew between the digits. The change is similar to a vestigial tail or a horse's hoof: the basic body plan—the skeletal feature—is still there.

A good engineer, in designing an engine or a machine, will try to make it as versatile as one can to meet all the demands of the environment in which it is supposed to function: within the environmental limits, of course. Similarly, if we think of an organism as a living machine, we expect it to be designed in such a way as to adapt to a changing environment; but again, within limits too. In engineering it is impossible to design something that will work in any environment; similarly, I would not expect living organisms to evolve (change) drastically in body shape; significantly, perhaps, but not drastically from one body structure to a completely different one. For example, I would not expect a frog to evolve to a crab: an endoskeleton to an exoskeleton; or a snake to evolve to an ape: from no limbs to growing hands and feet.

Molecular Clock

On hearing about the ability to determine the paternity of a child, following the discovery of DNA, it occurred to me that the human genome might not only have a history of human descent imprinted in it, but also a history of the common ancestor leading to the various phyla. As I already mentioned, I used to believe very strongly in macroevolution; I was so excited about the idea: I thought it would solve all the questions of evolution once and for all. However, my enthusiasm was short lived, to say the least!

Regarding molecular clocks, in his book *Darwin's Doubt*, Meyer informs us that the molecular clock hypothesis assumes the existence of a common ancestor. It also assumes that the mutation rates of organisms are constant and that they have remained relatively constant throughout geological time. There is nothing wrong with these assumptions, provided it could be shown that things added up.

Molecular clock studies take two similar genes in two or more animals and assume that the extent to which their sequences differ is a measure of the amount of time that passed since those animals, supposedly, parted from their common ancestor and continued to evolve separately: a small difference indicates a short time interval, while a big difference indicates a long time interval. [334] This is, of course, a very reasonable assumption to start with in one's research. Recall, however, from the last chapter on the "Origin of Life", that some, but not all, proteins can remain functional (perhaps even preserve a particular function) when a significant number of nucleotide bases are changed in their genes: so this linear relationship assumed above can easily be lost or confused.

For example, molecular-clock studies of birds and mammals assume a reptile to be their common ancestor. This is quite an assumption; the fossil record, in no uncertain terms, rejects the concept of transmutation of phyla, which we shall also confirm convincingly later on in this chapter. In any case, calibration is based on the age of this reptile, which can be dated accurately from the fossil record, somehow thought to be their most recent common ancestor. Then a number of similarly functioning genes in both the bird and the mammal in question are compared, and the differences in their sequences determined. From these molecular differences and the number of years elapsed from the assumed common ancestor (the reptile), a mutation rate is then calculated. Assuming again that different genealogies evolve at the same rate, this calculated mutation rate is used across the board to determine the time when any other pair of animals separated from each other in a presumed or preconceived evolutionary tree. [335]

---

[334]   Meyer, *Darwin's Doubt*, p. 102.
[335]   Meyer, *Darwin's Doubt*, pp. 102–3.

There is nothing wrong with the concept; the problem is which protein molecules would one select to calculate the mutation rate? In actual fact, they all give different rates: leading to inconsistent histories—of course, there should only be one true history. So which history does one choose: the one with preconceived notions based on apparent similarities? This goes directly against the grain of all scientific principles: that is, following wherever the data and the evidence lead. [336]

Universal Common Ancestor

Most evolutionary biologists believe in the *universal common ancestor* hypothesis; they think it is iron-clad, or unassailable. Let us now have a look at how unassailable it is. Advocates of this hypothesis tried to apply the molecular clock method to phyla (animals with different body-plans) that first appeared in the Cambrian period. By analyzing similar genes, RNA molecules, or proteins in pairs of animals, they tried to estimate how long before the Cambrian explosion the different phyla diverged from a presumed common Precambrian ancestor. Unfortunately the actual results were not promising at all.

In his book *Darwin's Doubt*, Meyer writes about a major 1990 such study performed by evolutionary biologists Gregory Wray, Jeffrey Levinton, and Leo Shapiro; its results were published in a 1996 paper. [337] Their conclusion was that the common ancestor of all the newly discovered Cambrian phyla existed about one billion two hundred million (1.2×10^9) years ago. This means that about seven hundred million (7×10^8) years elapsed for the various newly discovered Cambrian phyla to evolve from this common ancestor. Meanwhile, not a trace was found of any intermediate (connecting) ancestral branches for any of these phyla.

---

[336] Meyer, *Darwin's Doubt*, p. 103.
[337] Meyer, *Darwin's Doubt*, p. 103.

They tried to explain this strange phenomenon by postulating that the Precambrian ancestors had only soft body parts (no hard shells, bones, spikes, etc.) and that therefore it was unlikely for them to leave any kind of fossils behind. [338] [339] We have already seen, in the section on the "Fossil Record" above, that this explanation is implausible. As Meyer tells us, this argument does not hold water because we actually have numerous fossils of soft-bodied animals from the Cambrian period, as well as embryos and microorganisms from the Precambrian period. Thus the idea of an extensive period of undetected soft-bodied evolution is undermined: the evidence says there were *no* ancestors that could be fossilized! Moreover, to claim that the hard-bodied Cambrian animals evolved overnight, strictly from soft-bodied ancestors, is ludicrous and anatomically impossible. Fossils found in the Cambrian layer, for example, a brachiopod, which has hard shells on the upper and lower surfaces, cannot survive without its shell; or an arthropod, which is an invertebrate with an exoskeleton, a segmented body, and jointed limbs, cannot exist without its exoskeleton. A plausible ancestor of such animals would, in all probability, have left some fossils of hard or semi-hard body parts; yet none were found in the Precambrian period. [340]

A similar more recent molecular clock study by geo-biologist Douglas Erwin and his colleagues was performed on one hundred and thirteen (113) Metazoa. (Metazoa are multicellular animals with cells differentiated into *tissues* and *organs*. A tissue is made from a group of cells with a similar structure that work together to perform a particular function: for example, blood, lining of the lung or stomach, and muscle. An organ is made of a group of different tissues all working together to do a particular function: for example, heart, liver, lung, stomach, and brain.) They estimated that "the last common ancestor of all living animals arose nearly 800 million [8×10^8] years

---

[338]   Wray, Levinton, & Shapiro, "Molecular Evidence for Deep Precambrian Divergences among Metazoan Phyla".

[339]   Meyer, *Darwin's Doubt*, pp. 103–4.

[340]   Meyer, *Darwin's Doubt*, p. 105.

ago." [341] Similar studies come to the same conclusion, namely, a very ancient divergence of animal forms. You see, they have to conform to Darwin's affirmation that it must have taken a very long time "probably far longer than the whole interval from the Cambrian age to the present day" [342] if they believe in evolution from a common ancestor. Meanwhile, the fossil record shows that almost all animal phyla came into existence suddenly, in just five million ($5x10^6$) years, some five hundred and thirty million ($5.3x10^8$) years ago. [343]

Meyer concludes his treatment of molecular clock studies by pointing out that the results of these studies are so wide apart; yet everyone agrees that there should only be one true history of the common ancestor, if there was such an ancestor. So, assuming for a moment that there was such an ancestor, the fact that molecular clock studies do not agree, within experimental error, throws serious doubt on the concept as a scientific tool. One of the articles on such a study is called "Reading the Entrails of Chickens: Molecular Timescales of Evolution and the Illusion of Precision"; by its title alone one can tell how scientific it is. The other alternative is that there is no common ancestor; in which case, molecular clock studies between animal phyla are utterly irrelevant. [344]

Meyer then compares the various molecular clock results for a common ancestor of all animals; here's a summary of his comparisons. (1) There is a difference of about four hundred million ($4x10^8$) years between Wray's (and alias) and Erwin's (and alias). This kind of variance (+50% or -33%) in the results, using roughly the same principles, makes the method scientifically unacceptable, and most probably renders them both meaningless. (2) Results from many other studies also vary widely: they range from roughly six hundred million ($6x10^8$) to two billion ($2x10^9$) years ago. (3) Other results, mind-bogglingly, contend that the common ancestor of all animals

---

[341] Erwin, Laflamme, Tweedt, Sperling, Pisani, & Peterson, "The Cambrian Conundrum", p. 1092.

[342] Darwin, *The Origin of Species*, p. 438

[343] Meyer, *Darwin's Doubt*, p. 104.

[344] Meyer, *Darwin's Doubt*, p. 105.

lived after the Cambrian Explosion. (4) Occasionally, the same article (forthrightly) reports contradictory results. [345] (5) One study, claiming a ninety-five percent (95%) certainty in its results, oddly enough, concludes that the common ancestor for certain animals lived about 14.2 billion (1.2×10^{10}) years ago. This exceeds three times the Earth's age; obviously, this result cannot be taken seriously, but it also clearly shows how unreliable this, so called, scientific method is. [346] (6) Even studies of the same molecules disagreed significantly. [347] [348]

As we saw above, most evolutionary biologists, following Darwin's assertion just quoted, have preconceived ideas regarding divergence times from this presumed universal common ancestor. They know, as Darwin himself admitted, that it is fatal to the theory of evolution if they cannot show this; the whole basic idea of evolution between different phyla and possibly life emanating in a down-up fashion must be scrapped; leaving only room for a designer placing several types of organisms on earth. So they completely ignore certain molecules in their molecular clock studies because the results would grossly contradict their preconceived expectations of what the Precambrian tree of life should be like.

For example, they completely ignore *histones*, the proteins involved in packing DNA into chromosomes; different species exhibit hardly any difference in these proteins. Consequently, if they were used in molecular clock studies they would result in extremely short divergence times. Talk about scientific honesty! Such cherry-picking of evidence is conducive to bad science: research that adopts such philosophies is basically a loss of time and money. When preconceived ideas trump evidence, science becomes a faith or a religion. [349]

---

[345]   Bronham, Rambaut, Fortey, Cooper, & Penny, "Testing the Cambrian Explosion Hypothesis by Using a Molecular Dating Technique"
[346]   Graur & Martin, "Reading the Entrails of Chickens"
[347]   Ayala, Rzhetsky, & Ayala, "Origin of the Metazoan Phyla".
[348]   Meyer, *Darwin's Doubt*, pp. 105–7.
[349]   Meyer, *Darwin's Doubt*, pp. 107–8.

In their review article "Fossils, Molecules and Embryos", evolutionary biologist James Valentine, geophysicist David Jablonski, and paleo-biologist Douglas Erwin write: [350]

> Different genes in different clades [i.e., groups of organisms consisting of an ancestor and all its descendants] evolve at different rates, different parts of genes evolve at different rates and, most importantly, rates within clades have changed over time. [351]

So how can scientists set up a tree of life from the molecular clock hypothesis? They simply can't. The obvious conclusion is that the whole idea of a molecular clock hypothesis, unfortunately, has to be scrapped: different studies of different molecules generate widely divergent dates. The whole idea gives only the illusion of precision; if anything, it seems to prove that there was no common ancestor. Given the evidence from the Cambrian and the general geological fossil record, the molecular clock evidence seems to be corroborating evidence that the universal common ancestor is a figment of Darwinian imagination: just wishful thinking. If this is really the case, then similarities in organisms will only mean similar designs.

As we have already mentioned, most evolutionary biologists believe that the case for a universal common ancestor is close to unassailable: to them it is similar to a dogma, or blind faith statement one encounters in most religions. For example, evolutionary biologist Richard Dawkins, in his book *The Greatest Show on Earth*, asserts:

> And when we look comparatively at what is written in the [DNA] code—the actual genetic sequences in all these different creatures—we find the same kind of hierarchical tree of resemblance. We find the same *family tree*—albeit much more thoroughly and convincingly laid out—as we did with the vertebrate skeleton, the crustacean skeleton,

---

[350]   Meyer, *Darwin's Doubt*, p. 109.
[351]   Valentine, Jablonski, & Erwin, "Fossils, Molecules and Embryos", p. 856.

and indeed the whole pattern of anatomical resemblances throughout all the living kingdoms. [352]

Wow! A claim like that, made by a scientist across the board, must be really impressive to ordinary people. Having read the previous section, does the reader still think this is true? Does the analysis of both anatomical and genetic (DNA code) similarities converge on the same basic pattern of descent from a common universal ancestor?

After giving the above extensive evidence in support of the fact that different studies of different molecules generate widely divergent dates, Meyer concludes that, philosophically, there can only be one actual universal common ancestor for all the animals. If, on the other hand, widely varying divergence times result from gene comparative analyses, then obviously most, if not all, of them must be wrong; or that, most probably, a Precambrian animal divergence never happened. [353]

Given these conflicting points of view, how can we trust a scientist's statements? In general, nowadays, scientists claim that the spiritual and the supernatural are inexistent: that only the physical exists. Their agenda, therefore, is to eradicate all trace of anything spiritual or supernatural from science; consequently, in my humble opinion, they exaggerate actual data to achieve this. A few centuries, or even decades, ago a scientist did not care whether some naturally unexplainable phenomena implied God's intervention (we know so little about our universe anyway) so they presented data faithfully. This has been the case in the times of Copernicus (1473-1543), Galileo (1564-1642), Newton (1642-1726), Darwin (1809-1882) and even Einstein (1879-1955). These people were far from a hindrance to the advancement of science even though they still believed in God. Nowadays, any scientist who believes in a designer is ostracized and loses his job. As a result, actual data is manipulated to fit preconceived notions and beliefs. Science has become a matter

---

[352] Dawkins, *The Greatest Show on Earth*, p. 315 (emphasis in original)
[353] Meyer, *Darwin's Doubt*, p. 108.

of blind faith—a religion: the very same thing scientists hate so much about religious people.

Cladistics

Wikipedia defines *cladistics* as:

> [A]n approach to biological classification in which organisms are categorized based on shared derived characteristics that can be traced to a group's most recent common ancestor and are not present in more distant ancestors. [354]

Cladistics, therefore, again assumes a common ancestor to various organisms and a down-up tree-like Darwinian evolution. Cladistics picks various similar features, structures, or traits, shared by different types of organisms; it then concludes that, because of these similarities, the organisms must be related. The more similarities they have, of course, the more closely related it assumes they must be, which is not an unreasonable assumption to start with. It then generates a tree-like branching pattern, called *cladogram*, of relationships based upon an analysis of the various characters chosen.

For example, suppose five species exhibit the following characteristics. Species #1 has characteristic A; species #2 has both characteristics A and B; species #3 has characteristics A, B, and C; species #4 has characteristics A, B, C, and D; and finally species #5 has characteristics A, B, C, D, and E. Cladistics concludes that species #1 is the ancestor of the other four species #2, #3, #4, and #5; species #2 is the ancestor of the three species #3, #4, and #5; species #3 is the ancestor of the two species #4 and #5; and finally species #4 is the ancestor of the last species #5.

This looks simple enough; however, it is not as simple as it seems at first blush: life is never so simple. The problem is that,

---

[354]   https://en.wikipedia.org/wiki/Cladistics (2 October 2015)

in constructing cladograms, there are too many characteristics classification experts may consider while looking for similarities between species: such as, anatomy, genes, development, behavior, etc. Obviously, evolutionary biologists agree that the probability of characteristics suddenly appearing and disappearing numerous times, on separate evolutionary tree branches, should be minimal. However, no matter how hard they try, it simply cannot be avoided when considering actual species; that is, unless they cherry pick these characteristics. [355]

In his book *Darwin's Doubt*, Stephen Meyer explains that if we limit ourselves to analyzing a small group of traits in a given number of diverse animal species, we obtain an unquestionably probable history of the sequence of their presumed emergence from one another. The only problem is that if instead we consider another group of traits in the same set of animal species, we obtain a completely different unquestionable historical sequence of the presumed emergence of the species from one another. On the other hand, if we combine both of the above groups of traits, the result would be an embarrassingly complex sequence of "convergent evolution" and "loss of character": the ancestral tree would consist of many sudden appearances, disappearances, and reappearances of the various traits. [356] In other words, cladograms are inconsistent with one another when picking different characteristics. Alternatively, including all the characteristics at once and trying to explain all the instances of how traits moved in and out of the ancestral tree, through convergent evolution and loss of character in several cladogram branches, becomes somewhat embarrassing after a while. [357]

In the article "Meyer's Hopeless Monster, Part II", evolutionary biologist Nicholas Matzke criticized Stephen Meyer's book *Darwin's Doubt* for almost totally disregarding the explanatory powers of cladistics in the Cambrian fossil record. Matzke believed that if cladistics is applied to the Cambrian fossil record, an ancestral

[355]   Meyer, *Darwin's Doubt*, pp. 420–2.
[356]   Meyer, *Darwin's Doubt*, p. 422.
[357]   Meyer, *Darwin's Doubt*, p. 422.

pattern will emerge; effectively solving the mystery of their missing ancestral lineage. [358] Meyer countered by saying:

> [U]sing cladistics to infer such ancestral arthropods [animals with an exoskeleton, segmented bodies, and jointed limbs] requires postulating "ghost lineages" that imply the existence of still *more* missing fossils. The need to invoke hypothetical ghost lineages commonly arises when evolutionary biologists attempt to use cladistics to infer ancestors otherwise unattested by the fossil record. The reason for this is that the fossil record often reveals so-called stem groups [see next paragraph] contemporaneously, or even after, crown groups [see next paragraph]. [359]

To understand and appreciate the above quote fully, we need to define some technical terms: particularly *stem groups* and *crown groups*. A *clade* is a group of organisms which includes the inferred most recent common ancestor of all of its members and all of the descendants of that most recent common ancestor. A *crown group* consists of all the taxa (organisms) descended from the development of a new clade, the splitting of a single lineage into two distinct lineages: it is thus synonymous to a clade. It is recognized by possessing the *synapomorphy* (or characteristics) shared by all members of the clade. For example, arthropods are invertebrate animals with an exoskeleton, segmented body, and jointed limbs: that is, a clade with the common features listed. *Synapomorphy* is a character which is derived; and because it is shared by the taxa under consideration, it is used to infer common ancestry. A *stem group*, on the other hand, consists of all the taxa in a clade preceding the development of a new clade. These are often difficult to recognize because they may not possess all the synapomorpies found in the crown group. They have some, though not all, of the characteristics defining the crown

---

[358] Matzke, "Meyer's Hopeless Monster, Part II": http://pandasthumb.org/archives/2013/06/meyers-hopeless-2.html (23 January 2016).

[359] Meyer, *Darwin's Doubt*, pp. 424–5 (emphasis in original).

group. A *taxon* (plural *taxa*) is any group of organisms that is given a categorizing name, not necessarily a clade; and *taxonomy* is the science of classifying and naming organisms. [360]

It is of particular importance to note from the previous paragraph that, according to evolutionary theory, stem groups are supposed to *precede* crown groups, and not vice versa. If crown groups precede stem groups in the fossil record, it means that something went wrong, somewhere, with the inferences or assumptions taken.

A classic example is the fact that theropod (bipedal) dinosaurs are often thought to be the ancestors of birds. In fact, Wikipedia defines the archaeopteryx as "a genus of bird-like dinosaurs that is transitional between non-avian [non-flying] feathered dinosaurs and modern birds". [361] However, paleontologist and evolutionary biologist Robert Martin, in his book *Missing Links*, states that theropod dinosaurs "all occur in the fossil record after the *Archaeopteryx* and so cannot be directly ancestral". [362] In actual fact, theropod dinosaurs appear in the fossil record millions of years after the birds which are supposed to have evolved from them. [363]

Meyer continues answering Matzke's criticism:

> [M]any supposed members of stem [supposedly prior] group arthropods appear in the Cambrian fossil record *contemporaneously* with or *after* members of the crown [supposedly later] group arthropods that they supposedly preceded.
> The anomalocaridids (and other species) that, according to Matzke, represent Cambrian intermediates illustrate this kind of chronological problem. [364]

---

[360] http://www.ucmp.berkeley.edu/glossary/gloss1phylo.html (3 October 2015)

[361] https://en.wikipedia.org/wiki/Archaeopteryx (3 September 2015)

[362] Martin, *Missing Links*; and Swisher, Wang, Wang, Xu, & Wang, "Cretaceous Age for the Feathered Dinosaurs of Liaoning, China"

[363] Meyer, *Darwin's Doubt*, p. 425.

[364] Meyer, *Darwin's Doubt*, p. 425 (emphasis in original).

Anomalocaris or "abnormal shrimp" is an extinct genus of the anomalocaridid family; it was a lobster-like animal about the size of a human leg, and it was the largest animal—probably a predator—found among the Cambrian fossils. Now, one might ask: why are chronological inversions such a big problem? Meyer explains:

> For evolutionary biologists to produce phylogenic trees depicting evolutionary history consistent with cladistic analysis in cases involving inversions, they must draw back long branches representing lineages for which they lack fossil representatives. [365]

So, for Matzke's hypothesis to hold water, ancestors going very far back in time, connecting the entire set of stem (supposedly prior) and crown (supposedly later) groups of his alleged arthropod ancestors, should also exist in the fossil record. Meyer's bottom line quote goes:

> Indeed, *none* of the supposed *ancestral* stem [supposedly prior] arthropods, or their evolutionary histories, or related non-arthropod ancestral groups from which arthropods supposedly evolved are documented in the fossil record. [366]

Moreover, cladograms cannot be regarded as depicting real events in the evolutionary histories of organisms. The algorithms (formulae or procedures) used in cladistic analyses to reach the final result, presuppose rather than prove the existence of the common ancestry of the groups they analyze. Any method that presupposes the truth of a proposition needs to be supported by some other kind of evidence: it certainly cannot be used to prove the truth of that same proposition. Cladistics can be used to hypothesize an evolutionary proposition, as a starting point; but then it has to be substantiated in some other way by concrete evidence, such as a fossil record, say. [367]

---

[365] Meyer, *Darwin's Doubt*, p. 426.
[366] Meyer, *Darwin's Doubt*, pp. 426–7 (emphasis in original).
[367] Meyer, *Darwin's Doubt*, pp. 430–1.

Even more detrimental to the validity of the evolutionary history represented by cladistics is an intrinsic disadvantage it has. It was depicted graphically by historian of paleontology Keynyn Brysse in her 2008 paper "From Weird Wonders to Stem Lineages". She demonstrates, diagrammatically, how a simple cladogram consisting of just three characteristic traits is equally consistent with six evolutionary histories. It therefore follows that cladistic analysis cannot establish a definitive evolutionary history. [368] She explains the limitations of cladistics as follows: [369]

> Cladograms depict sister groups—taxa that are thought to be each other's closest relatives—but do not show ancestors or descendants. ... In other words, the very goal of evolutionary systematics—the determination of ancestor-descendent relationships—is on the cladistic method not just unattained but unattainable. [370]

So much for the evolutionary evidence from cladistics that Matzke proposed; such cladistics analysis seems to be an evolutionary biologists' wishful thinking. It does not seem to be evidence of evolution from common ancestors; it only seems to be evidence of similar design in organisms. Furthermore, as we have already mentioned above, an unbiased chronological unearthing of the Cambrian fossils shows only the top branches of the hypothetical Darwinian tree of life—no ancestors whatsoever.

---

[368] Meyer, *Darwin's Doubt*, pp. 434–6.
[369] Meyer, *Darwin's Doubt*, p. 436.
[370] Brysse, "From Weird Wonders to Stem Lineages", p. 306

Body Plans

In his book *Darwin's Black Box*, microbiologist Michael Behe writes:

> Consider this sequence: skateboard, toy wagon, bicycle, motorcycle, automobile, airplane, jet plane, space shuttle. It seems like a natural progression, both because it is a list of objects that can be used for transportation and also because they are lined up in order of complexity. ... But is, say, a bicycle a physical (and potentially Darwinian) precursor of a motorcycle? No, it is only a *conceptual* precursor. ... To be a precursor in Darwin's sense we must show that a motorcycle can be built from "numerous, successive, slight modifications" to a bicycle. [371]

While trying to accomplish the above, the bicycle has to remain a viable means of transportation, be always a better means of transportation than the model before it, and no big additions can be made to it: like installing an engine, because an engine has many integrated parts. Besides, an engine requires a completely different technology and energy source from pedaling. The engine's parts have to be added one by one, in random order, and minute changes with no foresight of a design of anything like an engine; not only that, every time a part is added, it is not just tolerated, but it must actually improve the system as a whole significantly: otherwise the change will not be adopted by natural selection. I'm sure the reader, by now, appreciates the inherent difficulty.

Similarly, it is difficult to imagine any way of transitioning between two completely different animal body plans such as one having an endoskeleton (an internal bone structure and soft exterior, like a frog) and another one having an exoskeleton (an external hard structure and soft interior, like a crab): the muscle and tendon attachments and operations are all reversed. In practice, this cannot

---

[371] Behe, *Darwin's Black Box*, pp. 43–4 (emphasis in original).

be done except by *re-designing* either one from scratch. In a later section in this chapter entitled "Macroevolution", we shall examine the current evidence as well as the possibility of transforming one body plan into another in greater detail.

Now, how could two such body plans appear for the first time together in the same fossil era, and at the same time have a common ancestor but with no intermediate predecessors in sight? Meyer summarizes the Cambrian explosion in the epilogue of his book *Darwin's Doubt* as follows:

> Darwin's Doubt makes its case for the reality of the Cambrian explosion chiefly, but not entirely, on the basis of the fossil record. Representatives of twenty-three of the roughly twenty-seven fossilized animal phyla (and of the roughly thirty-six total animal phyla) are present in the Cambrian fossil record. Twenty of these twenty-three major groups of animals make their first appearance in the Cambrian period with no discernable ancestral forms present in either earlier Cambrian or Precambrian strata. [372]

Meyer wrote a whole book of over five hundred (500) pages in support of the above paragraph; so I would like the reader to appreciate the full significance of this last quote.

Limiting our discussion to animals, we know of only about thirty-six (36) different phyla (body plans) that have ever existed over the whole history of the earth. Of these thirty-six (36) body plans, twenty-three (23) were found within the Cambrian explosion strata (430 to 425 million years ago). In all the strata prior to these Cambrian explosion strata, only three (3) of these twenty-three (23) animal body plans were found. This means that twenty (20), more than half of a total of the thirty-six (36) different body plans in the whole history of the world, appear suddenly in a relatively short interval of time (only 5 million years) with no Darwinian predecessors in sight. These twenty (20) new body plans found in the Cambrian explosion

---

[372] Meyer, *Darwin's Doubt*, pp. 417–8.

era seem to have just been *placed* there: they do *not* seem to have evolved from other organisms. There are only twenty-seven (27) fossilized animal phyla, of which twenty-three (23) were all found in this five million year period. Is it just me, or do you get the feeling that somebody is trying to tell us something? Do you recall that in an earlier chapter, entitled "Fine-tuning of the Universe", we saw that also the laws of physics are *discoverable*?

In his book *The Origin of Species*, Darwin himself admits the *downfall* of the extrapolation of his hypothesis—that of a universal common ancestor—if no fossils for the Cambrian explosion could be found! He writes:

> If numerous species belonging to the same genera or families have really started into life at once, the fact would be fatal to the theory of evolution through natural selection. [373]

None were found in over one hundred and fifty (150) years; despite numerous searches. Basically, Darwin extrapolated the evidence he had: the scientific evidence for microevolution (the straight line relationship, so to speak) too far, and stumbled in assuming macroevolution.

Finally, evolutionary biologist Richard Dawkins writes the following about genes in his book *The Selfish Gene*:

> A gene's influence ... is *always* indirect. The gene determines a protein sequence that influences X that influences Y that influences Z that eventually influences ... [374]

Given the fact, therefore, that a gene's influence is never direct, with all my experience in engineering, I simply cannot even imagine how a blind and random process (without any foresight or intelligence) could ever be able to construct a most complex system, like a body

---

[373] Darwin, *The Origin of Species*, p. 432
[374] Dawkins, *The Selfish Gene*, p. 240 (emphasis in original).

plan, without any end-vision of the body plan (design) in view; especially when one considers (as we have seen in the previous chapter) how difficult it is to come across just one functional protein sequence by chance alone.

Junk-DNA

If one wants to send a Morse-code signal using a light source or a buzzer, one cannot afford having big variations in the medium used for the signal: that is, the intensity of the light or the loudness of the buzzer; otherwise the integrity of the "long" and "short" signals making up the Morse-code alphabet will be compromised. In science, this is termed *noise to signal ratio*, which must be relatively low: typically less than ten percent (10%), say. What this means, in practice, is that the variations in the medium used to convey discrete (digital) information must be negligibly small in order not to affect the integrity (meaning) of the signal. Moreover, it is not desirable to have short spurious signals (spikes) come in between one's intended signals at random: they could confuse the original message.

We have already seen, in the previous chapter, that genes in the DNA code produce proteins. Well, in the human genome, it so happens that there are many more sections of DNA that do not produce proteins than sections that do. The sections that do not produce proteins were (and still are) called *junk-DNA*. Proponents of the theory of evolution pointed to this as obvious evidence of the random process of natural selection: since it, sort of, finds its way by a process of trial and error rather than directly—as in the case of an intended design. If something were designed, they argued, there would be no (or hardly any) junk-DNA; consequently there would not be a high noise to signal ratio (it would be typically less than 10%, say).

In 2001, right after the completion of the human genome project, scientists thought that about ninety-eight (98%) of the human genome was non-functional, or better, non-protein-coding: that is, junk-DNA. [375] Consequently, for quite a while, junk-DNA was considered to be strong evidence in favor of the theory of evolution.

Well, guess what, they were wrong—big time. In 2012, a group of "442 scientists in 32 labs around the world" determined that "some 80% of the human genome is biochemically active". [376] On the other hand, if DNA were actually designed, it is not inconceivable that there would still be a significant amount of noise embedded in it: because of the countless number of mutations over many years of microscopic evolution (for which we have enough evidence); but certainly not a very high percentage that would effectively compromise the intended function. In the human genome, as one would expect, there is evidence of mutational accumulation or degradation: it does display evidence of past viral insertions, deletions, transpositions, and the like; much as digital software, modified over and over, would accumulate imperfections. Nevertheless, the vast majority of base sequences serve an essential biological function. [377]

Contrary to evolutionary theorists' assertions, recent discoveries have proved that those sections of DNA that do not produce proteins happen to have many other important functions. As we have also seen in the previous chapter, proteins are the building blocks of the cell. Having produced a building block, the cell has to be told what to do with it, where to place it. Moreover, the production of these building blocks has to be controlled, somehow: there must be a feedback system so that there isn't a surplus or a deficit of these materials. As expected, it was discovered that the non-protein-producing sections of the genome manage the production of those RNA molecules that control the protein-producing regions. Furthermore, the cell has to be constantly maintained by editing code and defending against pathogens (infectious organisms). Not surprisingly, it was also discovered that these were other functions of the junk-DNA inside

---

[375] http://healthland.time.com/2012/09/06/junk-dna-not-so-useless-after-all/ (4 October 2015)

[376] http://healthland.time.com/2012/09/06/junk-dna-not-so-useless-after-all/ (4 October 2015)

[377] Meyer, *Signature in the Cell*, p. 461.

cells: in other words, junk-DNA sections behaved as a computer *operating system*—directing multiple operations. [378] In his book *Signature in the Cell*, Meyer gives the following list of functions performed by junk-DNA:

> [They] regulate DNA replication, ... regulate transcription, ... mark sites for programmed rearrangements of genetic material, ... influence the proper folding and maintenance of chromosomes, ... control the interactions of chromosomes with the nuclear membrane ..., ... control RNA processing, editing and splicing, ... modulate translation, ... regulate embryological development, ... repair DNA, and ... aid in immuno-defense [against infections] .... [379]

In his essay "Discovering Signs in the Genome by Thinking outside the BioLogos Box", evolutionary biologist Richard Sternberg writes:

> Now, the mouse and rat are estimated to have diverged 22 million years ago. ...
> *The mutational signal from mouse ... is equivalent to the mutational signal of rat ...*
> How strange that two independently acting degenerative processes—affecting mostly "junk DNA"—would lead to the same higher-order pattern. ...
> And this correlation occurs throughout both genomes. ...
> The strongest correlation between mouse and rat genomes is SINE [supposedly mutational insertions or junk] linear patterning.
> Though these SINE families have no sequence similarities, their placements are conserved.
> And they are concentrated in protein-coding genes. ...
> What *I am saying* is that we know a lot about the genome that is being glossed over in the popular works that the

---

[378] Meyer, *Signature in the Cell*, p. 407.
[379] Meyer, *Signature in the Cell*, p. 407.

> theistic evolutionists write. I am also saying that instead
> of finding nothing but disorder along our chromosomes,
> we are finding a high degree of order.
> Is this an anomaly? No. [380]

In the above quote Sternberg points out that what looks like junk DNA in the mouse and the rat is, strangely enough, faithfully preserved; and that this is not an anomaly—it is the norm in related species. There goes another strong piece of evidence in favor of Darwinian evolution: up in smoke!

Incidentally, I would not be surprised that the mouse and the rat are related, and that one evolved from the other or from their common ancestor: why, because they both have the same body plan.

In the early 1990s, when every mainstream scientist thought otherwise, Intelligent Design proponents predicted that the genetic signal in DNA should dwarf the noise element. They came to this conclusion because they were convinced that DNA was a designed system. Even though Intelligent Design proponents are often labelled as pseudoscientists, they were able to make scientific predictions, which turned out to be true, well ahead of their time to be accepted by the scientific community. To the scientific world, this is a sign of great genius; for example, when Einstein predicted gravitational lensing (that is, the fact that light bends when travelling beside a massive object), [381] or when Alpher and Herman predicted the existence of a cosmic microwave background. [382] This should serve as a lesson for scientific tolerance and mutual respect.

The headquarters of Intelligent Design proponents is also often criticized as being unscientific because there seems to be a shortage of laboratories for experimentation purposes; however, the historic discovery of DNA by Crick and Watson also teaches us a very

---

[380] Sternberg, "Discovering Signs in the Genome by Thinking outside the BioLogos Box", pp.78–81 (emphasis in original)

[381] https://en.wikipedia.org/wiki/Gravitational_lens (4 October 2015)

[382] https://en.wikipedia.org/wiki/Cosmic_microwave_background (4 October 2015)

important lesson: they collected all the information from everyone else (they did not find the information themselves) studied it, and came to the right conclusion regarding its structure. There are times when one has to stop and do some thinking; there is a place for science integrators and philosophers—there is so much scientific information, books and papers, to be read, examined, and evaluated correctly (preferably).

As if the above was not enough to highlight the importance of junk-DNA, Meyer gives us another parable that demonstrates better the design intricacies of DNA. He asks us to imagine a coded message, in the form of a letter, sent across enemy lines at a time of war. Read in the normal way, the letter describes everyday household challenges and solutions, the children's health and studies, the weather, etc.—all of it clear and accurate information and in perfect English prose. Yet embedded in the letter is an encoded military message regarding enemy strength, movements, supplies, and plans. In order to read the hidden message, one needs the key to decipher the message. This is what we actually find embedded in the DNA code. Needless to say, to create something like this is a sign of superior intelligence and design ability outside the reaches of chance. [383] Meyer quotes bioinformatician Wen-Yu Chung:

> As W.-Y. Chung, a bioinformatician ..., has noted, the existence of "dual coding" and overlapping protein-coding reading frames, just one of many cellular innovations for concentrating genomic information, is "virtually impossible by chance". [384] [385]

Random mutations may impart some advantages in adapting to a changing environment; however, there is no possible way that they can change an embedded code without compromising it. For example,

---

[383] Meyer, *Signature in the Cell*, p. 463.
[384] Chung, Wadhawan, Szklarczyk, Pond, & Nekrutenko, "A First Look at the ARFome".
[385] Meyer, *Signature in the Cell*, p. 464.

if the key to an embedded code is to pick out every third letter, random mutations in the main code are, most probably, going to compromise the embedded code.

This is probably one of the main reasons why I think that the second law of thermodynamics, the fact that everything deteriorates in nature, will trump evolution in the long run. As Chung affirms above, radically changing an embedded code while constructing a different body structure is virtually impossible for a Darwinian type of evolution to achieve. Because of the above and the experimental evidence we shall see in the next few sections, I doubt very much whether evolution between significantly different body plans is feasible.

Finally, Meyer also points out that the human genome is organized hierarchically, as we organize our personal computers: with files within various functional folders; this arrangement enables retrieving, manipulating, and expressing information more efficiently. Alternatively, we can compare the human genome to our use of language; the same way we use words ordered in sentences and sentences in paragraphs; nucleotide bases are ordered in genes and genes in gene clusters. These are all signs of significant intelligence, both of which are found in the DNA code. There is nothing like this kind of organization found elsewhere in nature; it is only available through us human beings, the only significantly intelligent beings. [386]

Macroevolution

In 1979, two geneticists, Christiane Nusslein-Volhard and Eric Wieschaus, generated thousands of mutations in the small subset of genes that regulate the embryonic development of fruit flies. These genes, known as *regulatory* genes, control numerous other genes that are responsible for the subsequent subdivision of the embryo into various anatomical parts constituting the body of the fly: its head, its thorax, its abdomen, etc. Regulatory genes tell the other genes

---

[386] Meyer, *Signature in the Cell*, p. 467.

when to start and when to stop, and so they indirectly control the growth and freeze the size of the various anatomical parts in the adult organism. These two geneticists mutated, one by one, all the genes of the fly's genome including the regulatory genes, and they figured out how and which gene controls, directly or indirectly, the growth of which part of the fly's body: the term normally used for such an inference procedure is commonly known as *reverse engineering.* [387]

In his book *From Molecular Patterns to Morphogenesis*, Eric Wieschaus describes how they reverse engineered the fruit fly's genome: [388]

> If transcription of a gene was essential for embryonic development, homozygous embryos [i.e., those missing both copies of the gene] should develop abnormally when that gene was eliminated. ... Based on these defects, it should be possible to reconstruct the normal role of each gene. [389]

In 1982, following his writing of a paper on the mutations he and his associate had induced in fruit flies, Weischaus made a presentation on the processes of macroevolution. In this presentation, he declared that, *without exception,* all the fruit fly mutants they had studied died as deformed larvae (a larva is a small, wormlike, early stage of animal development) long before achieving reproductive age: thus, they did not even have a chance to reproduce an organism with a novel body plan (for better or for worse). When asked by the audience about the implications of their findings on evolutionary theory, he replied:

> The problem is, we think we've hit all the genes required to specify the body plan of Drosophila [the fruit fly], and yet these results are obviously not promising as raw materials for macroevolution. The next question then, I guess, is

---

[387] Meyer, *Darwin's Doubt*, pp. 255–6.

[388] Meyer, *Darwin's Doubt*, p. 476, n. 1

[389] Weischaus, "From Molecular Patterns to Morphogenesis", p. 316

what *are*—or what *would be*—the right mutations for major evolutionary change? And we don't know the answer to that. [390]

To this day, more than thirty years later, both developmental and evolutionary biologists still have no clue as to the answer to this question. [391]

Genetic mutation experiments on fruit flies, nematodes (roundworms), mice, frogs, and sea urchins have left evolutionary biologists dumbfounded. In summary: mutating the genes that regulate the construction of the body plan end up in destroying the animal; then how are random mutations and natural selection supposed to create new body plans in a down-up tree of life evolutionary scenario? The hard evidence far from supports the hypothesis.

One concept that is often overlooked is the fact that proteins are only the building blocks of an organism: somewhat like the bricks, wood, drywall, etc., in a house. Hence, it is no use to have new building materials of a different shape, size, or quality unless there is also a higher level plan to determine where they should go. This is also essential in building a particular body plan; something must dispatch them to play their role in tissues, organs, etc. So now, besides changing the lower level code (proteins), the higher level code (organization) also needs to be changed simultaneously and in step. As we have seen in the parable of the coded letter above, in the opinion of many scientists, this is unreachable to the process of random mutation—since it has to work in isolation until natural selection kicks in. Recall that all the mutated embryos never reached reproductive adulthood, either; so the mutations produced never had a fighting chance—albeit they might have been monstrous-looking. [392]

---

[390] Quotes recorded in contemporaneous notes taken by biology philosopher Paul Nelson, who was in attendance at this lecture (emphasis in original notes).

[391] Meyer, *Darwin's Doubt*, pp. 256–7.

[392] Meyer, *Darwin's Doubt*, p. 257.

The Embryo

The process of organizing proteins while constructing an animal body plan, obviously enough, takes place during embryological development; so let us have a closer look at it.

(1) Embryology (Study of Embryos)

When Darwin wrote his book *The Origin of Species*, he collected all sorts of evidence in support of his theory of evolution from several fields of knowledge—embryology was one of them. There is hardly a more convincing piece of evidence for a common ancestor than the similarity of the embryos of different classes of animals. Regarding this similarity of embryos Darwin writes:

> [V]arious parts in the same individual which are exactly alike during an early embryonic period, become widely different and serve for widely different purposes in the adult state. ... [I]t has been shown that generally the embryos of the most distinct species belonging to the same class are closely similar, but become, when fully developed, widely dissimilar. A better proof of this latter fact cannot be given than the statement by [naturalist] Von Baer that [393] "the embryos of mammalia [mammals], of birds, lizards, and snakes, probably also of chelonia [sea-turtles] are in their earliest states exceedingly like one another, both as a whole and in the mode of development of their parts ...." [394]

But again, is this evidence of evolution from a common ancestor or of common design in the first stages of building an animal's body?

As a mathematician, I envisage this divergence of embryological development between species as straight lines diverging from a single

---

[393] Darwin, *The Origin of Species*, p. 587
[394] Von Baer, *Uber Entwickelungsgeschichte der Thiere (On the Development of Animals).*

point: some diverge at a small angle, resulting in a small difference in the body plan of the adult, others diverge at a large angle, resulting in a large difference in the body plan of the adult. This is the main reason why, in my opinion, embryos look so similar; because they all start from a single point (one cell) and in the same manner: namely, by the doubling of a single cell, followed by successive doubling of every cell. If you really think about it, this is direct evidence of similar designs. Embryos are similar not necessarily because they originate from a common ancestor; but because they originate from a common initial design. Conceptually, if one goes back far enough in this sort of design, embryos are definitely going to look alike at some place far enough backward, because they all look like a single cell at the beginning. Only the genetic code (information) in the nucleus of that cell is different.

(2) Ontogeny (Embryo Development)

In his book *Darwin's Doubt*, Meyer makes a very simple observation generalizing embryological development, namely, that all animal development has one goal: to grow into an adult and be able to reproduce.

In most animal species, a new offspring starts with a *zygote* (a single fertilized egg) and ends up with many different cells. Initially this single cell simply starts to double itself, over and over again, into daughter cells identical to itself. The next stage in turning into an adult form consists of this blob of identical cells producing many specialized cell types at the right time and in the right quantity, placing them in the correct positions; shortly afterwards, it turns into an *embryo*.[395] Developmental biologists have discovered that, during embryological development, the appropriate genes are turned on and off: ensuring the production of the correct type and quantity of proteins to achieve this. The proteins coordinating these roles are known as *transcription factors*. Transcription factors are themselves

---

[395] Meyer, *Darwin's Doubt*, p. 258.

controlled by circuits and signals of unbelievable complexity and precision, which are transmitted by other genes and proteins. [396]

Many protein and RNA molecules, that are required in the development of a particular animal body plan, transmit signals to neighboring cells that make the latter differentiate themselves. Moreover, these signals affect how cells interact with each other and organize themselves: they influence each other in such a way that they form networks of coordinated interaction: similar to *integrated circuits* on a printed circuit board. [397]

As early as 1969, although they did not have enough hard evidence, molecular biologist Roy Britten and developmental biologist Eric Davidson came to the following conclusion. Given that a significant number of specialized cell types arise during animal development and that every cell contains the same genome, some overall controlling system must determine which genes are activated, in different cells and at different times, to bring about the observed differentiation of cells: some system-wide regulatory logic must oversee and coordinate the genome expression. [398] [399]

Sure enough, during the last two decades, genomic research revealed that the non-protein-coding regions of the genome (commonly labeled as junk DNA) control and regulate the timing and the activation of the protein-coding regions of the genome. Davidson showed that the non-protein-coding regions of DNA and the protein-coding regions function together as interconnected networks, which look eerily like engineers' electronics schematics of remarkable complexity.

For example, the activation of the bio-mineralization genes (required for the structural proteins that constitute the skeleton or shell of animals) depends on other genes' being activated far upstream in development. Davidson named these apparent integrated circuits of protein and RNA-signaling molecules *developmental*

---

[396] Meyer, *Darwin's Doubt*, pp. 258–9.
[397] Meyer, *Darwin's Doubt*, p. 264.
[398] Britten & Davidson, "Gene Regulation for Higher Cells", p. 353
[399] Meyer, *Darwin's Doubt*, p. 265.

gene regulatory networks (or dGRNs); they are responsible for the differentiation and arrangement of the specialized cells that establish the body of the animal. [400]

I would like to make an obvious point here: electrical engineering blueprints, electronics wiring diagrams, and flow charts are complex products of intelligent design, similar to what only humans can produce; their complexity and their integration is obviously not the result of chance or blind mutation: they are the fruit of intelligence.

As we have seen above, in the section on "Irreducible Complexity", a complex coordinated system of more than three mutually integrated parameters has no real likelihood of emerging by random mutation, in combination with natural selection, even if we survey all the organisms of all the time the earth has existed. There is no way any of the above systems can make coordinated changes, as a group, to transform one animal into another; just because a few gene mutations occur; the system, as a whole, will not go anywhere. How these integrated systems came about in the first place is another question; as scientists we have to admit, at least for the time being, that we don't know the answer to this question; just as, for example, we have to admit that we don't know where all the matter in the universe has come from, yet.

Developmental regulatory genes resist mutational change because they are organized hierarchically: that is, they control various layers of other regulatory genes which subsequently control individual genes and proteins down the line. Central to this regulatory hierarchy are those genes that specify the development of the axis and global form of the animal body plan; these regulatory genes, in a special manner, cannot vary at all without resulting in a catastrophic degradation of the organism. [401]

In his book *Darwin's Doubt*, Meyer writes that biologist Eric Davidson's work emphasizes a very well-known principle in engineering, the *principle of constraints*: that is, the more intricately

---

[400] Meyer, *Darwin's Doubt*, pp. 265–7.

[401] Meyer, *Darwin's Doubt*, p. 268.

connected is a functional system, the harder it is to make significant changes to any portion of it without absolutely compromising its whole functionality. Neo-Darwinism contradicts this notion; it seems to think that replacing a jig-saw puzzle piece by any other random piece from any other picture, regardless of shape or content, will fit just as well. [402] Minor functional tweaking of an integrated system is possible in engineering; but when it comes to making flowchart or structural changes, say, which are dependent on one another, everything tends to crumble: unfortunately resulting in the system's not working any longer. In practice, it will probably be necessary, or perhaps much easier, to restart the design from scratch.

Actual experimental evidence supports this theoretical point of view. Nusslein-Volhard and Wieschaus, mentioned earlier in this chapter, discovered this phenomenon in their fruit fly experiments. Their experiments showed that when early acting molecules, affecting the body plan of an animal, are mutated, further development does not happen: it simply shuts down and the resulting embryos just die. [403] This was also found to be the case (without exception, as far as we know) in all other cases in which mutations occur early in the regulatory genes affecting body plan formation. As we have seen, regulatory genes are hierarchically interrelated, so an early change in the development of an animal will require several coordinated changes in other developmental processes downstream. [404]

Meyer then gives the following practical comparison for our better understanding. If an automaker changes a car's exterior paint color or the cloth material of the seat covers, no other modification—minor, significant, or major—needs to be made for the car to remain useful as a car—and even operate normally; because the core function of the car does not depend on these peripheral qualities. However, if a design engineer decides to modify the car's engine, say by changing

---

[402] Meyer, *Darwin's Doubt*, p. 269.
[403] Nusslein-Volhard & Weischaus, "Mutations Affecting Segment Number and Polarity in Drosophila";
Lawrence & Struhl, "Morphogens, Compartments and Pattern"
[404] Meyer, *Darwin's Doubt*, pp. 260–1.

the length of the piston rods by a substantial amount, he also has to modify the crankshaft accordingly: otherwise the engine will simply come to a grinding halt.

Meyer continues to explain that in order to produce significant changes in the form of an animal, timing is everything. Embryological development can be imagined as an expanding tree of decision-making where the earlier branches result in greater impact on body form. Mutating genes late in the development of an animal will only influence a few cells and, consequently, will not affect its structural features; because, by then, the basic body plan of the animal is already well-defined.

Mutations in early development, however, are the only ones that may affect enough cells that could eventually add up to produce significant changes in the animal's body plan. Therefore, the only realistic possibility of producing significant changes in the animal's body lies, exclusively, in mutations of the key developmental regulatory genes: thus, they are the only ones that could, conceivably, produce large scale evolutionary change—*macroevolution*. However, although many species have been subjected to such key developmental regulatory gene mutations, actual experimental results have shown that they, unequivocally, damage the organism beyond salvaging. [405]

Agreeing with the above concepts, in his article "Adaptation, Neo-Darwinian Tautology and Population Fitness", geneticist Bruce Wallace explains further why upstream mutations are exceedingly more likely to disrupt body development: normally, the earlier the mutation, the more widespread is the damage experienced by the organism. He writes: [406]

> The extreme difficulty encountered when attempting to transform one organism into another ... still functional one lies in the difficulty in resetting a number of the many

---

[405] Meyer, *Darwin's Doubt*, pp. 259–60.

[406] Meyer, *Darwin's Doubt*, p. 260.

controlling switches in a manner that still allows for the individual's orderly (somatic) development. [407]

Alternatively, one may again take my illustration and compare the different animal body plans to a number of straight lines diverging from a single point. Once the lines have separated, even minutely, if one happens to be on one of the lines, it is hard to imagine how one can go back smoothly onto another line, or to hop back and forth (zig-zag) from one line to the other.

In his article "The Molecular Basis of Adaptation", geneticist John Macdonald has called the above situation a "great Darwinian paradox". He draws attention to the fact that the variable genes observed in natural populations affect their form and function only minimally, while those genes that govern major changes do not vary at all; and when such major variations do actually occur, they are detrimental to the organism. [408] In other words, microevolution happens around us all the time; but macroevolution never happens beneficially. The kind of mutations that macroevolution would need to produce new body plans (namely, early beneficial regulatory changes) are never observed; whereas, the kind that macroevolution does not need (namely, genetic mutations late in development) do happen all the time. [409]

In his book *The Origin of Animal Body Plans*, evolutionary developmental biologist Arthur Wallace summarizes the mutagenesis experimental evidence from a wide variety of animal systems as follows: [410]

> Those genes that control key early developmental processes are involved in the establishment of the basic body plan. Mutations in these genes will usually be

---

[407] Wallace, "Adaptation, Neo-Darwinian Tautology, and Population Fitness", p. 70.
[408] McDonald, "The Molecular Basis of Adaptation", p. 93
[409] Meyer, *Darwin's Doubt*, p. 262.
[410] Meyer, *Darwin's Doubt*, p. 476, n. 11

> extremely disadvantageous, and it is conceivable that they
> are *always so.* [411]

Science philosopher Paul Nelson correctly points out that there is one noteworthy exception to this generalization made by Wallace: the loss of a structure. There are many well-documented cases of animals, birds, insects, and fish that show that animals will tolerate, and may even actually thrive, after losing certain traits to mutation; provided those functions are not essential for their survival in that special environment. Obviously, processes that result in a loss of a function or structure cannot be invoked to explain the origin of a function or structure: the former is an up-down process, while the latter is a down-up process. [412] A somewhat crude example would be: accidentally poking one's eye may result in one's losing one's sight; but no amount of poking a blind man's eye will enable him to see.

Even regulatory gene mutations rather late in the development of an animal's body formation (that is, after its axis and global form are completed) also seem to be always detrimental. Let us examine specifically actual mutations obtained by Nusslein-Volhard and Wieschaus in their fruit fly experiments.

A mutation in a particular regulatory gene produced an extra pair of wings on the fruit fly (which is normally two-winged). At first blush this may sound like a genuine down-up Darwinian improvement; in actual fact, however, it did not work out that way. This innovation resulted in a crippled insect, one that could not fly at all, because it lacked, among other things, the extra musculature necessary to support the use of its additional pair of wings. As explained in the automaker illustration above, because the seemingly beneficial mutation was not accompanied by several other coordinated changes that were necessary to make the wings useful, it ended up to be decidedly harmful. [413] Moreover, in this particular case, the tiny structures behind the wings that stabilize the fruit fly in flight, called

---

[411] Wallace, *The Origin of Animal Body Plans*, p. 14 (emphasis in original)
[412] Meyer, *Darwin's Doubt*, p. 477, n. 19
[413] Meyer, *Darwin's Doubt*, pp. 261–2.

balancers, were instead transformed into an extra set of wings; so the balancers were lost, and without the balancers the insect simply could not fly: needless to say, both pairs of wings ended up to be completely useless. [414]

Some people might still wonder, however (as I did, in fact), whether there would be cases where a four-winged fly, for example, would be able to find use of and have an advantage over a normal two-winged insect. While this probably has to remain an open question, I now doubt its feasibility. Utilizing Meyer's automotive example above, imagine somehow doubling the number of pistons in a four-stroke engine, without doubling the number of cylinders in the block, without simultaneously modifying the crankshaft significantly to accommodate eight arms instead of four, without modifying the camshaft to accommodate eight lobes instead of four, and without doubling the number of push-rods, spark plugs, and valves. What would the extra pistons do? Nothing: except come in the way.

Another one of their experimental mutations produced a fruit fly with legs in the place of its antennae; needless to say, it did not have any advantage over the rest of the fruit fly population. [415] Other mutations produced an eyeless insect, a short-winged mutant, and a curly-winged one; none of which could fly. In short, all such later developmental regulatory gene mutations ended up in disaster. [416]

Cis-regulatory elements (CREs) are segments of non-protein-coding DNA which regulate the transcription (copying) of genes in their immediate vicinity. (The Latin prefix "cis" translates to "on this side of"; hence, in the DNA sequence, one finds CREs situated very close to the gene, or genes, they regulate.) [417] Biologists and traditional Neo-Darwinists Hopi Hoekstra and Jerry Coyne note

---

[414] Meyer, *Darwin's Doubt*, p. 318.
[415] Meyer, *Darwin's Doubt*, p. 319.
[416] Meyer, *Darwin's Doubt*, p. 261.
[417] https://en.wikipedia.org/wiki/Cis-Regulatory_element (12 October 2015)

explicitly that the best examples of alleged CRE-induced mutations showed "losses of traits rather than the origin of new traits." [418] [419]

In his book *Foolish Faith*, Creationist author Judah Etinger, writes that during a televised interview, evolutionary biologist Richard Dawkins was asked whether he could point to any example in which a mutation has actually added new genetic information. Dawkins appeared extremely perplexed by this question and could not answer it on the spot. [420] Dawkins subsequently wrote an essay in response to this question, but in it he still did not point to any specific example. [421]

Apart from the bat, possibly acquiring the ability to fly; and the extinct archaeopteryx possibly acquiring a winged hand (or a fingered wing); I can't mention any other examples of animals *acquiring* a useful feature (assuming that is what actually happened, of course). It should, however, be the norm, not the exception, in a down-up evolutionary hypothesis: there should be innumerable instances of it: unless, of course, the hypothesis is false. Recall my illustration of the abundance of colored marbles in the large barrel. So much for macroevolution and common ancestor hypotheses! Notice that I call them hypotheses; in my opinion, we can only (still) call microevolution a theory.

In my younger days, I fully believed in macroevolution and the origin of all life from a common ancestor; particularly because of the vestigial (non-functional) parts in bodies: like our tailbone or the horse's hoof. Since, at that time, I was also a strong believer in a Creator (like Darwin), I used to compare God's creation of the evolutionary process—creating one organism from which all the others follow—to a master pool (or snooker) player who can put all the balls in the pockets with just one shot: every ball causing another ball to be pocketed until it ended up in a pocket itself. Of course, this attributes a greater ability to the pool player than one that pockets the balls one shot at a time: comparable to creating several organisms.

---

[418] Hoekstra & Coyne, "The Locus of Evolution", p. 1006

[419] Meyer, *Darwin's Doubt*, p. 481, n. 20

[420] Etinger, *Foolish Faith*, p. 48.

[421] Etinger, *Foolish Faith*, p. 81, n. 31

The interesting thing is that, at present, there are some Christian organizations (like *BioLogos*) that believe in macroevolution, in line with mainstream biologists. However, science philosopher Stephen Meyer and his Intelligent Design colleagues, including myself, oppose the stand of such well-meaning theistic organizations: for the simple reason that the scientific evidence does not point that way.

Epigenetic Information

There is a significant body of evidence that DNA is not the whole story in heredity; there seems to be hereditary information stored at a much higher level: the level of cell structures. *Epigenetic* information is stored in cell structures, not in DNA sequences. The Greek prefix "epi" translates to "above" or "beyond"; hence the term "epigenetic" means "beyond genes". The reader ought to realize that if this is the case, then Darwinian evolution is even more wrong than is commonly thought; because it postulates that mutations happen at the level of genes: that is, protein-producing DNA strands. Epigenetic information is stored at a much higher level. One may compare this situation to information comprising a whole encyclopedia, consisting of several, say, twenty-three (23) volumes (chromosomes), compared to information stored in a short paragraph in one of the chapters of one of these volumes. Following is some of the relevant scientific evidence.

(1) In 1924, scientists Hans Spemann and Hilde Mangold used microsurgery to cut a portion of a newt (lizard-like amphibian) embryo and transplanted it onto another newt embryo. The second embryo produced two bodies, each with a head and tail, joined at the belly: very much like Siamese twins. So, although the scientists did not change the second embryo's DNA, they managed to alter its anatomy drastically. [422] (2) In the 1930s and 1940s, biologist Ethel Harvey showed that sea urchin embryos continued to develop up to

---

[422] Spemann & Mangold, "Induction of Embryonic Primordia by Implantation of Organizers from a Different Species"

about 500 cells even after their nuclei were removed: that is, without any DNA. [423] (3) In the 1960s, scientists Jean Brachet, Helene Denis, and F. De Vitry chemically blocked the transcription (copying) of DNA into RNA in amphibian embryos; still, the embryos continued to develop until they consisted of several thousand cells. [424] (4) In the 1970s, biologists Y. Masui, A. Forer, and A. Zimmerman showed that a frog egg could continue its early development without its nucleus if just the cell division apparatus from a sea urchin was injected into the egg. [425]

In every one of the above cases, DNA was eventually required to complete development; however, these experimental observations suggest that DNA is not the whole story: that it represents only part of the picture. These experiments show that there are other sources of information playing an important role in animal development: at least in the early stages, if nothing more. [426] Surprisingly enough, in his book *Darwin's Doubt*, Meyer writes:

> Many biologists no longer believe that DNA directs virtually everything happening within the cell. Developmental biologists, in particular, are now discovering that crucial information for building body plans is imparted by the form and structure of embryonic cells, including information from both the unfertilized and fertilized egg. [427]

Meyer gives a couple of illustrations for our better understanding of the above quote.

---

[423] Harvey, "Parthenogenetic Merogony or Cleavage Without Nuclei in Arbacia Punctulata";
Harvey, "A Comparison of the Development of Nucleate and Non-nucleate Eggs of Arbacia Punctulata".

[424] Brachet, Denis, & De Vitry "The Effects of Actinomycin D and Puromycin on Morphogenesis in Amphibian Eggs and Acetabularia Mediterranea"

[425] Masui, Forer, & Zimmerman, "Induction of Cleavage in Nucleated and Enucleated Frog Eggs by Injection of Isolated Sea-Urchin Mitotic Apparatus"

[426] Meyer, *Darwin's Doubt*, pp. 271–2.

[427] Meyer, *Darwin's Doubt*, p. 275.

At a construction site, builders will use many materials: bricks, lumber, drywall, piping, wires, nails, etc.; however, these materials do not determine (are not the root cause of) what the overall plan of a house is going to look like, much less the neighborhood setting. Again, electronic circuits comprise many components: resistors, capacitors, inductors, transformers, transistors, etc.; however, these components do not determine their own arrangement in a circuit, nor the intended function of the circuit. Similarly, DNA services a given cell; however, it does not, by itself, direct how individual proteins are assembled into larger structures or systems: like cell types, tissues, organs, and body plans. [428]

Let us, therefore, look at some of the evidence for epigenetic information. As we have already seen in the section on the "Living Cell", eukaryotic cells (cells with a nucleus) have internal skeletons that give them shape. The cell's shape is supported by a structural framework called the cytoskeleton; it consists of three major types of structural materials: microtubules, microfilaments, and intermediate filaments. Recall that microtubules act as railroad tracks over which molecular motors can carry cargo to distant parts of the cell. The type of network and the positioning of the microtubules in the cytoskeleton result in the development of different types of embryos. These microtubule arrays (railroad tracks) within embryonic cells help to distribute the proper proteins used during development to specific locations in these cells. This network of microtubules in the cytoskeleton, therefore, plays a major role in developing different embryonic types and hence constitutes a form of precise and critical structural information.

These microtubule arrays are made of many identical proteins, called *tubulin*. Now, neither the tubulin proteins nor the genes that produce them can account for the differences in the shape of the microtubule arrays (network) that result in different kinds of embryos. [429] Conceivably, the quantity (number) of tubulin molecules

---

[428] Meyer, *Darwin's Doubt*, p. 277.

[429] Meyer, *Darwin's Doubt*, pp. 277–8.

may be regulated, but I do not think this is enough to determine the exact structure of a precise network: they can be placed anywhere. So where is this critical information that creates these different networks, resulting in different types of embryos, situated?

In an animal cell, there is a microscopic organelle situated close to its nucleus called the *centrosome*; it is the main location where the cell's microtubule arrays get organized. Oddly enough, the centrosome contains no DNA; yet, the microtubule arrays described in the previous paragraph originate from the centrosome. Centrosomes consist of two *centrioles* positioned perpendicular to each other and surrounded by a shapeless heap of protein molecules; each centriole looks like a cartwheel structure made up of twenty-seven (27) microtubules arranged in nine (9) linear triplets. [430] Although centrosomes are made of proteins, Meyer believes its structure is not determined by genes alone. [431] Although I normally agree with what Meyer has to say, I am not convinced by this last statement: I surmise the folding of the centrosome's apparently shapeless heap of proteins can determine its structure—I could be wrong.

We have already seen that after proteins are produced in embryonic cells, the cytoskeleton is used to transport them to the right locations; however, these tubulin tracks have no way of knowing where the right location might happen to be. So there seems to be some form of additional information on the cell walls defining the locations where these proteins are targeted to be deposited. Fruit fly developmental biologists have shown that these membrane targets precede the production of the embryo's regulatory molecules; these membrane targets are located on the inside surface of the egg cell, and they define the spatial coordinates where diverse regulatory molecules are to attach themselves. [432]

Ion channels are openings in the cell wall through which electrically charged particles (ions) can pass in either direction; these ion channels generate an electric field. Now experiments have

---

[430] https://en.wikipedia.org/wiki/Centrosome (16 October 2015)
[431] Meyer, *Darwin's Doubt*, pp. 278–9.
[432] Meyer, *Darwin's Doubt*, pp. 279–80.

shown that both electric and magnetic fields affect embryological development. Although the ion channels consist of proteins, their exact locations are again determined by preexisting cell membrane patterns or spatial targets. [433]

Sugar molecules can attach themselves to the fatty molecules constituting the cell wall or to the proteins embedded in it. Furthermore, biologist Ronald Schnaar explains that every sugar building block can assume several different orientations. [434] The various arrangements and orientations of several sugar building blocks outside cell membranes constitute what is commonly known as the *sugar code*: sequence-specific, information-rich structures, comparable to the DNA code. This sugar code is what dictates the arrangements of different cell types during development. Biochemist Hans-Joachim Gabius notes that:

> [Sugars provide a] high-density coding [system that is] essential to allow cells to communicate efficiently and swiftly through complex surface interactions. ... These [sugar] molecules surpass amino acids and nucleotides by far in information-storing capacity. [435]

Thus, apparently, the sugar code provides information over and above that stored in genes and DNA in general. [436] Epigenetic information, supposedly, resides in the structure of the maternal egg and is inherited directly from cell membrane to cell membrane independently of DNA. [437]

---

[433] Meyer, *Darwin's Doubt*, p. 280.
[434] Schnaar, "The Membrane is the Message".
[435] Schnaar, "The Membrane is the Message";
Gabius, "Biological Information Transfer Beyond the Genetic Code";
Gabius, Siebert Andre Jimenez-Barbero, & Rudiger, "Chemical Biology of the Sugar Code".
[436] Meyer, *Darwin's Doubt*, pp. 280–1.
[437] Meyer, *Darwin's Doubt*, p. 365.

Again, I'm not sure I agree that none of the above is determined by proteins or DNA. In his book *The Greatest Show on Earth,* Richard Dawkins shoots *epigenetics* down:

> 'Epigenetics' [is] a modish buzz-word now enjoying its fifteen minutes of fame in the biological community. Whatever 'epigenetics' might mean ... its enthusiasts cannot seem to agree with themselves, let alone with each other. [438]

Disagreement is not uncommon in any scientific field; that's not a real problem: in fact most string theorists cannot really define the theory.

I have included this section on epigenetic information for two reasons: first, to show that there is some evidence in favor of additional information independent of DNA; and secondly, in order that the reader may appreciate that even though I quoted Meyer extensively in the last two chapters, I do not adopt his convictions without weighing their validity.

Anyway, more to the point, Meyer concludes this subject by reminding us what Neo-Darwinism proposes: namely, that any new kind of information arises from random mutations at a very low (gene) level of the body's system; and yet, it seems that any significant body innovation depends upon changes at a much higher level. [439] So he writes:

> If DNA isn't wholly responsible for the way an embryo develops—for body-plan morphogenesis—then DNA sequences can mutate indefinitely and still not produce a new body plan, regardless of the amount of time and the number of mutational trials available to the evolutionary process. Genetic mutations are simply the wrong tool for the job at hand. [440]

---

[438]   Dawkins, *The Greatest Show on Earth,* p. 216
[439]   Meyer, *Darwin's Doubt,* p. 281.
[440]   Meyer, *Darwin's Doubt,* p. 281.

One may compare this to changing the values of the basic components: capacitors, resistors, inductors, transformers, and transistors, etc. slightly in an electronic circuit, without changing the whole circuit, or the mother board, of the electronic equipment. There is no way of producing different electronic equipment like a computer, a radio, a television, or a record player simply by changing the basic components. A modular, a whole board or circuit and possibly an interconnecting motherboard, change is required at some stage.

Although there is some evidence that (unexplainable) things do happen without DNA, the fact that DNA is eventually required to complete the animal development, does not preclude the possibility that DNA does, in fact, play a part in them—even initially. As evolutionary biologist Richard Dawkins affirms:

> A gene's influence ... is *always* indirect. The gene determines a protein sequence that influences X that influences Y that influences Z that eventually influences ... [441]

In fact, he also affirms that genes influence the environment outside the organism—the *extended phenotype*; [442] and I tend to agree with him when I observe the highways, dams, sewers, towns, and skyscrapers humans have constructed.

In my opinion, much more research is required in this area; and I have a feeling that we might discover new depths to the mystery of life: similar to the discovery of DNA itself inside the nucleus, which was, in my youth, thought to be a blob of simple organic materials like amino acids.

If, however, a higher level of information is actually discovered at the modular level, then the Darwinian mechanism of mutation at the much lower component (protein) level will be deemed impotent to make measureable changes at the modular level; simply because they are a world apart, many orders of magnitude apart. It is like

---

[441] Dawkins, *The Selfish Gene*, p. 240 (emphasis in original).
[442] Dawkins, *The Selfish Gene*, pp. 236–66

saying that a butterfly fluttering its wings can affect the weather significantly on the opposite side of the globe (although there are some scientists who actually believe this is the case); or else, that a change, or insertion, of a paragraph in an encyclopedia consisting of a number of volumes, is going to radically affect the encyclopedia as a whole.

Summary and Implications

As we have seen in this chapter, it seems that a down-up type of evolution does happen—but rarely. For example, Lenski's E-coli bacteria, under constantly varying conditions, developed the ability to assimilate citrate in the presence of oxygen; the bat may have developed the ability to fly; the archaeopteryx may have developed a winged hand; the tailbone in the human body may be evidence that we have evolved into more intelligent beings from the ape family. However, according to evolutionary theory, it should be the norm, not the exception.

The geological fossil record shows that typically most animal phyla (body-plans) appear out of nowhere on the scene; they thrive and mutate within rather strict limits for a while, and then they become extinct. Practically, geological time is referred to by the presence of their fossils in that era; the Cambrian explosion is just the most obvious case of this widely observed phenomenon.

Evidence from developmental gene mutations seems to show that evolution is incapable of radically changing a viable body plan; evolution seems to be a short term phenomenon in organisms: a survival tool for adapting to a changing environment. If the environment changes drastically, an organism becomes extinct.

So does this mean that God keeps placing new phyla on the earth, time and time again? This is a very good scientific question. From the scientific evidence to date, the answer to this question seems to be positive: it seems as if God wants life to continue on this planet, and that he intervenes from time to time. I must admit I do not know how or when or why he does this. Why God breaks his own laws is another good question; similar to why he performs miracles, at times. As we shall see in a later chapter on "Miracles", we have a significant body of direct evidence supporting their existence.

As a scientist, believe me, it goes directly against my grain to admit that, from time to time, God intervenes in nature and temporarily breaks nature's laws; and deep down, I wish I could tell the reader otherwise. But then, God seems to have created (started) the universe, anyway; the Cambrian explosion and the general geological fossil record could also be evidence along the lines of his direct intervention.

For all I know, it could have been some kind of extraterrestrial intelligence, rather than God, who just dumped the Cambrian and other phyla on earth. This may sound strange coming from a scientist; but that is what it looks like: as far as our current scientific knowledge goes. If we posit that the diverse animal phyla were placed on earth by extraterrestrials, we might only be pushing the scientific evidence one or two notches backward: we would also, eventually, need to provide some kind of hypothesis for the origin of these extraterrestrials. As one can easily foresee, this type of hypothesis for the origin of animal phyla (or life) might lead us to an eternal regress type of argument. But if one really believes in extraterrestrial intelligence and unidentified flying objects (UFOs) having landed on earth in the distant (or recent) past, this chapter could be positive evidence for its being the case.

Both religion and science have been of prime importance to me all my life: I was a member of the Society of Jesus (a Jesuit) for over six years, and I possess a Bachelor of Science degree in physics and mathematics. Questions about both the meaning and the origin of life took center stage all my life. Incidentally, I also discovered that the majority of scientists, engineers, and medical doctors I know personally are also very interested in this kind of questions.

Over the last couple of decades, however, I have decided to, systematically, base my faith on reason. This may sound like a contradiction in terms; in my humble opinion, faith disproved by evidence, or unsupported by reason, boils down to superstition—and superstition has a tendency of becoming dangerous. On the other hand, it is mind-boggling to me why most (mainstream) scientists lost their reason and the pursuit of evidence for their "faiths". The macroevolution hypothesis, that is, the supposed evolution of all diverse phyla from one common ancestor, is a typical case in point. Science is unfortunately becoming a religion: downplaying and manipulating evidence that does not conform to the mainstream scientists' "party line", and ostracizing any scientist who disagrees with them.

I may even add here that scientists still have no clue as to the origin of life, consciousness, matter, and the universe; or how and why the laws of physics and chemistry happen to be what they are; their explanations of the root cause, more often than not, gravitate towards chance as their final explanation: chance has become most scientists' "god of the gaps"—exactly what scientists say about believers.

There is nothing wrong with scientists' trying to find materialistic explanations for reality; but what I find objectionable is the fact that if a particular scientist believes in intelligent design (God), one is ostracized and ends up losing or being deprived of a good job. How can we have an unbiased search for truth if scientists are terrorized this way? History has shown that scientists who believed in God were still great scientists, and they contributed enormously to science: they were never a hindrance to science. What if, as the previous chapters indicate, intelligent design happens to be the truth about reality; where would today's science be leading our children?

I think the root cause of this problem is that various religions, both in the present and in the past, have preached war and violence in God's name; no wonder scientists simply cannot stomach religion and are determined to fight it at all costs. God is not on the side of this kind of behavior, of course, despite such teachings in almost every "holy" book. Religious fanatics are misled people, even though God may understand their good intensions. God does not need to be protected. Religions should preach love of God and neighbor, including those people with whom we disagree strongly.

Self-declared atheist Richard Dawkins deals with the dangers of religion at considerable length in his book *The God Delusion*. Dawkins is known for his outspokenness; but in this book he, most appropriately, writes:

> [S]uch hostility as I or other atheists occasionally voice towards religion is limited to words. I am not going to bomb anybody, behead them, burn them at the stake, crucify them, or fly planes into skyscrapers, just because of a theological disagreement. [443]

Scientific opinions are, normally, limited to words; that's true. But ostracizing someone from one's job is not: it means the quality of life for that person and his family—scientists nowadays *are* crossing the line!

Science is, consequently and unfortunately, becoming a religion too; I suppose scientists are fighting fire with fire, but I think it's disgusting. It is as unbearable as when a policeman, or a priest, turns sour: because we expect higher standards from them. I hope that, when the pendulum has swung fully to the other side, and this generation of scientists dies, science will again turn its eyes towards the truth. I have more faith in the methods of science (its history of correcting mistakes, flexibility, and basic philosophies), than in the methods of religion (its history of violence, dogmas, and infallible books or popes). Nowadays, with the internet readily available, nobody can lie or twist the truth too much any longer; hopefully both religion and science will, unequivocally, coordinate their efforts together and follow evidence and reason in their search for the truth.

---

[443] Dawkins, *The God Delusion*, p. 318

CHAPTER 8

MIND AND SOUL

Theseus' Ship

In the year 75CE, in his book on the life of *Theseus*, Greek historian, biographer, and essayist Plutarch wrote:

> The ship wherein Theseus and the youth of Athens returned from Crete ... was preserved by the Athenians: ... they took away the old planks as they decayed, putting in new and stronger timber in their places; ... this ship became a standing example among the philosophers, for the logical question of things that grow; one side holding that the ship remained the same, and the other contending that it was not the same. [444]

Philosophers in Plutarch's time thus questioned whether the ship would remain the same ship if it were, eventually, entirely replaced piece by piece by new planks. When I asked a friend this question, he answered, "Yes, because the information was preserved." I replied that it was a good answer; however Wikipedia adds that, centuries later, philosopher Thomas Hobbes introduced an additional scenario that complicates the situation even further. Hobbes asks, what if the original discarded planks were somehow located and assembled

[444] Plutarch, *Theseus*, http://classics.mit.edu/Plutarch/theseus.html (24 January 2016)

together into another (worn out) ship beside the maintained ship; which of the two ships would be considered Theseus' ship? [445]

In a parallel argument, in his book *The Self Illusion*, developmental psychologist Bruce Hood writes:

> Philosopher Derek Parfit uses these types of scenarios to challenge the reality of the self. [446] He asks us to imagine replacing a person cell by cell, so that the original person no longer contains any of the physical material before the process started. ... Using this logic, Parfit dismisses the notion of an essential self in the first place. [447]

So, regarding anything you remember from your childhood, that wasn't you: because there probably aren't any cells left in you from your childhood. In any case, your maternal fertilized egg must have died and been replaced by now; your original identity (the self) must have changed, or perhaps better it never existed: it is just an illusion.

This is probably where you start thinking that, somewhere along the syllogism, you have been hoodwinked. Most modern scientists are materialists: they want us to believe that only matter exists; so they trick us into believing their party line. They make no mention of your DNA, which is *information*, as my friend above pointed out, a great deal of it inside every single one of your cells: information that belongs to you, and to you alone, that has been transmitted all along the way preserving your personality (self) and your memories. If the DNA is identical, as perhaps in the case of twins, their identity is still preserved: for one thing, they do not have the same memories. Moreover, even two identical marbles are *qualitatively* the same, but not *quantitatively* the same: they still have their own identity, they are not one and the same; they have similar, but not the same, atoms constituting them.

---

[445]   https://en.wikipedia.org/wiki/Ship_of_Theseus (17 October 2015)

[446]   Parfit, *Reason and Persons*

[447]   Hood, *The Self Illusion*, p. 112

In the above riddle of Theseus' ship, if we introduce another parameter, time, then at any instant of time an object or person that is changing is uniquely and completely determined. Over time Theseus' ship was changing, but it was still Theseus' ship, with modifications, of course. In a sense, it was not exactly the same ship, because something had changed in it; but it existed at any instant of time: one can't say that Theseus' ship was an illusion, or even worse, that it never existed.

This is like saying that the book you are reading now is an illusion because I made so many changes to it. I could have saved every single one of my changes and called them "file1", "file2", "file3", etc.: they would all have been slightly different. The fact that I did not save every version does not mean that this book is (or was) an illusion, or even worse, that it does not (or did not) exist: it always existed, but in different versions.

Finally, if the original planks of Theseus' ship were gathered together and the original ship rebuilt, it will only be Theseus's ship number one thousand and one (1001), say, with the planks somewhat worn out, I suppose. It will exist contemporaneously with Theseus' ship number one thousand (1000), say, and there will be two items that can legitimately be called Theseus' ship, but for different reasons. The ship built from the original planks is not exactly the same as the original ship either: there is the time dimension mentioned above involved: the planks are worn out in fact. But it sure does not mean that Theseus' ship is just an illusion, or that it never existed, or even worse, that both ships don't exist now. This type of reasoning is ludicrous. So, the above is simply a philosophical trick by materialistic scientists to coast the general public into thinking according to their agenda. The reader's gut-feeling was right in the first place.

Materialistic scientists nowadays believe that only matter is reality; that's what physics says, they say. While there is a lot to be said in favor of this philosophy, I do not think I can agree with them. Let us think of a number, six, say. (Scientists deny the existence of thought too, but let's not get ahead of ourselves yet.) It could mean six doughnuts, six eggs, six pigs, six children, six dollars, six houses, six beaches, six countries, etc. In other words, a number is an *immaterial* concept. I agree that it would probably make more sense if associated with matter, say six doughnuts, but it does not necessarily have to be. So there *are* other "things", besides matter. (Please forgive my using the word "things" here, maybe the meticulous reader does not like the use of that word in this context, but you know what I mean.) Other examples of intangible things are thoughts, words, computer programs, a DNA sequence, etc.; they are all forms of information. I admit one usually needs a material medium to be able to express all the above; but I am sure the reader can appreciate that the information conveyed in all of the above examples is independent of the medium. For example, if I want to convey the idea that "it is sunny", I can say it in English, French, Italian, Chinese, etc., or I may not say it at all: I can write it down. So there are immaterial things: mainly those consisting of information.

What materialistic scientists would have us do is eradicate from our vocabulary anything that is spiritual or supernatural; and this is their first line of defense, their bastion, so to speak: their party line, "there is no reality except matter". Hence, for them, it follows that there cannot be a *self*, or a *soul*: it is not matter and therefore it does not exist. If the reader follows this kind of logic, one will end up with a lot of conclusions that go against the grain; the bottom line would be that the difficult questions one asks, pertaining to the mysteries of life, will remain unanswered.

Principle of Life

The word "animal" comes from the Latin word "anima" meaning "soul"; and it denotes, roughly speaking, anything that can move around of its own accord. It is the *principle of life* in a body: the *integrating factor* of the body as a whole; it relates to the entire body, not to one or a few organs. It is what distinguishes a living body from a body that has just died, irreversibly, by drowning or a heart attack, say.

In my language, Maltese, which is a Semitic language like Jewish, the word for "soul" is "ruh" (where the "h" is pronounced as in the word "house"); in Jewish, "ruah" is "life" or "vital breath". Again in Maltese the word for "wind" is "rih"; one can see that it has the same root (consonants) as "ruh". The soul is, therefore, associated with something intangible; as the wind was to primitive people. They knew that a wind existed: they could feel it, but they could not grasp it. Breathing created an effect similar to a small wind, and a living human breathed; they knew there was something different between life and inanimate matter: but they could not figure out what it was.

The interesting question is: does this soul have a separate existence from the body? Is it a separate entity? We know that there is such a thing as the *brain*, which is the stuff (matter) inside our skull; but when we speak of thoughts we refer to the *mind*, which is not synonymous to the brain. Thinking, presumably, happens in the brain; and without a brain, admittedly, one cannot think; one's logic is impaired, also, if one's brain is damaged: they are intricately connected. So again, is the mind a separate entity from the brain?

In his book *The Atheist's Guide to Reality*, philosopher Alex Rosenberg assures us that matter is simply atoms: there is no such thing as atoms that can convey other ideas, concepts, memories of locations, or thoughts: it is all just an illusion; he writes:

> Physics has ruled out the existence of clumps of matter of the required sort [i.e., conveying thoughts]. There are just fermions and bosons [see Appendix III] and combinations of them. None of that stuff is just, all by itself, *about* any other stuff. There is nothing in the whole universe—including, of course, all the neurons [explained presently] in your brain—that just by its nature or composition can do the job of being about some other clump of matter. So when consciousness assures us that we have thoughts *about* stuff, it has to be wrong. Therefore, consciousness

cannot retrieve thoughts *about* stuff. There are none to retrieve. So it can't have thoughts *about* stuff either. [448]

Neurons are cells that can be stimulated electrically by means of ions (electrically charged chemicals); thereby, they process and subsequently transmit information through electrical signals. [449]

I suggest the reader's not taking Rosenberg's quote too seriously; its author obviously cannot understand how inanimate matter, a complex aggregate of simple atoms, can possibly generate thoughts or retrieve memories in human beings—and neither can I. So for its author, ideas, thoughts, and memories cannot exist—problem solved! I'll let the reader make up one's own mind about that.

A living human being can generate thoughts, a human being that has just died irreversibly, presumably, cannot. So, what is the difference between them? They have the same cells; yet one can take care of itself one might loosely say indefinitely, while the other is doomed to constant decay.

The best analogy I can think of is a computer program. The mind is comparable to a program that controls the brain, while the soul is comparable to a program that controls the whole body. A computer is a machine made of atoms and molecules, but without an *executable* program (that is, a program that makes the computer run something: like a spreadsheet or a chess game) the computer will just sit there and do nothing. So an executable program is like part of the mind of a computer. Is a program a separate entity from the computer? Yes, of course, it is: it is intelligent information. Can it exist without the computer? Yes, it can exist apart from any computer: inside a disk, say; but it is not much use without a computer, or even inside a damaged computer.

In his book *The Physics of Immortality*, mathematical physicist Frank Tipler, writes the following about the similarity of the mind and the soul to a computer program:

---

[448]   Rosenberg, *The Atheist's Guide to Reality*, p. 179 (emphasis in original).

[449]   https://en.wikipedia.org/wiki/Neuron (17 October 2015)

> I claim that a "living being" is any entity which codes information (in the physics sense of the word) with the information coded being preserved by natural selection. Thus "life" is a form of information processing, and the human mind—and the human soul—is a very complex computer program. [450]

It is also my opinion that the mind, or the soul, is comparable to an ultra-complex computer executable program: one that can run any application. Notice, however, my repeated use of the word "comparable"; I will here explain exactly what I mean by it.

The mind, or the soul, is even more than a simple computer program. We know that the mind gives the animal, at least in the case of humans, *self-awareness* (or *consciousness*) of its own existence. If one looks at a mirror, one realizes that one is looking at oneself. If one places a computer in front of a mirror, the computer (as far as we know) will not realize that it is looking at itself; even if it has a camera. For example, when my chess program prints, "I resign" after it loses a game, it does so because its programmer designed it to print that phrase (or a similar phrase) whenever it loses a game; not because it realizes its own existence, or that it was playing a game, or that it just lost the game. If one places a computer, containing a camera, in the midst of other computers and in front of a mirror, that particular computer will not be aware of its own existence, or the existence of the other computers, or of anything else around it. Furthermore, it will not react, try to escape, or protect itself if it "sees" someone approaching it threateningly with a baseball bat, intending to destroy it. On the other hand, any animal will run away when it senses danger from a predator.

Setting aside consciousness, for the time being, the mind is still a fantastic program; it is obviously the result of intelligence. There is a common saying in computer programming: "Garbage in—garbage out". If a chess program can beat a chess-master, as some modern computers can, then its programmer, or conglomerate of

---

[450] Tipler, *The Physics of Immortality*, p. 124

programmers, must know the game of chess very well and, moreover, they must know how to make a machine help itself achieve a better strategic and material position that will eventually enable it to win the game. This is definitely the fruit of intelligence and design: it doesn't happen spontaneously, of its own accord, or by trial and error, or by chance. The next thing that current materialistic scientists need to do is to deny such things as *intelligence* and *design*; don't be surprised, they already deny such a thing as *thought*. Then, of course, their party line will follow, "there cannot be intelligent design".

We need to be aware, however, that there is such a thing as *artificial intelligence*. For example, a computer may have millions of games in its memory; it can therefore search through them all until it finds a winning game for any move played, or position reached, by its opponent. When one plays against such a computer, it appears that the computer is intelligent. Every time a game is played on such a computer, it is added to the database; thus the computer seems to learn how to play the game better over time. To date, this is possible in the game of checkers (or droughts) but not in the game of chess; because there are more possible moves in the game of chess than atoms in the observable universe. (According to mathematician, electronics engineer, and cryptographer Claude Shannon—the "father of information theory"—there are about 10^{120} possible moves in the game of chess, while there are only about 10^{80} atoms in the observable universe.)

When we make use of a search engine like Google (which will probably give us an answer better than any human being can) we do not attribute the answer we obtain to the intelligence of the computer itself, but to the intelligence of whoever placed the information and the indexing in the data base. When all is said and done, even a search program, by itself, is the product of high intelligence; no animal, besides a human being, can do it—let alone its happening spontaneously, through dumb luck.

But that is not how our mind works; it does not remember every game it played or every situation it encountered before; it just uses some form of logic, strategy, general rules, hints, clues, and only a little past experience for the current position. So the reader will appreciate the difference between the "intelligence" of a human being and that of a computer program.

Universal Wave Function

As I already pointed out above, there is a very important basic difference between a computer's executable program and the self of a living animal: and that is self-awareness. How does a living animal perceive danger? What makes an animal, a conglomerate of carbon molecules, act in such a way as to defend itself, to hide, or to run away?

In the introduction to his book *The Physics of Immortality*, Tipler writes:

> [Theologian] Wolfhart Pannenberg has suggested [in 1977 and 1981] that there may exist a previously undiscovered universal physical field ... which can be regarded as the source of all life, and which can be identified with the Holy Spirit. ... I shall argue ... that the *universal* wave function ... is a universal field with the essential features of Pannenberg's proposed new "energy" field. If this identification is made, it becomes reasonable as a matter of physics, to say God is in the world, everywhere, and is with us, standing beside us at all times. ... [S]uch Presence is a key property of the Christian God. (This does not mean, however, that God intervenes in human history in a supernatural way.) [451]

These are the words of a full-fledged scientist who endorses the idea of a theologian.

In the "Nicene Creed", formulated in the year 325, Christians profess as part of their faith: "I believe in the Holy Spirit, the Lord, the giver of Life ...." I am not sure, from a scientific point of view,

---

[451]   Tipler, *The Physics of Immortality*, pp. 13–4 (emphasis in original)

whether the Holy Spirit is another person distinct from God (the Father) as Christians believe; but I do believe that there is a close connection between God and life: giving living beings a self-awareness (or consciousness) among several other things, as we shall see later in the course of this chapter. I also believe that God is the only source of life: that all matter is intrinsically dead.

The bible, in "The Book of Genesis", depicts God as forming the first man, Adam, from mud (or clay) and then breathing life into it; throwing light on the fact that, in Semitic languages, the words "soul" and "wind" have the same root (recall: "ruh" and "rih", respectively, in Maltese):

> And the Lord God formed man of the dust of the ground,
> and breathed into his nostrils the breath of life; and man
> became a living soul. [452]

Some people, especially scientists, might think that I'm getting carried away, quoting the bible here: that is, in a book that is supposed to be scientific evidence for or against the existence of God. I used to believe in the infallibility of the bible; but not any more: I simply do not believe that the bible is the word of God. What I do believe is that the bible was written by people searching for God—like me; and therefore some of what they say may be significant. I further believe that although the bible is not literally inspired by God, like any other good book, some of the verses contain great wisdom and are inspired, or are inspiring, in the ordinary sense of the word. Personally, I always compare the bible to the sayings, or the wisdom, of a nation; I also admit that the bible may be biased in God's favor, and that it is very often unscientific. Finally, I believe that it has some ancient historical value like: Flavius Josephus' *Antiquities of the Jews*, Julius Caesar's *Gallic Wars*, and so on.

Let us get back to scientific evidence. Is there any evidence for the above claim, made by Frank Tipler, of a *universal wave function*, which is probably somewhat similar to a combination of a force

---

[452] *The Holy Bible*; Genesis: 2, 7.

field and *Schrödinger's wave equation*? This is not a subject that is developed or elaborated by scientists in general: they don't seem to make this connection, except for Tipler, of course; so I ask the reader to give me some latitude and cut me some slack here. (1) Well, there is *life* including: consciousness, free will, understanding, mental suffering, pain, pleasure, logic and reason, emotions (like love, empathy, hate, and anger), *qualia* (perceptions of color, taste, and smell—see the corresponding section below), and finally morality. (2) There is *dark energy*: recall, in the section on "Dark Matter and Dark Energy", about 73% of the universe consists of dark energy. What exactly is this dark energy? It seems to be a force field. No scientist knows, yet! (3) There is the *Quantum Vacuum* of free space: see the corresponding section above. (4) There is what scientists believe to be the nine or ten dimensions of free space, of which we only see three: height, width, and depth; the rest being "curled up very small" and out of our sight. (5) And finally, there is *String Theory*.

Although I disagree with most scientists on the last two items, scientists are not crazy: they must have had some evidence to suggest them; and they might even still be right. It is as a consequence of String Theory that scientists have postulated the nine or ten dimensions of free space and the existence of 10^{500} universes. String theory is more of a hypothesis rather than a theory. In fact, in his book *The Trouble with Physics*, theoretical physicist Lee Smolin, who had previously also contributed papers on string theory, writes:

> [S]tring theorists seem to have no problem believing that string theory must be right while acknowledging that they have no idea what it really is. [453]

In his book *The Universe in a Nutshell*, probably the most famous theoretical physicist and cosmologist Stephen Hawking explains:

> In M-theory, space has nine or ten dimensions, but it is thought that six or seven of the directions are curled up

---

[453]   Smolin, *The Trouble with Physics*, p. 270

very small, leaving three dimensions that are large and nearly flat. [454]

M-theory [is a] theory that unites all five string theories, as well as supergravity, within a single theoretical framework, but which is not yet fully understood. [455]

String [or superstring] theory [is a] theory of physics in which particles are described as waves on strings; [it] unites quantum mechanics and general relativity. [456]

Supergravity [is a] set of theories unifying general relativity and supersymmetry. [457]

Supersymmetry [is a] principle that relates the properties of particles of different spin. [458]

Although I'd rather, personally, call most of the above theories hypotheses (except for quantum mechanics and general relativity) physicists must have some reason to believe that there is enough evidence to call them theories. Physics is talking about empty space and matter here; why all this complexity? All these physics theories, or hypotheses, are models: they try to explain observations of the behavior of matter and energy in space. This type of complexity (which, in my opinion, is part-evidence for the universal wave function) can perhaps explain how an immaterial soul could possibly interact with (move) a material body: like a "ghost inside a machine", so to speak—which concept, incidentally, philosopher Gilbert Ryle rejects in his book *The Concept of Mind*. [459]

This universal wave function, if it actually exists, would be somewhat like the *BIOS*, or "Basic Input/Output System", which defines the basic rules linking a computer's microprocessor to its attached devices (it is also called "machine language"). I suppose the universal wave function transforms non-living matter into life. If God

---

[454] Hawking, *The Universe in a Nutshell*, p. 88

[455] Hawking, *The Universe in a Nutshell*, p. 205

[456] Hawking, *The Universe in a Nutshell*, p. 207

[457] Hawking, *The Universe in a Nutshell*, p. 208

[458] Hawking, *The Universe in a Nutshell*, p. 208

[459] Ryle, "Descartes' Myth."

is a reality (exists), then I believe that the God of the supernatural is the same God of the natural (matter); therefore, I think that there *must* be some kind of connection between the supernatural and the natural. Life is a mystery, a real mystery indeed; I admit I cannot explain everything, but the above is the most plausible explanation of life to me. The reader may or may not agree with me by the end of this chapter; in any case, let us continue examining some of the hypotheses of modern scientists and philosophers, in conjunction with some modern empirical (experiential) evidence.

Qualia

Qualia refer to an animal's subjective experiences: the way an individual animal perceives reality, or the way an animal's perception is connected to the reality outside it. For example, the way a human being sees colors, hears music, tastes wine, smells a rose, feels pain or pleasure, and so on. "Qualia" is the Latin neuter plural form of the adjective "qualis", meaning "what kind of".

If we were blindfolded and given an apple to taste, we would probably all agree that it was an apple; but unfortunately we do not know whether you taste an apple the way I taste a pear, say, or vice-versa. The same thing with odors, we can all agree that a gas station smells of gasoline; but we cannot tell whether you smell gasoline the way I smell turpentine, say, or vice-versa. Similar situations are the way we feel pain, a toothache as opposed to a headache, and pleasure. Most probably men and women enjoy sexual intercourse differently; but we will probably never know the exact difference, never mind the personal difference between individual males or females. We can all agree that a particular color is red, say; but we cannot know whether you see red the way I see red, or whether you see red the way I see blue, say, or vice-versa: they are subjective experiences. All these qualia, or first-person experiences, are connected to consciousness and life in general; but we don't know, and we probably cannot know, whether they are all connected the same way or differently in every individual.

Do I believe that color spectra, for example, are reversed in some human beings? No, I believe they are "wired" inside our brain the same way in all of us; as I shall show later, I see no special reason for complicating things and wiring them differently in every one of us. However, it is conceivable that they could be wired reversed in some of us; like traffic lights could be defined reversed: red for go and green for stop—and mounted reversed: green on top instead of red. If this were the case, I don't think that it would make any real difference to any one's life individually; in other words, I see no scientific reason why they should be wired in a specific manner rather than the other way around: I believe it is completely arbitrary. However, most scientists believe they are necessarily the direct result of physical states in the brain; we shall see in the course of the next few sections why I disagree with them, and why this disagreement is important—it is not trivial as one might think at first blush.

When it comes to qualia, science can only tell us something like: when light of wavelength 700nm falls on the retina of the human eye, that person experiences a red color; but it probably cannot tell us what red is like or what blue is like to one experiencing the colored light. Try and explain what red is like to a man blind form birth; admittedly, to some extent at least, this could be a shortcoming of our language: a certain inability to describe detail.

Qualia Differences

Do we all experience qualia identically or do some of us experience qualia differently? There is enough evidence to believe that some of us experience qualia differently; most definitely, different species of animals experience qualia differently from us humans.

The human ear can hear sound with frequencies ranging from 20Hz to 20KHz. Elderly people cannot hear the higher sound frequencies: say, between 15KHz and 20KHz. Occasionally, a customer (of the company I work for) complains of a buzzing or ticking noise, caused by a magnetic component; I have to ask the assistance of one of my younger work-colleagues to help me troubleshoot it. I do not say, "I cannot hear it, therefore it does not exist." Women, on average, seem to have a higher audible frequency range than men: they can hear hissing sounds that men cannot normally hear.

Some human beings experience a phenomenon known as *color blindness*: it is the complete inability to see any color, or a decreased ability to perceive color differences under normal lighting conditions. Wikipedia says that color blindness affects close to ten percent (10%) of the population. [460] I had a close friend who could not distinguish brown from red: while shooting snooker (a game similar to pool) he had to keep track of where the brown ball ended up—all the time. Drivers who are completely color blind can still drive; but only because the position of red, amber, and green lights are mounted the same way everywhere you go: top to bottom.

Moreover, practically everyone knows that dogs have a sense of smell much more sophisticated than us human beings. A lot of people also know that dogs and other animals can hear whistles that humans cannot hear: that scare these animals and can even drive them crazy. Examples of qualia we humans do not have are snakes' *heat-vision* and bats' *echolocation*; I shall discuss both of these in further detail later on in this chapter.

So summarizing, it looks like external reality is first perceived by the animal, these then cause *brain states* inside the animal, which finally induce the animal to act accordingly. Some individuals have below standard qualia (a deficient organ, leaving something to be desired) others experience better-than-expected performance from these brain states; presumably, approximately the same situation applies to animals of other species. This is what the great majority of scientists believe; and it is utterly reasonable: it is what I used to believe in my younger days, too. The problem is that life is never so simple: what we see is not always what we get.

---

[460] https://en.wikipedia.org/wiki/Color_blindness (20 February 2016)

Strange Qualia

What we see or what we feel, does not always correspond to the reality outside of us.

I have never been hit badly on the head so as to see stars or hear birds singing; but this is what children's comic magazines and cartoon movies portray all the time: there must be some truth to it—I'm not willing to prove it. In any case, according to Wikipedia, *phosphenes* are light flashes experienced by people as a result of pressure either inside the brain or on the eye. [461]

Floaters are another visual experience of something inexistent; they are small dark shadowy shapes of many sorts that seem to dart away when you try to look at them directly: there is no external reality attached to them. Obviously, what you see is not always what you get.

Phantom Qualia

Things get even more interesting when we realize that sometimes our brain tells us something that we know is a lie.

Phantom limb pain is a perception that an individual experiences relating to a limb that is not physically part of one's body any longer, usually the result of an amputation, even after it has completely healed. Phantom limb sensation is the perception of any feeling (excluding pain) at the location of an absent arm or leg, such as: apparent weight, itching, and thermal unpleasantness (feeling the limb hot or cold). According to Wikipedia, it is widely accepted that more than 80% of amputees experience various phantom sensations at least for some time in their lives; furthermore, some of them experience a level of phantom pain or feeling for the rest of their lives. [462]

---

[461] https://en.wikipedia.org/wiki/Qualia (24 February 2016)
[462] https://en.wikipedia.org/wiki/Phantom_pain (20 February 2016)

Can we tell these individuals that such pain or feeling cannot exist: that it is a figment of their imagination? Can we still say that such phenomena are the result of physical states in our brain? What is their brain trying to tell these people?

Reversed Qualia

I shall here mention just one of the many psychobiological experiments to show this very strange phenomenon concerning animal consciousness.

In 1957, neuropsychologist and neurobiologist Roger Sperry reported the following result of an experiment performed by his student Nancy Miner on a number of frogs. Basically, she cut a portion of skin from the back of the frog, and another same-sized portion of skin from the belly of the frog, swapped them and observed the results of this operation:

> [A] flap of skin was peeled off the trunk [belly] region of a frog, lifted, and cut completely free of all nerve and other connection, rotated 180 degrees, and re-implanted [on the back—and vice versa] ... The operation was done in early larval stages. When the tadpoles had grown up and undergone metamorphosis into the adult, we found that tickling these frogs on the back, within the graft region, caused them to scratch the belly with their foreleg. Conversely, stimulating them on the belly caused them to swipe at the back with the hind leg. [463]

I ask the reader here: is this an indication of the physical states of reality?

---

[463] Sperry, "Brain Mechanisms in Behavior." p. 26

Scientific Analysis

From several of his philosophical writings, theoretical physicist and quantum physics pioneer Erwin Schrödinger (of the Schrödinger wave equation fame) seems to believe that qualia are not physical. Regarding color perception by us human beings he writes: [464]

> The sensation of color cannot be accounted for by the physicist's objective picture of light waves. Could the physiologist account for it, if he had fuller knowledge than he has of the processes in the retina and the nervous processes set up by them in the optical nerve bundles and in the brain? I do not think so. [465]

Schrödinger adds that subjective experiences do not constitute a direct one-to-one correspondence with external stimuli. As an example he mentions the fact that light of about 590nm is experienced by us human beings as yellow; but we also experience the same yellow color with a combination of red light of wavelength 760nm and green light of 535nm. This evidence leads him to conclude that there is no "numerical connection with these physical, objective characteristics of the waves" and the physical sensations we experience. [466]

May I add my own observation here (and the reader will probably agree with me): it has always baffled me, personally, how the seven colors of the rainbow, every one of which looks "darker" than white (red, orange, yellow, green, blue, indigo, and violet) when combined together in a certain proportion give us an appearance of white light! Isn't that strange?

However I must also mention here, that most scientists disagree with Schrödinger and me; they believe that qualia are the direct result of physical states in the brain, and that science will one day be able to explain why we see red the way we see it, why we feel pain the way

---

[464]   https://en.wikipedia.org/wiki/Qualia (24 February 2016)

[465]   Schrödinger. *What is Life?* p.154

[466]   https://en.wikipedia.org/wiki/Qualia (24 February 2016)

we feel it, etc. These scientists contend that qualia simply do not exist at all, that only physical states in the brain exist, and that basically qualia are a figment of our imagination: an unnecessary addition to what is actually happening.

I honestly do not know how they can explain the evidence for phantom qualia and reversed qualia. Moreover, I do not know whether scientists will ever be able to explain pain, say, from a third person's point of view the same way one experiences it from a first person's point of view. How do simple ions (charged atoms) give different sensations of pain? One can tabulate the resulting sensations; but can one actually convey the same subjective feeling? Is it really just a case of unsophisticated language? Or is the subjective feeling *hardwired* to the organism?

Self

Most scientists contend that our self-awareness is also a figment of our imagination. In his essay "The Soul of the Matter", philosopher Charles Taliaferro summarizes Daniel Dennett's philosophy as follows:

> Dennett's case against subjective states of awareness is ... radical .... Dennett takes particular aim at our apparent awareness of ourselves as subjects. ... Dennett thinks there is nothing physical in the brain or the body as a whole that can play the role of such a substantial, individual subject. [467]

In his book *Consciousness Explained*, philosopher Daniel Dennett quite appropriately writes: "The trouble with brains is that when you look in them, you discover that there's nobody home." [468] I have to agree with Dennett: there is no "little person" inside our brain. However, I do believe that there is a connection, a "supernatural"

---

[467] Taliaferro, "The Soul of the Matter." p. 31
[468] Dennett, *Consciousness Explained*. p. 29

(in the sense of "above" the natural) "hardwiring", between the brain's physical states and the living organism as a whole entity. As I said before, matter is intrinsically dead; and it is this supernatural hardwiring that gives it life—the universal wave function. It is similar to telling the computer of a complex mechanical system which relay is number one relay, which proximity sensor is number three sensor, which pneumatic cylinder is number seven cylinder, etc.

A system architectural engineer could easily reverse the functions of relays number one and number seven, say; and as long as the operating system (program) knows which is which everything will function properly. (Recall what was said regarding reversing color spectra.) However, no design engineer, in his right mind, will construct every system differently and reprogram them accordingly; although it is utterly conceivable to do just so: one will probably design them all the same way. However, once a system is hardwired and the executable program finalized, if someone decides to reverse the connection of two relays, say, the wrong relay will be activated. (Recall Miner's frog experiments, where flaps of skin from the back and the belly were swapped.) The obvious question now crops up: in both our case and in the case of animals, how were our qualia hardwired?

Finally, as a philosophical observation, without scientists there would be no science; science consists of measurements taken, evidence collected, and conclusions reached by individual *selves* called scientists. Now, if the "self" (the little person inside our brain, so to speak) does not exist at all, in some form or another, how can scientists exist? Furthermore, how can we rely on science if the entities that discover its basic phenomena do not even exist? Isn't science undermining itself by taking such a distilled position? [469]

---

[469] Taliaferro, "The Soul of the Matter." pp. 33–4

Subjective Reality

True enough, our senses are not always to be trusted: they are sometimes misleading; but with the right instrumentation, in conjunction with our reasoning, our senses can be guided to the right path. However, still science is based on our senses!

For example, a piece of iron placed in an oven at eighty degrees Celsius (80C) will burn one's hand if touched; while a piece of wood at the same temperature (placed in the same oven for a very long time) does not: the iron "feels" hotter, in spite of their being at the same temperature. So, things aren't always what they feel or what they seem. But with the help of a thermometer and with the additional knowledge that iron conducts heat (and electricity) better than wood, we can come to the right conclusion; namely, that iron conducts heat faster into our hand than wood does: consequently, one feels pain, or at least discomfort, on touching the hot iron but not when touching the hot wood.

The point I am making here is that there is another (deeper) dimension to a subjective experience. [470] At the superficial level, there are the objective (physical) aspects of the phenomenon being observed: (1) the iron and the wood are at the same temperature, and (2) iron conducts heat faster than wood. But then we are also interested in answering the question of why one feels hotter than the other, even though they are at the same temperature; as human beings we are curious as to why things are experienced one way and not another. Our feeling the iron hotter is also an objective fact, so to speak, and it is still a scientific observation that cannot be completely excluded without compromising the whole picture of reality.

Another example will help us understand this concept a little better. When a (black) body is gradually heated up, it first appears (to us human beings) to be red-hot, then it appears white-hot, and finally it appears blue-hot. The reality is that in the first instance we (human beings) are approximately right: the red-hot body emits

---

[470]   Nagel, "What Is It Like to Be a Bat?" p. 443

overwhelmingly red light, or light of around 700nm, so it appears red. When it turns white, it is not emitting white light; it is emitting all light frequencies visible to us human beings, namely, from about 400nm to about 700nm; this proportional (right) mixture of colors, strangely enough as mentioned above, appears white to us. Finally when it turns blue, it does not stop radiating all the visible wavelengths of light; in fact, it radiates even more of them than before; but the intensity of blue light increases so much and so rapidly that it swamps all the other visible colors, and so it seems to emit blue light, or light of around 500nm. Of course, it also emits quite a bit of violet light (400nm) too, and also some of dangerous ultra violet (UV) radiation at this temperature; but the latter is invisible to us, and violet almost looks like blue, so the blue color still stands out. This represents the whole picture of the reality of our experience of viewing the heating up of a body: our sensations are a factor too.

Apparently snakes can only "see" infra-red (IR) radiation (heat, not red color): your body heat is like a lamp to a snake, especially at night when it is more conspicuous; that is why snakes tend to sneak up to sleeping humans. It is hard to imagine what a gradually heated object looks like to a snake, from our perspective; but if we imagine, for a moment, that it can only see red, say, it will only see various shades of red and perhaps an increase in intensity. As one can see the two experiences of the same physical phenomenon are completely different; to leave out the subjective portion (the human experience as opposed to the snake's experience) is to give an incomplete picture of reality and consciousness.

Bats, on the other hand, can see like us in daylight; but at night they make use of *echolocation*: a system of sound wave echoes of their own high-frequency shrieks reflected by objects nearby. Even if we were to show the various reflected sound waves on a television screen, say, we would never know how bats actually perceive reality around them: our brains would have to be wired like theirs for us to be able to experience echolocation the same way they do. [471] It is

---

[471]   Nagel, "What Is It Like to Be a Bat?" p. 439

something like trying to understand how one who has been blind, deaf, and dumb from birth experiences reality around oneself—but to a much greater difference, of course. [472]

Current Scientific Knowledge

At present, we seem to lack a threshold level of scientific knowledge to make a connection between matter and life. When certain scientists identify (propose they are one and the same thing) the physical states in our brain to qualia, say, none of them can explain how one actually translates to the other: nobody knows why the combined activity of a bunch of neurons in the brain results in a certain feeling. Why do we feel pain when we prick our finger? Why does the aroma of a flower (molecules evaporating from its petals) feel pleasant. Why does a red rose appear red? [473]

Nowadays most people can relate to a statement like, "mass is energy"; because most people know about Einstein's equation $E=mc^2$. But prior to Einstein, in the nineteenth century, say, very few people would have believed it; and to a pre-Socratic philosopher it would have seemed utter nonsense. [474] Ever since the discovery of the Higgs boson and consequently the Higgs field (which is comparable to an electromagnetic or a gravitational field) the concept of energy turning into mass has become even more understandable and tangible to us, especially to physicists.

This is the position we are in at present: we do not know how anything in the universe can turn to life or consciousness; yet we have ample evidence all around us that, somehow, matter turns to life. We can think of simulations of how things seem to come alive with complexity: examples would be movies, television, and holograms. The rapid display of still pictures from a movie strip seems to make it come alive. The innumerable electronic signals transmitted from a

---

[472] Nagel, "What Is It Like to Be a Bat?" p. 440

[473] Phillips, "The Ten Biggest Mysteries of Life" p. 30

[474] Nagel, "What Is It Like to Be a Bat?" p. 447

television station, unlike the movie stills, bear no resemblance to what we see on a television screen; but it surely looks alive. The natural tendency, therefore, is to conclude that with added complexity, things tend to come alive by themselves: as a matter of necessity. But I think this is the wrong conclusion; because, if you really think about it, movies, television, and holograms are not really alive, are they? Life only comes from life that already exists; we have, so far, failed to produce any kind of life.

Hardwiring

Science has discovered various physical fields: gravity, electromagnetism, strong nuclear, weak nuclear, and Higgs—all of which are invisible; it knows how they work, but it has no clue how they got to be that way.

However, mathematical physicist Frank Tipler does give us a clue: in his book *The Physics of Immortality*, he writes:

> [T]he universal wave function is the unique field which gives being to all other fields—the electroweak fields, the gluon fields, the quark fields, the lepton fields, indeed all the usual physical fields. ... So we have an all-pervasive physical field which gives being to all being, which gives life to all living things ... [475]

And even though he is a full-fledged scientist, he does not shy away from actually stating, later in the same book:

> In the biblical traditions, this life-giving power is the Holy Spirit. I am thus in effect proposing that we identify the universal wave function ... with the Holy Spirit. [476]

---

[475] Tipler, *The Physics of Immortality*; p. 183
[476] Tipler, *The Physics of Immortality*; p. 185

I concur with Tipler's proposal, and there are probably several other scientists who also do; but the subject is very controversial, and it sounds very unscientific and religiously biased.

But perhaps there is a hardwiring (connection) between the natural and the supernatural. Believe me; it goes against my grain, as a scientist, to say something like this. I do not claim to be a great scientist, by any means: I only have a Bachelor of Science degree; but I claim to be an evidence-oriented and results-oriented one. I also claim to be quite good at integrating scientific information (similar to what Francis Crick and James Watson did when they "discovered" the structure of DNA) and coming to the right conclusions this way.

I could be wrong, and I'm willing to accept my being wrong right now if I am so proved in the future: but I do not believe that consciousness or life could evolve from complexity; I believe that life comes from life, life originates from God alone—the giver of life—and that all matter is intrinsically dead. I shall continue to add all kinds of evidence along these lines throughout the rest of this chapter.

Free Will

In his book *The Atheist's Guide to Reality*, philosopher Alex Rosenberg makes reference to a series of experiments, boing back to about fifty years and replicated many times, which seem to show conclusively that our conscious decisions to do things do not actually do what we think they do. Our conscious decisions seem to happen too late and not to be involved in the choice process at all.

The most famous of these experiments was conducted in the late 1970s by neurologist Benjamin Libet. In this experiment, subjects were simply asked to push a button whenever they felt like it. Libet created a dot on the screen of an oscilloscope circulating like the hand of a clock, but moving much more rapidly: completing a rotation in one or two seconds, say. The subject was asked to visually note the position of the moving dot when one was aware of the conscious decision to move the finger or the wrist to push the button. At the same time the subject's brain was electrically wired to detect the motor cortex signal responsible for flexing the wrist and pushing the finger down. Normally it takes about two hundred (200) milliseconds (0.2s) from conscious willing to finger pressing; but the motor cortex signal starts about five hundred (500) milliseconds (0.5s) before the button is pressed. In other words, all that is needed to complete the finger pressing action has already taken place about three hundred (300) milliseconds (0.3s) before the subject is conscious of one's decision. Rosenberg concludes:

> The obvious implication: Consciously deciding to do something is not the cause of doing it. It's just a downstream effect, perhaps even a by-product, of some process that has already set the action in motion. A nonconscious event in the brain is the "real" decider. Maybe the real decision to act that takes place unconsciously really is a free choice. Or maybe, since your brain is just a purely physical system, you don't have any free will. [477]

While admittedly this is an interesting experiment, there are similar experiences with our vision, hearing, olfactory, and tactile functions: where what we think is happening "now" actually happened some short time earlier.

For example, we feel the pain of touching something that is very hot noticeably later in time. What we see is actually pre-filtered and re-interpreted by our brain, and we see it later in time. There could be reasons why vision is filtered by the brain; for example, one does not want to run away from the reflection of a predator in a pool, because that will mean that one is running towards the predator: so the brain takes some time to sort things out. But, just because we see something slightly later in time, it does not follow that we had no vision of the thing at all, or that what we see never happened. The three hundred (300) millisecond (0.3s) delay between the conscious and the subconscious just shows that there is a connection between

---

[477] Rosenberg, *The Atheist's Guide to Reality*, p. 153.

them that we still do not fully understand. True, we may not know enough about this filtering process, but it does not follow that we have no free will. Disposing of free will sounds more like a materialist's agenda: namely, eradicating all supernatural (meaning above what is natural) concepts.

It is not the first time someone decides, early in the morning, to go grocery shopping after work and actually does so; or one might change one's mind after work because it happens to be raining. A woman might decide to get pregnant; she asks her husband in the next few days to accommodate, and she actually does get pregnant; to say that she actually had no free will to make such a choice is ludicrous to me. Long term planning: like getting a university degree or planning for retirement; can we really say that we have no free will in embarking on such projects? A three hundred (300) millisecond (0.3s) delay is interesting, but not important enough in most situations in real life: it's splitting hairs. I think we are losing time and space here; let us move on.

Determinism

According to Newton's laws of motion, if we know accurately and completely the initial states (mass, position, velocity, direction, acceleration, temperature, impact-restitution factor, frictional coefficient, etc.) of two particles that interact (collide) with each other, we can predict their final states. Conversely, if we know their final states, we can tell what their previous states were. This also holds true for a system of particles, i.e. bodies, if we are able to know accurately and completely the state of every particle constituting the body.

If, for a moment, we forget about our inability to know the state of every particle accurately and completely, whatever those states might be; it follows, from the above reasoning, that the current state of any body is completely determined by its prior state, and that prior state is completely determined by an even prior state, and so on and on. It follows that any physical system, including our bodies and our brains, is completely determined by what occurred before; in other words, according to the laws of *classical physics*, we have no free will: Newton's laws are completely deterministic. This scientific conclusion, one must admit, is quite a formidable one. And this is exactly what Rosenberg proposes in his same book *The Atheist's Guide to Reality*:

> The mind is the brain, and the brain is a physical system, fantastically complex, but still operating according to the laws of physics—quantum or otherwise. ...
> When I make a choice—trivial or momentous—it's just another event in my brain locked into this network of processes going back to the beginning of the universe, long before I had the slightest "choice". Nothing was up to me. Everything—including my choice and my feeling that I can choose freely—was fixed by earlier states of the universe plus the laws of physics. End of story. No free will, just the feeling, the illusion in introspection, that my actions are decided by my conscious will. [478]

I'm sure most readers of this passage will feel something is wrong with the above argument, but where?

Feelings are not always to be trusted: it's usually science that has the last word. When we touch something hot or get electrocuted, it feels like something undesirable "flowed" into our bodies, in both cases. In the latter case we are correct: electrons flowed into our bodies. However, in the former case we are not; here's what actually happens. Heat is just motion of molecules; when we touch something hot, our skin cells are also set in motion by simple contact; the end result is that our skin cells are permanently damaged by this excessive motion; consequently, our nervous system sends a painful message to our brain, so that we don't do it again. No doubt, things aren't always what they feel or what they seem.

---

[478] Rosenberg, *The Atheist's Guide to Reality*, pp. 236–7.

It so happens, however, that the laws of physics are not as deterministic as described by Rosenberg above. When we speak of an electron orbiting around a nucleus, at first we imagine it orbiting in a two-dimensional circle or mild ellipse: perhaps like the moon or a satellite orbits round the earth, or like the planets orbit round the sun. But then, when you think about it, why should the electron settle in one plane; why shouldn't it move in a sphere all-around the nucleus? So all we can say is that the electron of every atom settles in some three-dimensional distance from the nucleus (in a spherical shell); at any instant of time, we cannot tell exactly where it is: we can only talk about its probability of its being in a certain location. Not only that, but even the distance of the electron from the nucleus is not defined precisely, it can deviate a little in a mild elliptical manner. One can plot a bell-shaped statistical distribution of the distance of the electron from the nucleus; it will be a very narrow distribution, but all the same we can only talk about the probability of an electron being a certain distance from the nucleus; therefore, we can only think of an electron as a negatively charged *cloud* when it is inside an atom.

It can be proved experimentally, as we shall presently see in this chapter, that matter, particularly electrons and light, can behave both like particles and like waves at different times. The theory of determinism originated from the behavior of particles alone; we don't even know if electrons and light are going to behave as particles or as waves at any given time: we can only talk about probabilities. So, by its very nature and at its very foundation, physics is probabilistic. Now, if you think about it, probability and chance cannot co-exist with determinism.

In his essay "The Measure of All Things: Quantum Mechanics and the Soul", science philosopher Hans Halvorson writes:

> "Classical Physics" is a catch-all phrase for a number of different theories developed roughly between the time of Galileo Galilei (1564-1642) and James Clerk Maxwell (1831-1879). Radically abstracting from the rich

detail of these theories, they are all based on two main assumptions: first, the state of each object in the world can be completely specified by assigning values to all of that object's quantitative properties (such as its position, its velocity, its mass, etc.). Second, there are laws of nature such that the state of each object at any future time is completely determined by the state of all objects at any previous time. [479]

The reader probably noticed the word "assumptions" in the above quote: it does *not* follow that there are such deterministic laws in nature.

Regarding determinism, in his book *The Physics of Immortality*, Tipler writes:

[T]he structure of the phase paths [histories of maximum probability] (more precisely their ultimate future) gives probability weights—guidance, so to speak, not rigid control—to all paths. The ultimate future guides all presents into itself. But this guidance is *not* determinism. [480]

In other words, the Holy Spirit gives us guidance, but he does not preclude our deciding to do what we want.

The reader may have heard about the strangeness of quantum physics (alternatively known as quantum mechanics), and may wonder what this strangeness is all about. There are many strange realities described in quantum physics, but I shall here limit myself to one extremely common phenomenon, the *duality* exhibited by particles and waves: that is, the fact that sometimes particles behave as if they were waves, and vice-versa, sometimes waves behave as if they were particles. And this happens with practically all basic (fundamental) matter! However, before we embark on this investigation, we must first appreciate the physical and experimental differences in the properties of particles and waves.

---

[479]   Halvorson, "The Measure of All Things", p. 140.

[480]   Tipler, *The Physics of Immortality*; p. 185 (emphasis in original)

Particles

Particles are confined to one specific location in space; they are not spread out in space as waves are: they are well-localized discrete objects.

In the absence of any force field (gravitational, magnetic, electric, etc.) they do not start to move on their own; but if they are already moving they will keep on moving in a straight line (they don't change direction) and with the same speed they already have. Therefore, in the absence of any force field, particles do not bend around corners or edges (as sound or water waves do) but keep on moving in a straight line; they can, of course, be *reflected* (bounced back) if they encounter obstacles. Particles may collide with (bump into) one another or they may coalesce (stick) together constituting a larger particle (or "body"). Particles are relatively simple to understand: they behave very much like miniature pool table balls.

Practically everyone knows that electrons are particles, not waves; but how do we know this? In the previous section on "Determinism", we described an electron orbiting around a nucleus as a negatively charged cloud; so how can we say that electrons are particles? This is the case when electrons are inside an atom; however, if we produce a beam of free electrons by using a thermionic (red-hot filament) electron gun, we can determine how free electrons behave. If we shoot such a beam through an opening in the shape of a Maltese (eight-pointed) cross onto a florescent screen, we obtain a very sharp image of the cross. In the absence of any force field, moving particles travel in straight lines; they do not bend round corners or edges as sound or water waves do. Had electrons consisted of waves rather than particles, they would leave a broad and diffused patch of light on the florescent screen, not a sharp detailed image of the eight-pointed cross opening.

Waves

Waves are a little bit more difficult to understand; but we all have enough everyday experience with sound and water waves to appreciate the differences between particles and waves. Waves are spread out in space; unlike particles, they are not confined to a particular (small) location, and they consist of a multitude of particles.

Waves do keep moving in a straight line in free space; but if they encounter a corner or an edge, they have the tendency of (sort of) clinging to and bending around them; this property of waves is termed *diffraction*.

For example, one can easily hear someone speak from around the corner of a building, even if the building is in the middle of nowhere where there is no possibility of any sound reflections. All the varieties of waves exhibit diffraction phenomena; the extent to which they bend around edges depends on the wavelength of the particular wave: the greater the wavelength, the greater the angle of diffraction.

Waves can reinforce one another if they are *in phase* (synchronized) or neutralize one another if they are oppositely synchronized. If an amusement park water-wave machine (alone) produces one-foot-high waves, and another water-wave machine (alone) produces two-foot-high waves; when working together they would produce three-foot-high waves if they are synchronized and one-foot-high waves if they are oppositely synchronized. In the first case the peaks and the troughs of the individual waves reinforce each other, while in the latter case the peaks and the troughs of the individual waves oppose each other. This property of waves is termed *superposition*. Particles, on the other hand, cannot be superposed like waves; they can only collide with one another or coalesce (stick) together.

Waves are said to interfere *constructively* (reinforce each other) if they arrive at a particular point in space synchronized: the peaks or troughs add together. On the other hand, waves are said to interfere *destructively* (cancel each other out completely or partially) if they arrive at a particular point in space in opposite phase (the troughs neutralize all or a portion of the peaks); the general term used for this phenomenon is *interference* of waves.

Practically everyone knows that light consists of waves, not particles. But wait a minute; light does leave a sharp image if one shines a laser beam through an eight-pointed cross opening onto a photographic plate; so how do we know that light does not consist of particles? If one observes the light shadow carefully, one does observe that the image of the cross is, in fact, very slightly diffused: this is evidence of the bending of light around edges. The shadow from an electron beam is much sharper; it is not diffused at all, not even minutely. Recall that the extent to which waves bend around edges depends on the wavelength of the particular wave. Light wavelengths are very small (4×10^{-7}m to 7×10^{-7}m) compared to sound wavelengths (1.7×10^{-2}m to 17m) hence the bending round edges in the case of light is not noticeable in everyday life; but in the lab, with proper equipment, it is quite obvious.

A final important phenomenon of waves is appreciated by looking at and examining the erosion of beaches by waves. Beaches are eroded over time by the successive beating of waves onto the beach; what the first wave cannot achieve, the next billion might: their effect is cumulative, but over time.

Quantum Physics (Quantum Mechanics)

We have already encountered a strange behavior of light waves. Recall that, in a dark room that is illuminated only by red light, a photographic film is not affected: no matter how intense the (red) light might be, or for how long it shines onto the film. Blue, green, or perhaps yellow light is required to produce an image; their higher frequency enables them to produce an image: the higher the frequency the higher the energy of a light train, or quantum. Recall that the energy of a light quantum is given by the equation $E=hf$, where "h" is a very small fixed number, known as Planck's constant, and "f" is its light frequency. Although there are many observations indicating that light consists of waves, at times it also behaves as particles—normally called photons. Unless these photons have a certain threshold (size) of energy, they will not produce any effect: the cumulative aspect of waves described in the last paragraph does not kick in. The red light in a photographic dark room is such an example.

As a second example of strange behavior in quantum physics, consider the following simple apparatus, placed on a horizontal table in a dark room. At one end of the table is a horizontal laser beam (light source), and at the other end is a vertical photographic plate perpendicular (at right angles) to the laser beam. In between is a screen, also vertical and perpendicular to the laser beam, with two very narrow parallel vertical slits (perpendicular to the table) and very close to each other.

Here is a diagram of the apparatus, viewed from above:

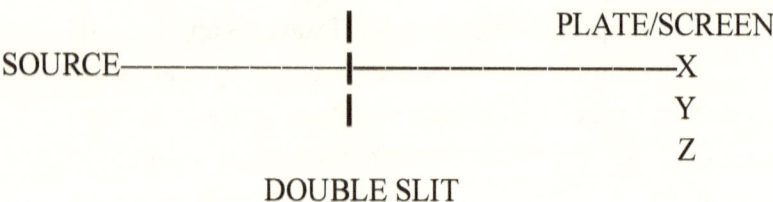

DOUBLE SLIT

Now a laser beam consists of monochromatic light: that is, it has just one color, and therefore one frequency and one wavelength; the light waves emitted by a laser beam are also all in phase (synchronized). Now imagine a line of symmetry, the perpendicular bisector of the horizontal distance between the two slits, extending from the source to the photographic plate; let's call this point on the photographic plate "X". Because of symmetry, the light beams exiting through both slits travel the same distance to X; therefore, having started synchronized, they will also arrive at X synchronized. Hence, they will interfere constructively and produce a bright vertical line, called a *fringe*, on the photographic plate at X. To either side of the center-line, on the photographic plate, the beam from one of the slits travels a longer distance than the beam from the other slit. So a *phase difference* starts to creep in: that is, the crests and troughs of the individual light beams start to become unsynchronized, and the two beams do not interfere constructively one hundred percent (100%) any longer; consequently, the intensity of the light starts to diminish away from the center-line. Please note that we are talking about very small distances here, because light has a very short wavelength. So, as we move further from the center-line on the photographic plate, we come to a location where one beam travels exactly one-half wavelength more than the other. At this point, let's call it "Y", the beams arrive oppositely phased and so interfere destructively: that is, trough cancels peak, because the two beams will be of roughly equal strength there. The result is that there is no illumination (light) at Y; so we end up with a dark fringe (vertical line) at Y. As we move further away from the center-line on the photographic plate, we come to a point, let's call it "Z", where one beam travels exactly one whole wavelength more than the other, and the two beams again interfere constructively, boosting each other up; consequently we again end up with a bright fringe at Z. If we carry on in this way, further away from the center-line on the photographic plate, we predict, and in fact obtain, a pattern of alternate bright and dark fringes on the photographic plate. (This is how sound and water waves behave too; that is why there are some locations in concert halls, or in churches, where one cannot hear properly—everything sounds muffled: that's destructive interference at that location.) In conclusion, since light waves interfere constructively and destructively, light must consist of waves.

Now, in the above apparatus, replace the light source by an electron beam (say from a thermionic electron gun or a mild beta-emitting radioactive source), and also replace the photographic plate by a florescent screen. Now block one slit; the result is a bright sharp image of the other slit on the florescent screen, slightly off-center in line with the other slit, of course. Now, instead of blocking the first slit, block the second slit; again we obtain a bright sharp image of the first slit, slightly off-center in line with the first slit, of course. If we do the same thing with light waves we do get similar slit images, in succession, but the images of the slits will not be as sharp, because light waves bend slightly round edges: the images will be slightly diffused. In conclusion, electrons are particles.

Generally, electrons do behave as particles: that's quite true; but look at the following strange behavior. If we leave both slits open for the electrons to go through either slit at will, we do not get two bright sharp fringes corresponding to the two slits, as we would expect: that is, the combined effect of the two separate (slit-blocking) experiments described above. Strangely enough we obtain an interference pattern similar to that of the laser beam (light waves)! That's strange, isn't it?

Even stranger is the fact that, if we leave both slits open but place two electron detectors one in each slit, we do not get an interference pattern any longer; we obtain two bright fringes: the images of each slit. Please note that the detectors do not interfere with the movement of the electrons through the slits: the electrons just pass through either detector as they choose. It is like our measurement is affecting, or "spooking", the behavior of the electrons: sometimes they act like particles and sometimes like waves, depending on whether we are "looking" at them or not. This experiment has been repeated hundreds of times, giving the same result; it is not a fluke, it is reality; it is not a small miracle, it is physics: it is just one of the many strange phenomena of quantum physics.

Consequently, in his book *The Spiritual Brain*, neuroscientist Mario Beauregard writes:

> This area of physics, quantum physics, is the study of the behavior of matter and energy at the subatomic level of our universe. Briefly, the synapses, the spaces between the neurons of the brain, conduct signals using parts of atoms called ions [charged particles]. The ions function according to the rules of quantum physics, not of classical physics.
>
> What difference does it make if quantum physics governs the brain? Well, one thing we can dispose of right away is

determinism, the idea that everything in the universe has been or can be predetermined. [481]

There you have it; our brains function at the subatomic level and therefore obey the laws of quantum physics, not of classical physics. (I shall be using concepts and ideas from Beauregard's book extensively in this chapter.) In quantum physics there is a significant amount of indeterminism, as the above experiment shows. So much, therefore, for the materialistic scientists' view that free will does not exist because the laws of physics and chemistry in the universe are deterministic. Basically, it's not the whole truth.

As an aside, I don't think that the law-enforcement and legal systems would like to concede that none of us have free will; I don't know how we can fare in society without taking responsibility for our actions. No free will has the stench of a wrong conclusion.

Turing Test

We now come back to the question of the so-called intelligence of computers. The *Turing test* is a test proposed by computer scientist Alan Turing; it is intended to determine whether a computer is able to simulate intelligent behavior equivalent to, or rather indistinguishable from, that of a human being. It is not intended to test its ability to give the correct answers to questions, because human beings do not always give the right answers anyway: it depends on how knowledgeable they are in that particular subject; it is only intended to evaluate intelligent, or rather relevant, answers.

Wikipedia describes the test as consisting of three participants: a human evaluator converses naturally with another human being and a machine (computer), placed in separate rooms isolated from each other and also from the evaluator. The evaluator is aware of the fact that one of the other two entities answering questions is a human while the other is a machine. The answers are given to the evaluator

---

[481] Beauregard & O'Leary, *The Spiritual Brain*, p. 32

in the form of text on a screen through a keyboard, so that the quality of speech is eliminated as a factor. From the answers received, the evaluator is supposed to judge whether the responses to natural human conversation were given by a normal human being or by a machine purposely designed to conduct human-like conversation. If the evaluator cannot distinguish reliably the human from the machine, the latter is deemed to have passed the Turing test. [482]

According to some scientists, for example mathematical physicist Frank Tipler, already mentioned above, such a machine can be considered a "person". In fact, in his book *The Physics of Immortality*, regarding super-machines that pass the Turing test comfortably, Tipler writes:

> Of course, it would not be wise for humans to try to enslave intelligent machines, or to murder them. [483]

I can't agree with Tipler in this case, even though I admire him immensely: I don't think that, in the future, anybody will be charged for murder because one smashed one's own computer, no matter how intelligent it might have seemed. Neither do I believe, as some scientists actually do, that automobiles are alive: simply because we can't help "reproducing" them. We shall come back to this idea of murdering an intelligent machine after the next few sections: in the section entitled "Consciousness".

Chinese Room

The *Chinese room* is a thought experiment originally conceived by philosopher John Searle.

In this thought experiment, he *assumes* the existence of a computer program that can pass the Turing test described above, but in the Chinese language: that is, it can carry on an intelligent,

---

[482]   https://en.wikipedia.org/wiki/Turing_test (29 April 2015)
[483]   Tipler, *The Physics of Immortality*, p. 44

or rather a human-like, conversation in Chinese. He then supposes that an English-speaking person, who knows no Chinese at all—either written or spoken—is locked in a room with a set of rules corresponding to the operations of the above computer program—but written in plain English. These enable an English-speaking person to correlate one set of formal symbols (Chinese characters constituting the "question", so to speak) with another set of formal symbols (different Chinese characters constituting the "answer", so to speak).

Consequently, if pieces of paper written in Chinese are slipped under the door, by hand-simulating the above program, the English-speaking person can carry on an intelligent conversation in Chinese: without understanding a single word! The process will, admittedly, be very lengthy and much slower than that of a computer; but nevertheless, conceivably at least, he can. Therefore, Searle concludes that, if there were such a program that allows a computer to carry on an intelligent conversation in any given language, the computer executing the program would not *understand* the conversation either. [484]

Now, it is worth noting here that Tipler, in his book *The Physics of Immortality*, regarding the Chinese room thought experiment writes:

> A human being could no more hand-simulate a program
> that could pass the Turing Test than she could jump to the
> Moon. [485]

And again a bit later he insists:

> As I said, a man can no more hand-simulate a Turing Test-
> passing program than he can jump to the Moon. In fact it
> is far more difficult. [486]

He also gives convincing mathematical calculations to this effect; however, I'm not sure I agree with his initial assumptions in his

---

[484] https://en.wikipedia.org/wiki/Chinese_room (03 May 2015)
[485] Tipler, *The Physics of Immortality*, p. 39
[486] Tipler, *The Physics of Immortality*, p. 40

calculations: I still think that if the person in the room is given a manual corresponding to the program, one or a dozen non-Chinese-speaking people can hand-simulate a program.

In any case, I won't go into the assumptions and calculations because it's academic to my argument; let's stick to the concept here, and suppose for a moment that it *is* possible. I only wanted to point out, for the sake of fairness, that a great scientist, Tipler, thinks that hand-simulating a computer that passes the Turing test is physically impossible: just keep that in mind for now.

Understanding

One of the things I realized, since my youth, is that machines can do things much better and much faster than the human beings that invented them: a car runs much faster than a human being; a crane can lift loads much heavier than a human; a sewing or weaving machine can do the job much faster and much better than a woman; and so on.

So, is there anything that makes us, their creators, "better" than machines? We have creative powers, and they possibly cannot carry out an intelligent conversation, that's true; machines don't have that … yet; but we don't know what will happen in the future! Truly we are an efficient package; but our own machines seem to be out-running us—by far.

I was told when I was still very young, that we have "intelligence", which machines don't have. But then this last statement has always been baffling to me. A ten-dollar ($10.00) calculator can do calculations much better and much faster than I can, even though I have a Bachelor of Science degree in mathematics: and I surmise that puts me in the top ten-percentile (10%) of humanity. So how intelligent does a machine have to be before it is considered human? Talking about intelligence, chess is a game of both intelligence and strategy: nobody can deny that; I'm a very good chess player, I'm not a master, but I pride myself on being able to play a good game of chess; again I surmise I'm in the top ten-percentile (10%) of humanity in my ability to play a good game of chess. Yet, I can hardly beat my computer program without taking some moves back. I don't have the least doubt that I could mistake my computer program for a human being if I gave it a Turing test in chess, rather than in English. So exactly how "intelligent" does a machine have to be before we consider it human?

So for a while, until the recent past, I leaned towards scientists like Frank Tipler who believe that machines can eventually be considered persons. However, deep down, I somehow felt something was wrong in my way of thinking. I think I resolved the puzzle only lately, during my research for this book, when I came across the Chinese room thought experiment and realized that, as Tipler points out in his book *The Physics of Immortality*:

> Searle's central point in the Chinese Room Experiment is "A computer has syntax, but no semantics." That is, all the program does is manipulate symbols according to certain formal rules (syntax). It has no *understanding* of what the symbols *mean* (semantics). True enough, symbol manipulation per se gives no understanding. ... The meaning in the symbols comes from how the symbols in the program are connected through the computer hardware to the environment, not from the manipulation of the symbols themselves. [487]

Manipulations are carried out depending on the exact position (first, second, third, etc. location) in a line, or *string*, of characters. That's *not* intelligence; that's a mechanistic process: typical of a machine. The machine does not *understand* anything it is doing!

---

[487] Tipler, *The Physics of Immortality*, p. 42 (emphasis in original).

Consciousness (Self Awareness)

We have already talked a little about consciousness (or self-awareness) above, and pointed out that computers are not conscious of themselves or anything around them. However, nowadays, some scientists (including Tipler) think that, as a system becomes more and more complex, somehow, it automatically develops consciousness: somewhat like a pattern emerges from the natural arrangement of a large number of basic entities. They then conclude that our brains became conscious in a similar fashion: automatically, as a result of its complexity. I do believe that, to some extent, the whole is greater than the sum of its parts: that a whole system can accomplish feats that the individual (separate) parts cannot. So the above claim is not entirely without merit.

A computational model, such as a weather-forecasting model, contains numerous variables that characterize the system being modeled; it uses mathematics, physics, and computer science to study the behavior of complex systems by computer simulation. Because of such a system's many connections to the outside world, some scientists nowadays think that, somehow, it becomes conscious automatically. The *computational theory of mind* holds that the mind is a computation that arises from the brain acting as a computing machine: that is, the brain is a computer and the mind is the result of the program that the brain runs. In his book *Consciousness and Language*, philosopher John Searle writes the following regarding consciousness emerging from complexity or computation:

> [C]omputational models of consciousness are not sufficient by themselves for consciousness. The computational model for consciousness stands to consciousness the same way the computational model of anything stands to the domain being modeled. Nobody supposes that the computational model of rainstorms in London will leave us all wet. But they make the same mistake of supposing

that the computational model of consciousness is somehow conscious. [488]

Consciousness and understanding are what, I believe, makes us superior to the machines we create; and this is not just wishful thinking on my part: I honestly believe it. They are what make us living beings, capable of being hurt or murdered.

How, exactly, are our brains connected to the environment to give us consciousness and understanding? Science cannot explain this yet; I have no intention of cheering (adoring) the gaps in science: I will keep an open mind. However, presently, I do not see a light at the end of the tunnel explaining these two phenomena, universally recognized by every human being; except, of course, the *universal wave function* discussed above, that is, if it really exists. We have the power to create any machine superior to us in every functional respect, but not to give it life, understanding, or consciousness.

As I hinted above, there is other evidence of our soul being hard-wired to God: (1) the first is reason, or logic; (2) the second is morality: a sense of right and wrong, basically, that we should treat others as we would like to be treated ourselves; (3) the third is emotions: love and empathy, particularly for other people, especially conjugal love; and (4) the fourth is qualia: the way every one of us perceives reality. I shall not go into further details in these areas; I shall only point out that although animals are alive, like us, they probably don't have all or most of these qualities; or if they do, they have them to a much lesser degree. Qualia are probably an exception: for example, a dog has a much better smelling ability than a human; cats, unlike us, can see in the dark; bats "see" by echolocation at night, etc.

But one might still not be completely convinced of this universal wave function because it does present a disconnection from mainstream science. It seems to bring into the picture a ghost or an Omnipotent Being to explain what science cannot explain yet.

---

[488]   Searle, "The Problem of Consciousness", p. 16

Mainstream science supposes that "the mind *is* the brain" [489] and that the mind emerges from the complexity of the brain. But if we could *separate* the two, the mind from the rest of the body—the program from the machine—then we would know that the mind is a separate entity from the brain; and consequently, that the soul is a separate entity from the body! The soul and the mind are synonymous concepts, because the brain supposedly controls the whole body. So, is there any tangible, positive evidence for the possibility of the soul's existence completely separate from the body: similar to the manner in which an executable computer program can be stored on a disk?

There is such evidence; however at this point in time, it possibly lacks the rigorous element of normal science; mainly because the great majority of scientists refuse to consider evaluating such evidence seriously: they exclude it a priori.

Near Death Experiences (NDEs)

The last interesting question about the soul, or principle of life (that is, the integrating factor of the body's functions), is whether it can exist separate from the body; as a computer program can exist on a disk, say: completely separate from the computer it normally runs on.

I shall only give a few instances of the vast body of empirical (experiential) evidence that comes to us through *near death experiences* (NDEs) reported by survivors of a *clinical death* occurrence. Before I start, I ask the reader to appreciate that there is a multitude of such experiences; they are also increasing in number at a faster rate with the advance in medicine: corresponding to the increase in the survival rate of people experiencing clinical death. There are, of course, many other private and undocumented similar experiences; in fact, I shall also mention one such case later on in this chapter.

---

[489] Rosenberg, *The Atheist's Guide to Reality*, p. 236.

In his book *The Spiritual Brain*, neuroscientist Mario Beauregard gives the following *well documented* account of a particular near death experience. The reader can refer to the internet for confirmation of what follows and possibly for more details.

In 1991, thirty-five year old singer and song writer Pam Reynolds of Atlanta, Georgia, USA, started to feel dizziness, lose her speech, and experience general difficulty of movement. So she had a *CAT scan* done. (The acronym "CAT" stands for "computerized axial tomography"—a CAT scan is a special X-ray-type test that produces cross-sectional images of the body on a computer.) Her CAT scan spelled very bad news: she had a grossly swollen blood vessel in the brain stem that was inoperable. Attempting to drain and repair it would most probably kill her anyway; her doctor told her that she had no chance of survival if conventional procedures were used. [490]

But Pam's mother had heard of neurosurgeon Robert Spetzler, who was a specialist and pioneer in a dangerous, but sometimes necessary, procedure called *hypothermic cardiac arrest*. The technique basically consists of cooling the body down to a very low temperature (60F): so low that the body is essentially dead; but then the body is brought back to normal body temperature (98.6F) in a timely manner, before the brain has had enough time to suffer irreversible damage. The swollen blood vessels that would easily burst at normal body temperatures become soft and operable with less risk at the low temperature used in the procedure. There is also another added advantage. Since the brain is non-functional in this cooled state, it uses much less oxygen. As a result, it can last much longer without oxygen before irreversible damage sets in. Pam consented to this procedure because she realized she didn't have much of a choice.

So, for all practical purposes, Pam was actually *dead* during her surgery: in fact, her heart stopped, and her *EEG* was "flat" (a horizontal straight-line). (The acronym "EEG" stands for "electroencephalogram"—it is a test that detects electrical activity

---

[490] Beauregard & O'Leary, *The Spiritual Brain*, p. 153

in the brain using electrodes, or small flat metal discs, attached to the scalp.)

Beauregard continues his account by giving several details of Pam's NDE, which I shall simply quote for the sake of authenticity: [491]

> When all of Reynold's vital signs were stopped, the surgeon began to cut through her skull with a surgical saw. At that point, she reported that she felt herself "pop" outside her body and hover above the operating table. From her out-of-body position, she could see the doctors working on her lifeless body. ... She described with considerable accuracy for a person who knew nothing of surgical practice, the ... bone saw used to open skulls. Reynolds also heard and reported later what was happening during the operation and what the nurses in the operating room had said. ... [S]he became conscious of floating out of the operating room and travelling down a tunnel with a light. Deceased relatives and friends were waiting at the end of this tunnel, including her long-dead grand-mother. She entered the presence of a brilliant, wonderfully warm and loving Light and sensed her soul was part of God and that everything in existence was created from the Light (the breathing of God). This extraordinary experience ended when Reynolds' deceased uncle led her back to her body. She compared entering her body to "plunging into a pool of ice". [492]

Possibly, she felt so cold because her body was still at a very low temperature when she came to again.

The reader is here asked to notice the sentence above: "She entered the presence of a brilliant wonderful warm and loving Light and sensed her soul was part of God and that everything in existence was created from the Light (the breathing of God)." Recall Tipler's concept of a *universal wave function*, and the bible's (Genesis')

---

[491] Beauregard & O'Leary, *The Spiritual Brain*, p. 154

[492] Beauregard & O'Leary, *The Spiritual Brain*, p. 154–5

concept of God *breathing* life into every living being; I think they fit Pam's experience like a glove.

Many near death experiences (NDEs) have been reported; of course, not all of them are equally credible. However, Pam Reynolds' case is unique for several reasons: she had the experience at a time while she was *fully instrumented*; she was under observation by the *medical profession* and known to be *clinically dead*; furthermore, she was able to *recall verifiable facts* that happened *while* she was clinically dead: things she could not have known if she were not somehow *conscious* during her surgery.

Clinical death is the state in which all vital signs have ceased; the medical profession can tell that someone is clinically dead by the following observations. (1) The heart is in *ventricular fibrillation*: that is, the muscle that normally contracts (squeezes) the ventricles (chambers) to pump the blood out of the heart does so in an uncoordinated manner, making them quiver rather than contract properly. (2) Brain-stem activity is abolished: characterized by loss of corneal (blinking) reflex, fixed and dilated pupils, and loss of the gag reflex. (3) There is a total lack of electrical activity on the cortex (outer layer) of the brain (that is, the EEG is flat): during a cardiac arrest, the brain's electrical activity vanishes after 10 to 20 seconds. [493]

Neuroscientist Mario Beauregard concludes his account of Pam Reynolds' NDE with the following two statements:

> Pam Reynolds's case strongly suggests that: … mind, consciousness, and self can continue to exist when the brain is no longer functional and clinical criteria of death have been reached; and … RSMEs [religious, spiritual, and mystical experiences] can occur when the brain is not functioning. In other words, this case seriously challenges the materialist view that mind, consciousness, and self are simply by-products of electrochemical brain processes, and RSMEs are delusions created by a defective brain.

---

[493]   Beauregard & O'Leary, *The Spiritual Brain*, p. 155

Such a view is based on metaphysical belief, not on scientifically demonstrated facts. [494]

I would like to add a few experts' opinion regarding consciousness. In his book *God and the Folly of Faith*, particle physicist, philosopher, and self-declared atheist Victor Stenger admits:

> The one major area where we do not yet have a plausible physical model that satisfies a consensus of experts in the field is the question of the nature of consciousness. [495]

I can't say that I agree with Stenger that everything in our universe can be explained physically, but it is worth noting the exception he makes.

In his book *The Taboo of Subjectivity*, researcher and author Alan Wallace writes the following: [496]

> "Mainstream neuroscience … insists that individual consciousness vanishes with the death of the body. However, given its ignorance of the origins and nature of consciousness and its inability to detect the presence or absence of consciousness in any organism, living or dead, neuroscience does not seem to be in a position to back up that conviction with empirical scientific evidence." [497]

In other words, Alan Wallace is saying here (and many scientists agree) that mainstream neuroscience has absolutely no clue as to what consciousness might be all about. As we have seen above, mainstream science downplays consciousness: explaining it away as the direct result of complexity and culminating in an illusion of self. In actual fact, current empirical (experiential) evidence from NDEs seems to support the opposite view: namely, that consciousness can

---

[494] Beauregard & O'Leary, *The Spiritual Brain*, p. 155

[495] Stenger, *God and the Folly of Faith*, p. 44.

[496] Beauregard & O'Leary, *The Spiritual Brain*, p. 153

[497] Wallace, *The Taboo of Subjectivity*, p. 5

still have an existence separate from its body; hence neuroscience has no standing argument in declaring that consciousness disappears at the organism's death.

Pam Reynolds' NDE account impressed me a lot when I first read it; but it had a much greater impact on me, when out of the blue over dinner, a friend of our family recounted a similar experience of a close female friend of hers who had just had a heart attack and was resuscitated. I don't remember the exact details of her account, so I will not try to make up a coherent story of what our friend told us; I will only write down what I remember. Her friend mentioned seeing a light in a tunnel, being met by deceased relatives, and being led back to life by them. Her friend was very emotional about the whole experience and didn't want to talk too much about it; she was confiding in our friend only because the latter was a very close friend of hers. I asked our friend whether her friend was a reliable person (that is, one not prone to confabulations) and whether she believed what her friend had said to her. The answer to both questions was, "yes". Then I gave our friend the above account to read; she too was impressed by the similarities; she also told me that she had never read anything like it before.

It is also noticeable that Pam Reynolds had what is commonly known as an *out-of-body experience* (OBE). Such experiences are quite common during NDEs; Dutch cardiologist, researcher, and author Pim van Lommel describes them as follows: [498]

> "*Out-of-body experience (OBE)* ... is an experience of floating outside one' own body, while retaining one's identity and a very clear consciousness. Most patients report looking down from above. As we have seen, in some cases, patients have reported information that was later verified." [499]

---

[498]   Beauregard & O'Leary, *The Spiritual Brain*, p. 157
[499]   Van Lommel, "About the Continuity of Our Consciousness", pp. 120–3.

Beauregard then mentions the following statistical results. In 1988, van Lommel started a study consisting of interviewing heart-attack survivors in less than a week of their being resuscitated from clinical death; he interviewed three hundred and forty-four (344) patients. Medical science has no doubt that a person in a state of clinical death is not aware of anything happening around him or her; yet, sixty-two (62), or about eighteen percent (18%), of the above subjects reported experiences of varying intensity during the exact time they were clinically dead; twenty-four (24) of them, or about seven percent (7%) of the total interviewed, reported a very deep such experience. These results showed no connection to the patients' educational or religious background. [500] Similar studies resulted in roughly the same percentage of very deep experiences: in an American study (Greyson, 2003) the rate was ten percent (10%), and in a British study (Parnia et al., 2001), the rate was over six and three-tenths percent (6.3%). [501]

In my opinion, besides the fact that a cardiologist can be considered a scientist, I think these numbers are significant enough to the point of triggering a rigorous future investigation by the scientific community. With the advance of emergency medicine, recent heart-attack survivors lend themselves to better NDE studies; obviously, they are a favorite group for such studies, because medical records are available. These records can confirm whether they were actually clinically dead and unconscious for some time after their cardiac arrest; they can also confirm whether their brains were abnormally low in oxygen: such people would have died from irreversible damage had they not been resuscitated within five to ten minutes. [502]

We now move on to some more general details reported in NDEs. The evidence from several NDEs made me think deeper regarding our experiences in life and try to answer some very persistent questions we all have. You will see what I mean in a short while.

---

[500] Van Lommel, "About the Continuity of Our Consciousness"
[501] Beauregard & O'Leary, *The Spiritual Brain*, pp. 156–7
[502] Beauregard & O'Leary, *The Spiritual Brain*, p. 156

An interesting phenomenon, revealed in NDEs, is that people who were blind from birth sometimes reported NDEs in which they could see: in other words, these NDEs were hard to explain otherwise. People blind from birth often learn the world within their reach fairly accurately: through touch rather than sight. However, they cannot detect color, background, or changes in the position of objects situated at some distance away from them. For example, blind persons cannot feel a tunnel as we see a tunnel: they can only feel its walls locally; nor can they see a light at the end of a tunnel, or see anything happening at a distance: like someone smiling or opening one's arms to greet them. Given these basic differences, between normal and blind people, one can probably tell whether a blind person's NDE was visual or not. [503]

Of special interest, in my opinion, is the NDE account of a heart-attack survivor, which is *typical* of many similar NDEs:

> All my life up till the present seemed to be placed before me in a kind of panoramic three-dimensional review, and each event seemed to be accompanied by a consciousness of good or evil or with an insight into cause or effect. Not only did I perceive everything from my own viewpoint, but I also knew the thoughts of everyone involved in the event, as if I had their thoughts within me. This meant that I perceived not only what I had done or thought, but even in what way it had influcnced others. [504]

In his article "Who's Afraid of Life after Death?" philosopher Neal Grossman gives the following interesting *summary* from NDEs in general:

> The evidence from the NDE ... suggests that God is not vengeful, does not judge us or condemn us and is not angry at us for our "sins"; there is judgement, to be sure, but the reports appear to be in agreement that all judgement

[503] Beauregard & O'Leary, *The Spiritual Brain*, p. 158
[504] Beauregard & O'Leary, *The Spiritual Brain*, p. 157

comes from within the individual, and not from the Being
of Light. It seems, in fact, that all God is capable of giving
us is unconditional love. [505]

Many modern scientists are inclined to write off NDEs as
hallucinations of a dying brain. However, experts believe that one's
culture seems to have a significant impact on one's hallucinations:
what they believe in is what they imagine. In Christian teaching, God
is overwhelmingly represented as a strict, unrelenting judge who
prepared an eternal fiery hell for sinners, and who sacrificed his son
because he can't tolerate sin, not as a benevolent father capable only
of unconditional love. Why would the *majority* of people near death
conjure up something so clearly revolutionary and unconventional?

Natural Explanations

To start with, there is, of course, the possibility that survivors
reporting NDEs may embroider or even fantasize their accounts: just
to enjoy their fifteen minutes of fame.

Also, in most cases, we, as observers, can't know for sure exactly
when their NDEs happened: whether it was before, during, or after the
time they were clinically dead. We probably can't even tell whether it
was during the recovery period: during which time their mind may
have been hallucinating or sort of half-dreaming. Very often we just
assume that these experiences happened during the time the survivors
were clinically dead; unless, of course, verifiable details are given
that happened during the time the patient was actually clinically
dead. But even then, proximity is a factor: there is a lot we still don't
understand about the subconscious mind. One must appreciate here,
that it is practically impossible to obtain exactly timed scientific
evidence when one is trying frantically to save someone's life. The
bottom line is that there are still significant doubts regarding these

---

505 Grossman, "Who's Afraid of Life after Death?", p. 14

experiences. Hopefully, better documented evidence will be collected in the future.

Finally, regarding the light seen at the end of a tunnel, which is so commonly experienced in NDEs, most scientists believe that this phenomenon is characteristic of the brain shutting down; similar to what happens when one turns off an older type television set: all the screen turns white before it goes dark.

Proof Requirements

On the other hand, if it could be established, without leaving any reasonable doubt, that in *some* (a few) of the NDEs, patients were actually able to relate unknowable and verifiable facts, and that these facts happened at the *exact* time they were clinically dead, and preferably at a significant *distance* away from the location of the patient, then science will have to reconsider the existence of the self or the soul. The evidence could be something as simple as one nurse relating a private matter or a work situation to another at the exact time the patient had a flat EEG; even better, if it happened at a considerable distance away from the operating room, to preclude subconscious hearing. One has to look at the details in such cases and evaluate them honestly and scientifically.

Pam Reynold's case, described above, seems to satisfy most, if not all, of these requirements. In his article "Who's Afraid of Life after Death?" philosopher Neal Grossman writes the following regarding this case:

> Perhaps the "smoking gun" case is the one recently described by [cardiologist] Michael Sabom (1988). In this case, the patient had her NDE while her body temperature was lowered to 60 degrees [Fahrenheit], and all the blood was drained from her body: "her electroencephalogram was silent, her brain-stem response was absent, and no

blood flowed through her brain (Sabom, 1988, p.49). [506] A brain in this state cannot create any kind of experience. Yet the patient experienced a profound NDE which included detailed veridical [coinciding with reality] perception of the operation. [507]

The reader probably noticed the author's use of the phrase "smoking gun": meaning, that there is hardly any doubt as to its authenticity.

---

[506] Sabom, *Life and Death*, p. 49.
[507] Grossman, "Who's Afraid of Life after Death?", p. 6

CHAPTER 9

MIRACLES

I would like to start this chapter by apologizing to those readers who do not profess my religion: I am only familiar with miracles pertaining to the Roman Catholic Church. This section is not intended to be an exhaustive list of miracles; only a short list will be given. More detailed accounts of the miracles mentioned here are, of course, available on the internet should the reader like to confirm details or numbers I mention here. I shall only go into details in one or two accounts I give in this chapter; I delve deeper into the details of a few miracles simply to answer questions the reader might have regarding miracles in general.

In his book *The God Delusion*, self-declared atheist Richard Dawkins defines miracles as follows: "Miracles by definition violate the principles of science." [508] I have to say that I agree with this statement. Further on in the same book he writes: "What is the use of a God who does no miracles and answers no prayers?" [509] I can't say I agree with him completely here, but he makes a good point for one who finds it hard to admit any evidence for the supernatural. But is this really the case?

In this chapter, I shall try to show that this is *most probably* not the case; unless, of course, one absolutely shuts off one's mind to the possibility that the evidence given here might be the truth. I think that God does, on very rare occasions, perform miracles and

---

[508] Dawkins, *The God Delusion*, p. 83
[509] Dawkins, *The God Delusion*, p. 84

answer prayers; but normally he does not interfere at all: allowing us to carry our own weight, deal with one another the way we decide to, and enjoy or suffer the consequences of our actions. What I said about near death experiences (NDEs) in the previous chapter I repeat here: there is a significant amount of empirical (data-based) evidence in support of miracles; however at the present day, it may lack the rigorous element of normal science; again, mainly because the great majority of scientists refuse to consider investigating any evidence of a supernatural nature seriously: they exclude it a priori.

A word of caution before proceeding further: there are many magicians who can perform startling spectacles that look like miracles; but everyone knows these are normally illusions rather than violations of the laws of nature. I'm certainly not qualified enough to distinguish between the two; experts in the field of illusions might have to be called upon in determining, rigorously whether an observed phenomenon might be an illusion or not. In what follows I am assuming there is no foul play by illusionists; I'll let the reader discriminate for oneself!

Fatima

Wikipedia gives the following account of a *well-documented* and *amply testified* event, commonly known as the *Miracle of the Sun*: an event of extraordinary solar activity which lasted about ten minutes. It happened in a field called "Cova da Iria" ("Irene's Cove") near Fatima in Portugal just after midday on Sunday, 13 October 1917. It was witnessed by a very large crowd: estimated between thirty thousand (30,000) and one hundred thousand (100,000) people. Avelino de Almeida was a newspaper reporter for the masonic and anticlerical Portuguese newspaper *O Seculo* who was present at the place and time of the event; he estimated the crowd there to be between thirty thousand (30,000) and forty thousand (40,000) people. Dr. Joseph Garret who was professor of natural sciences at the University of Coimbra was also present there on that day during the event; he estimated the crowd present at between ninety thousand (90,000) and one hundred thousand (100,000) people. To reconcile these two reports, by supposedly reliable sources, a crowd of seventy thousand (70,000) people, the mean, is often quoted.

The people gathered there because the three children (aged nine, eight, and six), to whom the lady (Our Lady of Fatima) was appearing, promised that on that day she would provide a miracle "so that all may believe". The event place, date, and time were predicted by Our Lady three times, namely, during her apparitions to the three children on 13 July, 19 August, and 13 September, 1917. It rained all morning prior to the event and consequently both the crowd and the ground were soaked wet. Several newspaper reporters, including reporters from *hostile* newspapers of the time, were present. They reported what they saw personally, and also took testimony from many people present who witnessed the event. [510]

Here is a summary report in Wikipedia of what actually happened:

> According to many witnesses, after a period of rain, the dark clouds broke and the sun appeared as an opaque spinning disc in the sky. It was said to be significantly duller than normal, and to cast multicolored lights across the landscape, the people, and the surrounding clouds. The sun was then reported to have careened [rushed] towards the earth before zig-zagging back to its normal position. Witnesses reported that their previously wet clothes became [511] "suddenly and completely dry as well as the wet and muddy ground that had been previously soaked because of the rain that had been falling". [512]

I'm sure the reader will appreciate the *fact*, here, that the event was *predicted* by Our Lady on three separate occasions: that is why so many people present were there. The *exact* place, date, and time

[510] https://en.wikipedia.org/wiki/Miracle_of_the_Sun (28 October 2015)
[511] https://en.wikipedia.org/wiki/Miracle_of_the_Sun (12 June 2016)
[512] De Marchi, *The Immaculate Heart*, b:150

were predicted: so it is improbable that it was just a random cosmic occurrence.

As I said at the beginning of this section, this event was very well documented. Italian Catholic priest and researcher John De Marchi spent seven years in Fatima (from 1943 to 1950) conducting his own research by interviewing witnesses. In 1952, he published his research in a book entitled *The Immaculate Heart*, consisting of several volumes, from which most descriptions of the Fatima events and statements are normally cited. [513] De Marchi reports the following:

> [The observers'] ranks included believers and non-believers, pious old ladies and scoffing young men. Hundreds, from these mixed categories, have given formal testimony. Reports do vary; impressions are in minor details confused, but none to our knowledge has directly denied [check another later quote, however] the visible prodigy of the sun. [514]

Although many reported basically the same experience, the event is not easily attributable to a mass hysteria or mass hallucination: given the mixture of characters present. Nevertheless, most materialistic scientists resort to this dubious, if not infantile, explanation, as we shall see later on in this section.

De Marchi quotes several newspaper reports of the time, most of which were anticlerical, pro-government, and hostile to the Catholic religion because the then current government was masonic: Freemasons and Catholics have been enemies at odds with each other for centuries. The newspapers' main aim there was to ridicule and belie both the people present and the alleged apparitions reported by the three little children. [515]

---

[513]   https://en.wikipedia.org/wiki/Miracle_of_the_Sun (28 October 2015)

[514]   De Marchi, *The Immaculate Heart*, b:143

[515]   De Marchi, *The Immaculate Heart*, a:174

Avelino de Almeida was chief editor for Lisbon's (Portugal's capital city) most popular and influential newspaper *O Seculo*; he was there to belittle whatever happened there. One can see this from his hugely underestimated size of the crowd (between 30,000 and 40,000) compared to that of the Coimbra University professor (between 90,000 and 100,000). His previous articles were a mockery of the circulating reports; [516] yet, regarding the events of that day he reported:

> Before the astonished eyes of the crowd ... the sun trembled, made sudden incredible movements outside all cosmic laws—the sun 'danced' according to the typical expression of the people. [517]

He was vehemently criticized for his report of that day's event by the anticlerical press, but he re-iterated his experience about two weeks later:

> Attacked violently by all the anticlerical press, Avelino de Almeida renewed his testimony, 15 days later, in his review, l'"Ilustraçao Portuguesa". This time he illustrated his account with a dozen photographs of the huge ecstatic crowd, and repeated as a refrain throughout his article: "I saw ... I saw ... I saw." And he concluded fortuitously: "Miracle, as the people shouted? Natural phenomenon; as the experts say? For the moment, that does not concern me; I am only saying what I saw ... The rest is a matter for Science and the Church". (Article of October 29, 1917) [518]

Other then-current newspaper reports confirmed Avelino de Almeida's report above, adding further details; for example, Lisbon's newspaper *O Dia* reported the following:

---

[516] De Marchi, *The Immaculate Heart*, a:174
[517] De Marchi, *The Immaculate Heart*, b:144
[518] http://www.fatima.org/crusader/crintro/crintropg16.asp (28 October 2015)

> The silver sun, enveloped in the same gauzy grey light,
> was seen to whirl and turn in the circle of broken clouds ....
> The light turned a beautiful blue ... and spread itself over
> the people who knelt with outstretched hands ... wept and
> prayed ... in the presence of a miracle they had awaited." [519]

Notice that the newspaper reporter expressly mentions that the people who were there "awaited" the "miracle" to happen.

Dr. Domingos Pinto Coelho reported the following in the Catholic newspaper *Ordem*:

> The sun, at one moment surrounded with scarlet flame,
> at another aureoled in yellow and deep purple, seemed
> to be in an exceedingly swift and whirling movement, at
> times appearing to be loosened from the sky and to be
> approaching the earth, strongly radiating heat. [520]

Notice that Dr. Coelho described the sun as falling towards the earth and radiating more heat than normal. This is no small suspension of the laws of physics and probably impossible to explain scientifically. But then, that's exactly what a miracle is supposed to be: a suspension of the natural laws, the likes of which only an omnipotent God can perform.

Following are a couple of reports by highly educated people who were present at the field of apparitions (Cova da Iria) also given in De Marchi's accounts. The first quote comes from Dr. Almeida Garret, who was then the professor of natural sciences at the University of Coimbra (the same one who estimated the crowd at between 90,000 and 100,000 people):

> The sun's disc did not remain immobile. This was not the
> sparkling of a heavenly body, for it spun round on itself in
> a mad whirl, when suddenly a clamor was heard from all
> the people. The sun, whirling, seemed to loosen itself from

---

[519] De Marchi, *The Immaculate Heart*, b:143
[520] De Marchi, *The Immaculate Heart*, b:147

the firmament and advance threateningly upon the earth
as if to crush us with its huge fiery weight. [521]

The second quote comes from a priest, who was then a professor at the
seminary of Santarem, Dr. Manuel Formigao. He was also present at
the field of the apparitions during the previous (September) apparition
of Our Lady. During that time, he thoroughly and meticulously
interrogated the three children, not just once but several times,[522]

> [L]ike a bolt from the blue, the clouds were wrenched
> apart, and the sun ... appeared .... It began to revolve
> vertiginously [dizzyingly] on its axis, like the most
> magnificent fire-wheel ... taking on all the colors of
> the rainbow and sending forth multicolored flashes of
> light .... This sublime and incomparable spectacle, which
> was repeated three distinct times, lasted for about ten
> minutes. [523]

The next quote comes from another priest, Rev. Joaquim Lourenço,
who was still a boy at the time. He saw the phenomenon of the
sun from eighteen (18) kilometers (11 miles) away, in the village of
Alburitel.

> I looked fixedly at the sun, which seemed pale and did
> not hurt my eyes. Looking like a ball of snow, revolving
> on itself, it suddenly seemed to come down in a zigzag,
> menacing the earth. Terrified, I ran and hid myself among
> the people, who were weeping and expecting the end of
> the world .... [524]

This account shows a spontaneous reaction of a child as a consequence
of what he saw. Our final quote comes from a poet, Alfonso Lopes

---

[521] De Marchi, *The Immaculate Heart*, b:146
[522] De Marchi, *The Immaculate Heart*, b:146
[523] De Marchi, *The Immaculate Heart*, b:146
[524] De Marchi, *The Immaculate Heart*, b:149

Vieira, who also was away from the field of apparitions because he happened to forget all about the children's miracle prediction.

> [W]ithout remembering the predictions of the children,
> I was enchanted by a remarkable spectacle in the sky
> of a kind I had never seen before. I saw it from this
> veranda .... [525]

The reader will probably agree that this account does not support the possibility of a mass hysteria or a mass hallucination.

De Marchi finally adds the conclusions of engineers' studies. There are newspaper pictures showing the whole crowd of people holding their umbrellas over their heads, proving that it was raining in the morning of that day. Engineers' calculations seem to indicate that a tremendous amount of heat must have been required to dry up the pools of rain water on the ground and the people's clothes in about ten or fifteen minutes. Yet, this sudden blast of heat did not harm any of the people present in any way. Truly remarkable! [526] Examining this statement from a scientific point of view, the event could not possibly have been the result of a mass hallucination: from what we know about hallucinations, they cannot dry up muddy pools (or clothes) instantaneously.

Wikipedia then mentions the following observations, given by De Marchi himself, which effectively debunk criticisms of the event's not being a miracle: (1) the alleged miracle was predicted, but nobody knew beforehand what it would consist of; (2) it both started and stopped suddenly; (3) the observers comprised a wide spectrum of religious backgrounds; (4) a great multitude of people saw the same thing; (5) scientists could not come up with any known underlying natural cause; (6) the event was also visible eighteen (18) kilometers (11 miles) away. All this evidence leads one to conclude that both mass hysteria and mass hallucination were out of the question

---

[525] De Marchi, *The Immaculate Heart*, b:148–9
[526] De Marchi, *The Immaculate Heart*, b:150

as viable explanations; in other words, it seemed to be a genuine miracle. [527] [528] [529]

The following report by a Benedictine priest, physicist, and historian Stanley L. Jaki, again taken from De Marchi's accounts, is very interesting. This is usually the source document that most atheists and scientists normally cite to question and debunk this alleged miracle:

> [N]ot all witnesses reported seeing the sun "dance". Some people only saw the radiant colors. Others, including some believers, saw nothing at all. [530] [531] [532]

While admitting that this is rather odd, one must keep in mind that when a miracle happens, God does not necessarily allow everyone around to witness it; for some strange reason, some people are not allowed to see anything at all. This somewhat odd phenomenon is noticeable in many accounts of other miracles.

To prove this point, for example, while Our Lady was appearing and speaking to the three children atop the little holm oak tree, no one else was able to see her. From this, one cannot conclude that Our Lady was not appearing to the children at all; the fact that the children predicted the miracle of the sun proves otherwise. We shall see this being the case, over and over again, in the ensuing accounts of miracles described in this chapter.

If you think about it, oddly enough and in a weird sort of way, this report precludes the possibility of a mass hallucination; and probably it would also stump a mass hysteria because people around those that saw nothing would not have been swept along with the flow.

---

[527] De Marchi, *The Immaculate Heart*, b:150, esp. pp. 278–82

[528] https://en.wikipedia.org/wiki/Miracle_of_the_Sun (8 June 2015)

[529] http://www.liquisearch.com/miracle_of_the_sun/critical_evaluation_of_the_event (9 June, 2016)

[530] De Marchi, *The Immaculate Heart*, b:150, esp. pp. 278–82

[531] http://www.liquisearch.com/miracle_of_the_sun/critical_evaluation_of_the_event. (9 June, 2016)

[532] Jaki, *God and the Sun at Fatima*

Moreover, for those who believe in unidentified flying objects (UFOs), had an alien UFO been hovering above and playing tricks in the clouds, these few people that saw nothing would have seen the UFO as well.

Regarding the possibility of its being a natural solar or astronomical occurrence, De Marchi also mentions the fact that there were no extraordinary scientific reports of that nature at that time, or any eyewitness reports, outside a radius of sixty-four (64) kilometers (40 miles) of the field of apparitions. [533] [534]

Kevin McClure is an atheist and author of several books researching unusual phenomena; he also admirably tackled the Fatima event head on: this is quite uncommon among atheists, as I shall show shortly.

> Kevin McClure claims that the crowd at Cova da Iria may have been expecting to see signs in the sun, since similar phenomena had been reported in the weeks leading up to the miracle. On this basis, he believes that the *crowd saw what it wanted to see*. [535] However, none of the previous phenomena had to do with the sun; the focus, for the most part, was on the little [holm oak] tree where the lady was said to appear. Kevin McClure stated that he had never seen such a collection of contradictory accounts of a case in any of the research that he had done in the previous ten years, although he has not explicitly stated what these contradictions were. [536]

Even McClure, however, seemed to have sifted through hundreds of testimonies and cherry-picked a handful to make his case. He fails to address the much greater number of consistent reports that disagreed

---

[533]   De Marchi, *The Immaculate Heart*, b:150, esp. pp. 278–82
[534]   http://www.liquisearch.com/miracle of the sun/ critical evaluation of the event. (9 June, 2016)
[535]   McClure, *The Evidence for Visions of the Virgin Mary* (emphasis in original)
[536]   https://en.wikipedia.org/wiki/Miracle of the Sun (28 October 2015)

with his hypothesis. The website "Arguing with Atheists" makes the following observation:

> The preposterous retort by atheist Kevin McClure that he had "never seen such a collection of contradictory accounts" flies in the face of the striking uniformity of the actual witness accounts. [Author John] Haffert's 200 signed statements from real witnesses as well as the nearly 300 documented interviews by Fr. John De Marchi are virtually identical. Fatima supporters contend that McClure never interviewed one witness, but spent months sifting through countless witness statements cherry-picking four slightly different than the rest, then founded his case on that. [537]

To me, it sounds like McClure suffers from extreme skepticism and excessive bias.

Most atheists do not even address the miracle at Fatima, because, I contend, they think it is a losing battle. For example, I was rather disappointed to notice that particle physicist Victor Stenger does not even mention this event in his book *God and the Folly of Faith*. On the other hand, I admire self-declared atheist Richard Dawkins for, at least, addressing this event in his book *The God Delusion* and trying to explain it away; however, I was somewhat disillusioned by his analysis of the event. Here's what he writes:

> On the face of it mass visions, such as the report that seventy thousand pilgrims at Fatima in Portugal in 1917 saw the sun 'tear itself from the heavens and come crashing down upon the multitude', are harder to write off. It is not easy to explain how seventy thousand people could share the same hallucination. [538]

---

[537]   http://www.arguingwithatheists.com/pages/Arguments.htm
[538]   Dawkins, *The God Delusion*, p. 116

Why does Dawkins want to "write off" a witnessed event? And why does he assume it is a hallucination? Chances are it really did happen given the above testimonies. In any case, Dawkins continues:

> But it is even harder to accept that it really happened without the rest of the world, outside Fatima, seeing it too—and not just seeing it, but feeling it as the catastrophic destruction of the solar system, including acceleration forces sufficient to hurl everybody into space. [539]

This definitely would be true *if* the laws of physics were obeyed; but not if they were, somehow, temporarily suspended by an omnipotent God. I suppose that if God can establish the laws of physics, he can also suspend them, change them, and even eliminate them altogether. This is exactly what a miracle is! Didn't Dawkins himself define a miracle this way a few pages back; let me repeat it here: "Miracles by definition violate the principles of science." [540] So, basically, Dawkins is admitting that it must be a miracle (according to his own definition) but at the same time he wants to "write off" the miracle. Why? Because miracles *never* happen; right? But I think that, in this instance, God proved otherwise here, rather convincingly: he proved that he can perform *great* miracles, even by Dawkins' definition.

As to whether the event was a miracle or not, De Marchi quotes Jesuit, mathematician, and astronomer Pio Scatizzi: [541]

> The ... solar phenomena were not observed in any observatory. Impossible that they should escape notice of so many astronomers and indeed the other inhabitants of the hemisphere ... there is no question of an astronomical or meteorological event phenomenon .... Either all the observers in Fatima were collectively deceived and erred

---

[539] Dawkins, *The God Delusion*, p. 116

[540] Dawkins, *The God Delusion*, p. 83

[541] https://en.wikipedia.org/wiki/Miracle_of_the_Sun (28 October 2015)

in their testimony, or we must suppose an extra-natural intervention. [542]

I think this should be enough on this event or miracle, whatever the reader would like to call it. One either believes evidence, the testimony of many people, or one does not. But I think it is the epitome of arrogance to just say, categorically, that miracles never happen; especially knowing about empirical evidence such as the above and the medical evidence from "Lourdes", which is treated in a later section. The least one can do is to keep an open mind; definitely not to try to "write it off".

I don't think it is enough to simply quote David Hume, as Dawkins does:

> David Hume's pithy [concise and forceful] test for a miracle ... [543] "No testimony is sufficient to establish a miracle, unless the testimony be of such a kind, that its falsehood would be more miraculous than the fact which it endeavors to establish." [544]

Given that a miracle is, *by definition*, an event that violates the principles of science; the above quote, in my opinion, argues in circles.

The above account is certainly not the only well documented miracle that ever happened; the complete list would literally be overwhelming. Because of lack of space I shall only give a brief description of a very short list of miracles; one can also research their details further on the internet, if one so desires.

---

[542] De Marchi, *The Immaculate Heart*, b:150, esp. pp. 278–82

[543] Dawkins, *The God Delusion*, p. 116–7

[544] Hume, *An Enquiry Concerning Human Understanding, Section 10: Of Miracles Pt. 1.*

Medjugorje

According to Wikipedia, in the late afternoon on 24 June 1981, sixteen-year-old Mirjana Dragicevic and fifteen-year-old Ivanka Ivankovic reported seeing an apparition of Our Lady in a village of a region of what was at the time called Yugoslavia, nowadays known as Bosnia and Herzegovina. Another vision was reported the next day by four other young people: seventeen-year-old Vicka Ivankovic, sixteen-year-old Marija Pavlovic, sixteen-year-old Ivan Dragicevic, and ten-year-old Jakov Colo. The youngsters claimed that they saw an apparition consisting of a white form of a young lady with a child in her arms on Mount Podbrdo, which is part of Mount Crnica.

The six visionaries have also reported seeing daily apparitions of Our Lady for several years; some reports say that Our Lady has been appearing daily to three of these visionaries ever since; other reports say that they have stopped having daily apparitions; nevertheless, it seems that Our Lady is still appearing to the six of them from time to time. [545] The reader should note here the length of time Our Lady has been appearing to these visionaries, practically every day: for over thirty-five years; simply amazing! The length of time of these apparitions is a unique characteristic in the history of Marian apparitions.

Another unique feature of the apparitions in Medjugorje is the sheer number and diversity of the apparitions: there were no large crowds that reported seeing the same thing, as in the case of the Fatima event, but only small groups reporting seeing different supernatural and paranormal (beyond scientific explanation) phenomena. This way, I suppose, no one could conclude that they were cases of mass hallucination, or false reporting of things not really witnessed: the outcome of some kind of mass hysteria or peer pressure. One either believed the accounts or one didn't; however the numbers of reported supernatural and paranormal phenomena in Medjugorje are staggering. In his book *The Queen of Peace Visits Medugorje*,

---

[545]   https://en.wikipedia.org/wiki/Our_Lady_of_Medjugorje (13 August 2016)

Augustinian priest and author of several books on Marian apparitions Joseph Pelletier writes:

> It is difficult to give precise dates for many of the signs [i.e., miracles] that have been seen at Medugorje. This is not surprising, given their great number and the long period of time over which they have been witnessed. [546]

Wikipedia also reports diverse and numerous supernatural and paranormal phenomena being seen or experienced in Medjugorje. For example, several groups of people, at various times, reported seeing the sun spinning on its axis, moving around (dancing) in the sky, changing colors, and being comfortable to look at (pale and not harmful): similar to what people saw in Fatima, Portugal in 1917; but the groups of people reporting these phenomena at any one time were not as numerous as in Fatima. At other times, several groups of people reported seeing the sun surrounded by hearts or crosses. People also reported the disappearance and reappearance of the tall and heavy cement cross, which weighs tons; it was erected by the local people on the top of Mount Podbrdo, the location of many of the reported apparitions. Alternately, people occasionally saw the cement cross lit, as if by light bulbs; even though there is no electrical power lines going up the mountain. Typically, no large crowds but rather small groups of people reported seeing different paranormal phenomena at different times. In addition, there were also some reports of miraculous personal healings. [547]

Having the advantage of modern-day science, technology, recording, and reporting facilities, the Franciscan priests at Saint James Parish in Medjugorje decided to perform some tests on the seers at the very moment they were experiencing apparitions. In the same book mentioned above Pelletier reports the following incidents:

---

[546]   Pelletier, *The Queen of Peace Visits Medugorje*, pp. 72–3
[547]   https://en.wikipedia.org/wiki/Our_Lady_of_Medjugorje (13 August 2016)

[O]ne of the Franciscan priests put some pressure on Vicka's arm. She manifested no awareness of this …. Then he lifted little Jakov. He appeared heavy but there was no evident reaction on his part …. When the apparition had ended, the priest questioned the two seers, who indicated that they felt nothing while the tests were going on. On another occasion Ivan's brain activity during an apparition was measured. The encephalogram [EEG] indicated normal brain activity. There was no sign of dreaming or hallucination. Once, a member of the Bishop's Commission of Investigation stuck a needle in Vicka's back, shoulder and arm. Though blood was visibly drawn, she manifested no sign of pain. [548]

These incidents seem to indicate that there is some kind of connection between the natural and the supernatural: the seers' qualia were, somehow, compromised during their mystical experiences. These incidents, if we believe them, constitute further corroborating evidence of the existence of a universal wave function permeating the whole of space.

In the same book, Pelletier reports another interesting incident to the same effect; he quotes from a written report of a mystical experience one of the seers, Ivanka, had on 7 May 1985:

Our Lady asked me [Ivanka] if I had some wish. I told her I would like to see my earthly [dead] mother. She smiled, nodded her head, and my mother immediately appeared. She was smiling. Our Lady told me to stand up. I did and my mother embraced and kissed me and said, "My child, I am so proud of you." Then she kissed me and disappeared. [549]

Notice the physical connection between mother and daughter: between the supernatural and the natural. Could it be a hallucination?

---

[548] Pelletier, *The Queen of Peace Visits Medugorje*, p. 69
[549] Pelletier, *The Queen of Peace Visits Medugorje*, p. 239

Maybe; but recall from above that Ivan's brain activity showed no evidence of hallucination or dreaming during an apparition.

Pelletier gives us another detail regarding physical contact between the natural (several people) and the supernatural (Our Lady): "Those who had touched her [Our Lady] said that they felt a numbing of their hands." [550] This is evidence of a force field.

Elsewhere in the book, he quotes one of the seers, Vicka: "She [Vicka] said, 'Yes, I touched her [Our Lady's] dress. But it resists like metal.'" [551] Again, this description indicates what seems like a physical force at the interface of the natural and the supernatural.

Following is a description of the phenomena preceding Our Lady's apparitions also given by Pelletier:

> Before Our Lady appeared [on 25 June 1981], a bright light "something like lightning" was seen by those who were on Mount Podbrdo and also by people in the [nearby] village who happened to look toward Mount Podbrdo. ... The six seers have all asserted that Our Lady came each day preceded by three flashes of light. ... [552]
>
> At the beginning of the first apparitions, the seers used to see three brilliant lights or flashes of light, and then Our Lady. After a few days, Our Lady was preceded by only one brilliant light. Our Lady appeared coming out of this light. She stood on a cloud that grew wider the longer she stayed. [553]

Notice the electromagnetic (light) and the chemical or physical (cloud) interaction between the supernatural and the natural. If we decide to believe these accounts, they constitute more evidence for the existence of a universal wave function; unfortunately, I cannot explain any of it in scientific detail: we know too little about this wave function, if it really exists.

---

[550] Pelletier, *The Queen of Peace Visits Medugorje*, p. 54
[551] Pelletier, *The Queen of Peace Visits Medugorje*, p. 33
[552] Pelletier, *The Queen of Peace Visits Medugorje*, p. 17
[553] Pelletier, *The Queen of Peace Visits Medugorje*, p. 69

Neither can we, however, really explain the other force fields we are so familiar with in our everyday experiences. When we deal with forces as a result of fields we cannot see anything in the space between. An object on earth will fall towards the ground if released; but we do not see the earth pulling the object down by means of a string or an elastic band. A magnet is pulled towards a refrigerator door; but there is nothing visible on either the magnet or the refrigerator door to explain this attraction. Unlike charges attract one another while like charges repel one another at any distance: but there is no interconnecting string between them. Most of matter is empty space: the size of a nucleus inside an atom is comparable to a pea in the middle of a race track; nevertheless, we cannot push our way through matter, because the atoms are bound together by the strong nuclear force. The bottom line is that we cannot explain the mechanism of any of these forces and fields; but they have been proved to exist, scientifically.

Pelletier also reports a curious physical phenomenon that happened during several of the apparitions of Our Lady at Medjugorje:

> A most unusual facet of this apparition [on 27 June 1981] was Our Lady's departure in two separate instances when people stepped on her veil. A thing like this had never happened during any other Marian apparition. It will be repeated in subsequent apparitions. [554]

A strange phenomenon indeed: why it bothered Our Lady so much as to suddenly disappear when the crowd accidentally stepped on her veil. What is also very interesting is that in both occasions Our Lady came back: it seems that she did not disappear willingly. I wonder whether we shall one day have a scientific explanation for it.

We shall now address the appearance of Our Lady during these apparitions; that is, what she looked like to the six seers. According to Wikipedia, Franciscan Fr. Janko Bubalo asked the Medjugorje visionaries to describe what Mary looked like: her physical appearance

---

[554] Pelletier, *The Queen of Peace Visits Medugorje*, p. 27

as they saw it. They described her as "18 to 20 years old, slender and around 165 centimeters (5 ft. 5 in) tall." [555]

Here is a similar account of what Our Lady looked like to the six seers, given by Joseph Pelletier in his book *The Queen of Peace Visits Medugorje*:

> She appeared young, nineteen or twenty. She was strikingly beautiful, with a beauty beyond anything of this world. Her eyes were particularly beautiful and expressive. They were light blue and so tender and kind. [556]

The reader should note here that, if we are to believe the youngsters as to what Our Lady looked like, people in heaven seem to be, and remain, young (that is, at the prime of the human life cycle); even though, according to several New Testament books, Our Lady survived middle age (that is, about 50 years of age) on earth, as we shall see later on in this chapter.

Lourdes

While the main characteristic of the events in Fatima and in Medjugorje seems to be of a paranormal (difficult to explain scientifically) nature, the main characteristic of the events in Lourdes, France, seems to be of a healing nature. There is obviously a time-related factor regarding miracles: what science, medicine, and technology have achieved nowadays, do not preclude miracles being so at the time they happened. For example, healing blindness and leprosy some two thousand years ago, as Jesus is claimed to have done, cannot be judged by modern advances in medicine.

According to Wikipedia, on 11 February 1858, a lady dressed in white is said to have appeared for the first time and spoken to a fourteen-year-old peasant girl called Bernadette Soubirous. This first appearance happened in a cave about a mile from the town of

---

[555] https://en.wikipedia.org/wiki/Our_Lady_of_Medjugorje (13 October 2016)
[556] Pelletier, *The Queen of Peace Visits Medugorje*, p. 15

Massabielle in the vicinity of Lourdes in France while Bernadette was gathering firewood with her sister and a friend. Bernadette reported seventeen (17) similar apparitions by the lady that year. [557]

During an apparition on 25 February 1858, the lady asked Bernadette to dig the ground at a certain location; the latter exposed a spring there, and the lady asked her to drink from it. As word of the apparitions spread, this spring water was given to all sorts of medical patients many of whom reported miraculous cures. Seven (7) of them were declared as medically unexplainable by Professor Verges in 1860. [558]

Since then, many thousands of pilgrims have visited Lourdes and followed Our Lady's instructions to "drink at the spring and wash in it". This practice was never encouraged by the Catholic Church; however, Lourdes water was still considered by believers a tangible item of devotion to Our Lady of Lourdes, but also as an infrequent miracle potion. It is still provided free of charge, to this very day, to anyone who might ask for it.

The spring water was analyzed in 1858 by a professor in Toulouse, who determined that it was drinkable, essentially quite pure, and inert: in other words, he found nothing special with the water. Nevertheless, many people claimed to have been cured by drinking or bathing in the spring water. [559]

According to Wikipedia, both Roman Catholics as well as Protestants believe Our Lady's reported apparitions at Lourdes to be genuine: being substantiated repeatedly by many testimonies of healings claimed to have occurred at the spring. [560]

The Lourdes Medical Bureau (Bureau Medical) was established at the request of Pope Pius X; however, it has always been run completely under medical not ecclesiastical supervision. The Bureau's aim is to ensure that reports of miraculous healings are examined professionally and scientifically by a medical team: thus

---

[557] https://en.wikipedia.org/wiki/Our_Lady_of_Lourdes (1 November 2015)
[558] https://en.wikipedia.org/wiki/Our_Lady_of_Lourdes (1 November 2015)
[559] https://en.wikipedia.org/wiki/Our_Lady_of_Lourdes (1 November 2015)
[560] https://en.wikipedia.org/wiki/Our_Lady_of_Lourdes (1 November 2015)

protecting the town from rash or fraudulent claims of miracles. Since the Lourdes apparitions, there have been about seven thousand (7000) people who submitted their case for examination by the Bureau in the hope of having it confirmed as a miracle; however, only sixty-eight (68, just under 1%) were officially declared to be "beyond medical explanation". Nevertheless, there were a few of these medically unexplainable cases that may not have been real miracles after all, because the illness reappeared in later years; but in my opinion, a medical relapse does not preclude a temporary miracle. A crude example would be: if someone is raised from the dead, we shouldn't expect him or her to live forever in order to consider it a miracle. However, the greater majority of the Bureau's judgements of cures declared as miraculous were upheld whenever questioned later. [561] Wikipedia concludes:

> The officially recognized miracle cures in Lourdes are among the least controversial in the Catholic world, because Lourdes from the very beginning was subject to intense medical investigation from skeptical doctors around the world. All medical doctors with the appropriate specialization in the area of the cure have unlimited access to the files and documents of the Lourdes Medical Bureau (Bureau Medical), which also contains all approved and disapproved miracles. Most officially recognized cures in Lourdes were openly discussed and reported on in the media at the time. [562]

Now, for someone, unqualified in medical affairs, to go against these medical doctors and the open documentation provided, simply because one believes that miracles cannot happen, is equivalent to denying empirical scientific evidence, to say the least.

---

[561]   https://en.wikipedia.org/wiki/Our_Lady_of_Lourdes (1 November 2015)
[562]   https://en.wikipedia.org/wiki/Our_Lady_of_Lourdes (1 November 2015)

Garabandal

According to Wikipedia, the apparitions of Garabandal in the Cantabrian Mountains of northern Spain are claimed to have occurred from 18 June 1961 to 18 June 1965. [563] They were apparitions of the Our Lady and Saint Michael the Archangel to four young schoolgirls in the village of San Sebastian de Garabandal. The girls experiencing the visions were twelve-year-old Conchita Gonzalez, twelve-year-old Jacinta Gonzalez, twelve-year-old Mari Loli Mazon, and ten-year-old Mari Cruz Gonzalez.

Many of the thousands of the supposed apparitions, consisting mainly of paranormal (beyond scientific explanation) phenomena, were allegedly filmed or photographed in the presence of thousands of witnesses. Skeptics claim they are illusionary tricks, the likes of which are performed by ordinary magicians. [564]

I have included this account because the alleged visions of Saint Michael the Archangel, if true, may be positive (empirical) evidence in favor of the existence of angels.

Paray-le-Monial

According to Wikipedia, in the monastery of Paray-le-Monial in Burgundy, France, a Roman Catholic nun, Margaret Mary Alacoque, received several private revelations from Jesus Christ. The first happened on 27 December 1673, and the final one eighteen (18) months later.

Notably, on 27 December 1673, Margaret Mary reported that Jesus permitted her to rest her head upon his heart, then he told her that he had chosen her to promote devotion to his Sacred Heart. [565] The reader is asked to note here that, if we are to believe that Margaret

---

[563] http://www.piercedhearts.org/treasures/shrines/garabandal.htm (3 November 2015)

[564] https://en.wikipedia.org/wiki/Garabandal_apparitions (3 November 2015)

[565] https://en.wikipedia.org/wiki/Margaret_Mary_Alacoque (3 November 2015)

Mary did actually rest her head on Jesus' breast, then it is actually possible, somehow, to have physical contact between a physical being (Margaret Mary) and a supernatural being (Jesus—who lived some two thousand years ago). To put it in another way: a ghost can drive a machine; that is, of course, unless it was a hallucination of Margaret Mary's rather than a miracle.

Besides this alleged miracle constituting positive (empirical) evidence for a natural to supernatural interface, I have included this account in order to point out that besides the "usual" apparitions of Our Lady, visions of Jesus have also been reported.

Guadalupe

According to Wikipedia, on 9 December 1531, a recently baptized fifty-seven-year-old convert, a peasant named Juan Diego, saw a vision of a beautiful young lady surrounded by light at a place called Tepeyac Hill near Mexico City, later known as Guadalupe. She told Juan to ask the local bishop to build a church on that hill.

The bishop initially rebuffed Juan and asked for a sign from heaven before he would comply. Mary first gave Juan a personal sign (miracle) by healing his uncle who was dying; she then asked him to gather some out-of-season flowers from the top of the hill. Juan complied anyway, and he did in fact find such flowers, some of which he carried in his cloak. When Juan opened his cloak before the bishop, on the fabric was the image of Our Lady of Guadalupe. [566]

I included this account to show that miracles have probably been happening for centuries; it is also noticeable that Europe may not be the only continent where miracles have supposedly happened.

---

[566] https://en.wikipedia.org/wiki/Our_Lady_of_Guadalupe (3 November 2015)

Personal

Personally, I *never* experienced a single miracle or even a deep spiritual experience in my entire life (so far).

However, I had a very interesting experience. A person I know very well claimed having a deep spiritual experience that was divulged to me at the time: predicting a certain event at a certain time. The timing and the natural conditions of the situation absolutely precluded the possibility of such foreknowledge. At the time I was told I didn't pay much attention to this person's alleged spiritual experience, and basically mentally dismissed it as a hallucination of some sort. But when the event actually happened in the predicted time frame, even though the odds were stacked heavily against its happening, I wondered. I'm afraid I'm not at liberty to say any more; believe it or not.

Jesus' Lifetime Miracles

In this section I shall be using the bible, particularly the New Testament, not as supernaturally inspired or infallible information, but as ordinary (human) ancient historical literature.

The first question we need to deal with is: Did Jesus Christ really exist; is he a historical figure? Or is he just a mythical figure? There is hardly any doubt about Jesus' existence and historicity; just as there is hardly any doubt about the existence and historicity of the Roman statesman, general, and author Julius Caesar, say. New Testament scholar and self-declared atheist Bart D. Ehrman wrote a whole book entitled *Did Jesus Exist?* answering this question. He used current *historical-scientific* methods concluding that Jesus is a historical figure. Let me quote his conclusion:

> We have considered substantial and powerful arguments showing that Jesus really existed (chapters 2-5 above). Many of the arguments made by the mythicists, by contrast, are irrelevant to the question (chapter 6); many of

the others are relevant but unsubstantial or, quite frankly, wrong (this chapter). There was a historical Jesus, a Jewish teacher of first-century Palestine who was crucified by the Roman prefect Pontius Pilate. [567]

And again, towards the end of the same book, Ehrman simply and categorically declares: "Jesus did exist. He simply was not the person that most modern believers today think he was." [568]

The second question relevant to this section is, obviously: did Jesus, really, perform many miracles? The four canonical (official) Christian gospels record (only) thirty-seven (37) specific miracles; however, in several of them there are statements like:

> [H]e healed many that were sick with diverse diseases .... [569]
> [W]heresoever he entered, into villages, or into cities, or into the country, they laid the sick in the marketplaces ... and as many as touched him were made whole. [570]
> Jesus went about all the cities and the villages ... healing all manner of disease and ... sickness. [571]
> [H]e came forth, and saw a great multitude, and he had compassion on them, and healed their sick. [572]
> [T]here came unto him great multitudes, having with them the lame, blind, dumb, maimed, and many others, and they cast them down at his feet; and he healed them. [573]
> [S]o much the more went abroad the report concerning him: and great multitudes came together to hear, and to be healed of their infirmities. [574]
> [A] great multitude of his disciples, and a great number of people from all Judaea and Jerusalem, and the sea coast of

---

[567] Ehrman, *Did Jesus Exist?* p. 263
[568] Ehrman, *Did Jesus Exist?* p. 336
[569] *The Holy Bible*; Mark: 1, 34.
[570] *The Holy Bible*; Mark: 6, 56.
[571] *The Holy Bible*; Matthew: 9, 35.
[572] *The Holy Bible*; Matthew: 14, 14.
[573] *The Holy Bible*; Matthew: 15, 30.
[574] *The Holy Bible*; Luke: 5, 15.

Tyre and Sidon, which came to hear him, and to be healed
of their diseases; ... were healed. [575]
In that hour he cured many of diseases and plagues ...; and
on many that were blind he bestowed sight. [576]
[W]hen he was in Jerusalem ... many believed on his
name beholding his signs [i.e., miracles] which he did. [577]
[A] great multitude followed him, because they beheld
the signs [i.e., miracles] which he did on them that were
sick. [578]

Hence Jesus' lifetime miracles, according to the accounts in the four
gospels, probably numbered in the hundreds: most of them consisting
of healings, triggered by his compassion for human suffering.

But I'm not going to consider any of this as scientific evidence
for the occurrence of miracles. Jesus' lifetime miracles, alleged in
the gospels, are almost completely omitted in this chapter because,
from a historian's point of view, we certainly lack corroborating
documentation from outside (non-Christian) sources: that is,
contemporaneous writings from non-believers in Jesus. Taking
evidence from believers in Jesus is questionable, to say the least,
because they would obviously be biased in his favor.

However, first century Jewish (non-Christian) historian Flavius
Josephus, in the eighteenth book of his history of the Jewish people
entitled *Antiquities of the Jews* writes:

> At this time there appeared Jesus, a wise man, if indeed one
> should call him a man. For he was a doer of startling deeds,
> a teacher of people who receive the truth with pleasure.
> And he gained a following both among many Jews and
> among many of Greek origin. He was the messiah. And
> when Pilate, because of an accusation made by the leading
> men among us, condemned him to the cross, those who

---

[575]   *The Holy Bible*; Luke: 6, 17–8.
[576]   *The Holy Bible*; Luke: 7, 21.
[577]   *The Holy Bible*; John: 2, 23.
[578]   *The Holy Bible*; John: 6, 2.

had loved him previously did not cease to do so. For he appeared to them on the third day, living again, just as the divine prophets had spoken of these and countless other wondrous things about him. And up until this very day the tribe of Christians, named after him, has not died out. [579]

In his book *Did Jesus Exist?* Ehrman writes the following, regarding this paragraph of Josephus:

> The problem with this passage should be obvious to anyone with even a casual knowledge of Josephus. We know a good deal about him, both from the autobiography that he produced and from other self-references in his writings. He was thoroughly and ineluctably Jewish and certainly never converted to be a follower of Jesus. But this passage contains comments that only a Christian would make: that Jesus was more than a man, that he was the messiah, and that he rose from the dead in fulfillment of the scriptures. In the judgement of most scholars, there is simply no way Josephus the Jew would or could have written such things. So how did these comments get into his writings? [580]

By his own admission, Josephus turned traitor in the Jewish war against Rome, and the Jews considered him as such; consequently, he was not a popular author among his people. The reader should realize here, that no original first century papyrus, parchment, or any other form of paper writing has survived to the present time. We have some small, possibly original, scraps from the second century; but all our other first century manuscripts are copies, of copies, of copies, etc.! During the Middle Ages Josephus' writings, however, were handed down to us this way by Christians rather than by Jews (because he was considered a traitor by his people). This is how we can explain why the above core Christian claims about Jesus ended up in Josephus' writings: when Christian scribes copied his text, they

---

[579] Josephus, *Antiquities of the Jews*, bk. 18, chap. 3 para. 3.

[580] Ehrman, *Did Jesus Exist?* p. 60

added a few phrases here and there in order to make known their beliefs to the reader.

So the obvious question arises: did the Christian scribe only add a few key phrases here and there, or did he make up the whole paragraph and insert it in an appropriate place? Ehrman answers this question thus: [581]

> The majority of scholars of early Judaism, and experts of Josephus, think that it was the former—that one or more Christian scribes "touched up" the passage a bit. If one takes out the obviously Christian comments, the passage may have been rather innocuous, reading something like this: [582]
>
> "At this time there appeared Jesus, a wise man. He was a doer of startling deeds, a teacher of people who receive the truth with pleasure. And he gained a following both among many Jews and among many of Greek origin. When Pilate, because of an accusation made by the leading men among us, condemned him to the cross, those who had loved him previously did not cease to do so. And up until this very day the tribe of Christians, named after him, has not died out." [583]

If this was the original text, as the majority of scholars agree, then Josephus, an unbiased first century author, is telling us here that Jesus was a well-known personality. Jesus was known for his wisdom, teaching, and love of truth; he was also thought to have done "startling deeds", and had numerous followers. Josephus also confirms that the Roman prefect Pontius Pilate condemned Jesus to death by the cross because of Jewish accusations brought against him. Finally, Josephus tells us that Jesus continued to have followers among the Christians even after his death. [584]

---

[581]   Ehrman, *Did Jesus Exist?* p. 60

[582]   Ehrman, *Did Jesus Exist?* p. 60

[583]   Josephus, *Antiquities of the Jews*, bk. 18, chap. 3 para. 3.

[584]   Ehrman, *Did Jesus Exist?* pp. 60–1

One might ask, here, how do scholars know what to include in and what to leave out of a passage that has been touched up? The answer is from the style, vocabulary, anachronisms, context, background information, etc.

The point I want to make here is that even atheistic scholars left the phrases: "He was a doer of startling deeds ..." in the passage; which means that Jesus had a reputation of performing miracles, according to a first century author. Now most people do not obtain this kind of reputation for no reason at all.

In the same book *Did Jesus Exist?* Ehrman addresses another passage: this time in the twentieth book of Josephus' *Antiquities of the Jews*:

> Here, Josephus is referring to an incident that happened in 62CE, before the Jewish uprising, when the local civic and religious leader in Jerusalem, the high priest Ananus, misused his power. The Roman governor had been withdrawn, and in his absence, we are told, Ananus unlawfully put to death a man named James, whom Josephus identifies as "the brother of Jesus, who is called the messiah." [585] [586]

Now "messiah" is the Jewish equivalent of the word "Christ". "*Christ*" is not Jesus' last name. "Christ" comes from the Greek word "christos" which in English translates to "anointed", and the Jewish word for "anointed" is "messiah".

In the Jewish tradition, when someone was sent on a special mission, or given a special position of authority, he was previously anointed with some perfumed ointment by a person of superior authority. The Jews were long awaiting a deliverer from the oppression of the Roman Empire: a deliverer sent by God himself, the anointed of God: *the* messiah. (So technically we should say "Jesus *the* Christ" not "Jesus Christ".) The fact that Jesus had this reputation

---

[585] Josephus, *Antiquities of the Jews*, bk. 20, chap. 9 para. 1.

[586] Ehrman, *Did Jesus Exist?* p. 59

shows that he must have done something special ("startling") in his lifetime. Since we do not have supporting documentation from outside Christian sources regarding his so-called miracles, I shall not try to show which miracles really happened and which did not. It just seems that he made some miracles: otherwise he would not have had such a reputation. So, I shall not examine Jesus' life-time miracles here, because they are beyond the scope of this chapter and the book.

Jesus' Resurrection

Jesus' alleged final resurrection from the dead, however, is another matter.

It might surprise many Christians that, unfortunately, we do not know who any of the authors of the four gospels might be; moreover, we do not know who most of the authors of the twenty-seven "books" of the New Testament are, either. There's one exception: Paul, originally called Saul from Tarsus in Turkey. And even then, not all thirteen "books", or better letters, attributed to him were written by him. In another book entitled *How Jesus Became God*, New Testament scholar Bart Ehrman writes the following about Paul's letters:

> Of the thirteen letters that are under Paul's name in the New Testament, critical scholars are reasonably sure that Paul actually wrote seven of them—Romans, 1 and 2 Corinthians, Galatians, Philippians, 1 Thessalonians, and Philemon ...; these are called the *undisputed Pauline letters*, since almost no one disputes that Paul was their author. [587]

The other six letters were written posthumously in Paul's name by his followers, apparently after circumstances had changed somewhat. All of the authentic Pauline letters were written in the 50s CE: they are our earliest, and therefore most reliable, surviving documents

---

[587] Ehrman, *How Jesus Became God*, p. 215 (emphasis in original)

written by a very early Christian we actually know by his real name, autobiography, and also someone else's ("Luke's"?) biographical accounts of him. [588]

Paul was initially a dedicated Jewish Pharisee (similar to a political priest) who persecuted the Christians because of their belief in Jesus' being the Christ (messiah); but then he converted to Christianity around 33CE. Here is Paul's autobiography and sworn testimony, written around 50CE, in one of these undisputed Pauline letters (that's some twenty odd years after Jesus' death, and some twenty odd years before the first of the four gospels was written):

> [Y]ou heard of my former way of life in Judaism, how I persecuted the church of God beyond measure and tried to destroy it, and progressed in Judaism beyond many of my contemporaries among my race, since I was even more a zealot for my ancestral traditions. But when (God) ... was pleased to reveal his Son [Jesus] to me, ... I did not immediately consult flesh and blood, nor did I go up to Jerusalem to those who were apostles before me; rather, I went into Arabia and then returned to Damascus.
> Then after three years I went up to Jerusalem to confer with Cephas [the leading apostle Peter] and remained with him for fifteen days. But I did not see any other apostles, only James the brother of the Lord [Jesus]. (As to what I am writing to you, behold, before God, I am not lying.) [589]

In another of the authentic letters, written around 56CE, Paul testifies, *personally* to Jesus' resurrection, and he adds the testimony of more than five hundred 500 people, most of whom were still alive at the time and could still be consulted if one so desired. Obviously this is significant evidence of Jesus' resurrection from the grave by someone who originally persecuted his followers.

---

[588]   Ehrman, *How Jesus Became God*, p. 215
[589]   *New American Bible*; Galatians: 1, 13–20.

> I handed on to you as of first importance what I also received: that Christ died for our sins in accordance with the scriptures; that he was buried; that he was raised on the third day in accordance with the scriptures; that he appeared to Cephas [Peter], then to the Twelve [i.e., the initial apostles]. After that he appeared to more than five hundred brothers [i.e., Christians] at once, most of whom are still living, though some have fallen asleep. After that he appeared to James [Jesus' brother], then to all the apostles. Last of all, as one born abnormally, he appeared to me. [590]

Later, in the same chapter of this letter, Paul writes:

> [I]f Christ has not been raised, then empty ... is our preaching; empty, too, your faith. Then we are false witnesses to God, because we testified ... that he raised Christ, whom he did not raise .... For if the dead are not raised, (then) neither has Christ been raised, and if Christ has not been raised, your faith is in vain .... If for this life only we have hoped in Christ, we are the most pitiable people of all. [591]

Why would the Christian faith be empty; and why would Christians be the most pitiable people? Because Christ was, in fact, crucified: he underwent the most ignoble death of his time (similar to the electric chair of our time); not much to put into the resume of a Christian!

Finally, according to tradition, Paul died beheaded for what he believed in. This is basically why I think there is significant evidence supporting Christ's resurrection from the dead: Paul's own sworn testimony of his witnessing it, his inclusion of about 500 witnesses most of whom could still be consulted, coupled with his extraordinary conversion and culminating in his final testimony, his martyrdom.

---

[590] *New American Bible*; 1 Corinthians: 15, 3–8.
[591] *New American Bible*; 1 Corinthians: 15, 14–19.

People do not usually die for a lie; he was convinced of its truth (have another look at the first and last verses of the last quote).

I have one last question to the reader: would you believe someone today telling you what happened two thousand years ago, or would you believe a contemporaneous person, together with other witnesses, telling you what happened then? There are certain facts that are not repeatable.

The Empty Tomb

The account of the discovery of Jesus' empty tomb, on the other hand, is not a very convincing argument for Jesus' resurrection.

First of all there is serious doubt as to whether Jesus' dead body was ever buried. Normally, when the Romans crucified a political criminal, they first stripped him naked in order to completely humiliate him; after his death, his body was left hanging on the cross to decompose for days on end: their body usually ended up torn apart and practically consumed by scavenging birds and animals: such a sight made a horrible public spectacle, and it was an obvious deterrent to others to partake in revolutionary activity. It was not sufficient for the Romans to simply kill revolutionaries, after inflicting them with excruciating pain; they also deprived them of a decent burial, which was extremely significant to most contemporary nations—especially the Jews. [592] Such a custom leaves serious doubt as to whether Jesus was actually given a decent burial: why should the Romans give special treatment to our Jesus? He was considered an enemy of the state: a revolutionary, the worst kind of criminal. In his book *How Jesus became God*, Ehrman writes to this effect:

> It is absolutely true that as far as we can tell from all surviving evidence, what *normally* happened to a criminal's body is that it was left to decompose and serve as food for scavenging animals. Crucifixion was meant to

---

[592] Ehrman, *How Jesus Became God*, p. 157

be a public disincentive to engage in politically subversive activities .... [593]

Moreover, as Ehrman points out in this same book, if we read carefully the gospel written first chronologically (allegedly Mark's) of the four canonical (official) gospels, we find a subtle contradiction regarding Jesus' burial that is not so easy to spot. Here is Mark's burial account:

> Joseph of Arimathea ... was a respected member of the Council, who was waiting for the coming of the Kingdom of God. ... Joseph went boldly into the presence of Pilate and asked him for the body of Jesus. Pilate was surprised to hear that Jesus was already dead. He called the army officer [centurion] and asked him if Jesus had been dead a long time. After hearing the officer's report, Pilate told Joseph he could have the body. Joseph bought a linen sheet, took the body down, wrapped it in the sheet and placed it in a tomb which had been dug out of solid rock. Then he rolled a large stone across the entrance of the tomb. [594]

Crucifixion was supposed to be a slow death; that's why Pilate was surprised that Jesus was already dead.

The "Council" in this passage refers to the *Sanhedrin*: which was the Jewish ruling body made of "chief priests, the elders, and the teachers of the Law (scribes)", as Mark explicitly tells us. [595] Here is Mark's trial account on the eve of Jesus' crucifixion:

> The chief priests and the whole Council tried to find some evidence against Jesus in order to put him to death, but they could not find any. [596]

---

[593] Ehrman, *How Jesus Became God*, p. 157 (emphasis in original)

[594] *New Testament and Psalms*; Mark: 15, 43–6.

[595] *New Testament and Psalms*; Mark: 14, 53.

[596] *New Testament and Psalms*; Mark: 14, 55.

[T]he high priest questioned him [Jesus], "Are you the Messiah, the son of the Blessed God?" "I am," answered Jesus .... The high priest tore his robes and said, "We don't need any more witnesses! You heard his blasphemy. What is your decision?" They all voted against him: he was guilty and should be put to death. [597]

Tearing one's robes was a visible sign of outrage at something said—a blasphemy.

In the same book Ehrman points out:

Mark himself said that at Jesus' trial, which took place the previous evening, the "whole council" of the Sanhedrin (not just some or most of them—but *all* of them) tried to find evidence "against Jesus to put him to death" ([Mark:] 14, 55). At the end of this trial, because of Jesus's statement that he was the Son of God ([Mark:] 14, 62), "they all condemned him as deserving death" ([Mark:] 14, 64). In other words, according to Mark, this unknown person [never mentioned previously in his gospel], Joseph, was one of the people who had called for Jesus's death just the night before he was crucified. ... My hunch is that the trial narrative and the burial narrative come from different sets of traditions inherited by Mark. Or did Mark simply invent one of the two traditions himself and overlook the apparent discrepancy? [598]

The only question mark I have about this argument is the fact that Joseph of Arimathea is mentioned in all four gospels. I can understand him being in the three synoptic gospels, Mark, Matthew, and Luke; but why is he also in John's gospel?

In spite of all the above evidence against Jesus' burial, I personally still think that he was properly buried; not as a matter of faith or devotion, but because of a single consistency in all four gospels: all

---

four written about a decade apart, and originating from separated communities of early Christianity.

First, it is just possible, in my opinion, that Pilate was a little concerned about Jesus' popularity and wanted to avoid an uprising during Passover, at which time many Jews visited Jerusalem. Passover is a feast recalling the liberation of the Jews from Egyptian slavery; it was only one small step away, for revolutionaries, from the concept of liberation from Roman oppression. Consequently, making an exception, Pilate might have given Jesus' followers the body of Jesus to be decently buried, something which is very important to Jews even to this day, in order to avoid an uprising by his admirers.

Secondly, and more importantly, all four gospels agree that the tomb was found empty by Mary Magdalene. The testimony of a woman did not amount to much in first century Palestine; yet the four gospels are consistent in this respect. I also believe that all four gospels tried to add other witnesses—a second witness, at least—to give the discovery of the empty tomb more weight; but none of the other witnesses mentioned in the four gospels agree. Nowadays, however, we give as much weight to a woman's testimony as that of a man.

Moreover, as we saw above in Paul's letter, written in the 50s CE, Christians already believed and formulated that Jesus was buried.

> I handed on to you as of first importance what I also received: that Christ died for our sins in accordance with the scriptures; that he was buried; that he was raised on the third day in accordance with the scriptures; that he appeared to Cephas [Peter], then to the Twelve [i.e., the initial apostles]. [599]

The shroud (burial cloth) of Turin, Italy, does not seem to provide scientific evidence in support of Jesus' burial: it seems to be a *fake* produced around 1300CE, from radiocarbon (^{14}C) dating. This is

---

[599] *New American Bible*; 1 Corinthians: 15, 3–5.

roughly around the time it appeared on the scene: confirming modern scientific evidence.

Jesus Seen Alive Again

In his book *How Jesus Became God*, Ehrman writes the following regarding Jesus' resurrection:

> There can be no doubt, historically, that some of Jesus's followers came to believe he was raised from the dead — no doubt whatsoever. This is how Christianity started. If no one had thought Jesus had been raised, he would have been lost in the mists of Jewish antiquity and would be known today only as another failed prophet. But Jesus's followers—or at least some of them—came to believe that God had done a great miracle and restored Jesus to life. This was not a mere resuscitation, a kind of near-death experience. For Jesus's disciples, Jesus was raised into an immortal body and exalted to heaven where he currently lives and reigns with God Almighty. [600]

It was not Jesus' teaching that resulted in Christianity's flourishing; his teachings were good, of course, but they were by no means original: other wise men taught the same charitable principles before him; yet very few of us know about these people: hardly any record of them was left for us. Furthermore, it was his teachings that resulted in his becoming a victim of Church and State. In actual fact, it was Jesus' first followers' belief in his resurrection that resulted in such a numerous subsequent following of Jesus, nothing else. Further on in the same book, Ehrman writes:

> It is indisputable that some of the followers of Jesus came to think that he had been raised from the dead, and something had to have happened to make them think so. Our earliest records are consistent on this point, and I

---

[600] Ehrman, *How Jesus Became God*, p. 174

think they provide us with historically reliable information
in one key respect: the disciples' belief in the resurrection
was based on visionary experiences. [601]

So the followers of Jesus believed that he rose from the dead, not
because of Mary Magdalene's finding the tomb empty, but because
a number of them had visions of him alive after his crucifixion—
and the Romans made sure they did a thorough job of it. This is
one of the reasons why, I believe, God suffered Jesus to die such an
ignominious and painful death, not to be able to forgive our sins:
but because it was a *public* death. This way everybody knew that he
really died: otherwise there would be serious doubt about his eventual
resurrection had he died on his bed in a private room, say. In fact
people question his resurrection all the time, but hardly anyone ever
questions his death on a cross.

Regarding these alleged visions, detractors of Jesus' resurrection
say that his disciples were simply hallucinating; because he was so
beloved to them, and he was taken away from them so suddenly: they
simply imagined him still alive. There might be some truth in this;
however, this is certainly not the case with Paul: Jesus was nothing
to Paul. Paul hated Jesus and wanted to kill all those who followed
him. But God, somehow, "revealed his Son" to Paul.

Finally, according to New Testament scholar Bart Ehrman, it is
historically certain that at least some of Jesus' followers believed
that Jesus rose from the grave because they had visions of him alive
again. If you were to judge, who would you tend to believe more: a
contemporary skeptic of Jesus (like Paul), who came to believe that
Jesus rose from the grave because he himself saw him alive again; or
a twenty-first century skeptic who refuses to believe in the possibility
of miracles?

---

[601]   Ehrman, *How Jesus Became God*, pp. 183–4

Resurrection of the Body

So how exactly did this resurrected body of Jesus appear to Paul? The answer to this question is in his writings. This is how Ehrman, in his book *How Jesus Became God*, summarizes Paul's writings regarding Jesus' resurrection:

> The body Jesus had when he was raised was not simply his resuscitated corpse brought back to life. It was an astoundingly immortal body, a "spiritual" body. A body, yes. A material body, yes. A body intimately connected to the body that died and was buried, yes. But a transformed body that could not experience pain, misery, or death. [602]

This throws a monkey-wrench into the cogwheels of the idea of physical suffering in hell; but in any case, this is how Paul tries to explain things in the above undisputedly authentic letter to the Corinthians:

> You fool! When you sow a seed in the ground it does not sprout to life unless it dies. And what you sow is a bare seed, perhaps a grain of wheat, or some other grain; not the full-bodied plant that will later grow up. God provides that seed with the body he wishes; he gives each seed its own proper body. [603]

A seed does not die, of course, prior to growing into a plant; but that was the common "wisdom" of the time: the seed was buried under the ground and consequently they thought it died before it grew into a plant or tree. Furthermore, the difference between a seed and its plant or tree is an astronomical metamorphosis; this is the idea Paul wanted to convey to the early Christians in Corinth. Moreover, I think, the fact that different kinds of seeds grow accordingly into

---

[602] Ehrman, *How Jesus Became God*, p. 177
[603] *New Testament and Psalms*; 1 Corinthians: 15, 36–8.

different kinds of plants or trees implies that there will be levels of resurrected bodies—not just one type.

So Jesus' resurrected body was nothing like Lazarus' resurrected body that Jesus, according to the John's gospel, brought back to life after it was dead for four days. [604] Lazarus, I guess, died again—a second time; that's why no one ever talks about him nowadays. People may have talked about him in his time (assuming the account is true) while he was alive; but not so much after he (supposedly) died a second time. No, Jesus' resurrected body was different in some obvious way.

In his book *The Physics of Christianity*, theoretical physicist Frank Tipler proposes that Jesus' resurrected body was made of neutrinos and antineutrinos [605] rather than the carbon atoms which normally constitute our human bodies. As we have seen, neutrinos are extremely minute uncharged particles about one-ten-thousandth (10^{-4}) of the size of an electron; they can pass through matter unhindered and practically undetected; do you get the clue? Our identity, which we have likened to a detachable executable computer program, can conceivably be changed in format and loaded onto another type of medium—the neutrino-antineutrino.

People often ask me what our body will look like in heaven. I do not pretend to know the answer; but if we believe what people who experienced apparitions say, we can get an idea (I shall also add some of my own speculations). The visionaries in Medjugorje (mentioned in this chapter on "Miracles") describe Our Lady's (Mary's) appearance as being "18 to 20 years old". The children of Fatima described Mary as "a most beautiful lady". Now, according to "The Gospel According to John" [606] and "The Acts of the Apostles" [607] Jesus' mother, Mary, was still alive after Jesus' death. Since Jesus died at about thirty-three years of age, she would have been close to fifty years of age, at least:

---

[604]   *New American Bible*; John: 11, 38–44.

[605]   Tipler, *The Physics of Christianity*

[606]   *New American Bible*; John: 19, 25.

[607]   *New American Bible*; Acts: 1, 14.

she was not exactly a woman in her prime when she died; the human body reaches its peak between twenty and thirty years of age.

Because of this seers' testimony, my theory is that the bodies of all those in heaven will be at their physical peak, and there will be no handicapped people in heaven. Aborted fetuses, babies, and children that died young will also be fully grown adults. This might seem strange to some people: seeing your children, parents, and grandparents the same age as you. However, people are still recognizable if you look closely at pictures of their youth: I suppose we'll get used to it in no time.

I also think that we shall all be naked in heaven; the undeniable evidence is that clothes are left behind in the coffin. Conceivably, light will emanate from our bodies; if we take Jesus' transfiguration account in Matthew's gospel literally, assuming it is a true account of what really happened, "his face did shine as the sun". [608] This may be of some consolation for shy people; this light will probably cover some of our nakedness by blinding the full view. I believe that wearing clothes is an unnatural but practical human proposal intended to curb or control our inordinately strong sexual appetite here on earth. In fact, "The Book of Genesis" depicts Adam and Eve, supposedly our first parents, as initially naked; yet, they were not ashamed of each other. [609] That is how they were supposedly created by God, so why would it be different in heaven? We shall not have such uncontrollable desires in heaven; therefore we shall not need to wear any clothes. There will probably be no sexual mating in heaven, although some researchers of the afterlife believe there will be some. There will be no need for reproduction since we shall be immortal there; our role in God's creation-participation project (see the next chapter) would have been completed.

There is no time in the afterlife, as there probably is no time here: it seems to be our own creation because everything deteriorates in our universe. We shall be so happy that "time" flies without our noticing

---

[608]   *The Holy Bible*; Matthew: 17, 2.
[609]   *New American Bible*; Genesis: 2, 25.

it. Our existence will probably consist if a harmonious, care-free, and painless "living"; never getting bored or tired: communicating with God and everyone else, getting to know them intimately and building satisfying relationships; learning about the mysteries and beauty of reality; indulging in our favorite pastimes; participating in banquets and dances; etc. What a "life"!

CHAPTER 10

Philosophy, Morality, and Metaphysics

The main object of this last chapter is to tie up a few loose ends left as a byproduct of the conclusions drawn or suggested in this book. Naturally, it cannot be an exhaustive treatise: there is enough material for another entire book; I can only skim the surface.

Summarizing the first seven chapters of this book, we have compelling scientific evidence in our universe pointing to the existence of an intelligent creator: the existence of matter and energy, the expansion and winding down of our universe, the exquisitely fine-tuned laws of physics and chemistry, and ultimately the mind-boggling genetic code and variety of life. This creator's ability seems to surpass, by far, our human limitations; indeed it seems to be above anything we know of in the universe: he seems to possess a supernatural ability.

However, this is only the scientific evidence; which is the strongest type of evidence: as in a court of law. But besides forensic evidence, our law courts also heed eyewitness evidence—with a pinch of salt. Admittedly eyewitness evidence, more often than not, is less reliable than forensic evidence; but occasionally it could be more conclusive. In this book, therefore, I presented experiential evidence: what every one of us feels subjectively. Finally, I also presented empirical (data-based) evidence from people who had near death experiences or witnessed miracles.

The scientific community does not even want to touch this last category of evidence with a ten foot pole; in fact, many of my scientifically oriented friends, who previewed my book, advised me to leave these sections out because it would distract readers from the main argument. But you see, despite the corroborating scientific evidence pointing to an intelligent creator, people still ask: Can we ever see or experience this creator in a more direct manner? Is there any smoking-gun sort of evidence for the existence of this creator? Is it possible to find direct physical evidence for such a creator: like the gravity effects implying dark matter and dark energy in spite of their not being visible to us? Or is he completely elusive to our instruments and senses: like the concept of the multiverse? Does his existence fall into the realm of science or of pseudoscience, where our imagination takes over.

This is the main reason why I decided to include the previous chapter on "Miracles". If the body of evidence from reported miracles could somehow be scientifically authenticated (that is, confirmed that they actually happened, and barred of any possibility of illusions by entertainers such as magicians), then the temporary suspension of the laws of nature would most probably imply a physical intervention by the creator of these laws.

I feel that there is such a great body of empirical evidence; if I left it out, the evidence we humans have of the supernatural would be incomplete: consequently I would be doing my readers a disservice. So, even at the risk of compromising the main argument of this book I feel I must include it as well. Notwithstanding the opinion of most scientists, I believe that empirical evidence is also a form of scientific evidence, especially if it shows a pattern: in which case it usually leads scientists to the right conclusions—the medical profession has made use of empirical data since time immemorial.

Currently science has no clue as to what our consciousness (mind) is all about; however, there is no reasonable way of denying it. Near death experiences seem to point to the existence of an out of body experience (soul) that is separable from the body. Moreover, both near death experiences and miracles seem to point to some form of existence of people who are known to have died, but somehow still exist in a form quite different from earthly beings. So the obvious question is where are these souls? How come we can never, or hardly ever, see them?

A parallel question would be: if (as the book shows) scientific evidence implies the existence of a supernatural creator, who is not confined to the universe, where is he to be found? Is he outside the universe? If so, how can our prayers reach him in time to be of any benefit to us? Even at the maximum speed possible, the speed of light, a message takes about fourteen billion years to reach the end of our universe. This is why a little metaphysics is included in this chapter.

Near death experiences also seem to support evidence of a benevolent creator; yet from the theory of evolution—survival of the fittest, in particular—the whole world is built on violence: predators killing and eating prey. Moreover, from our every-day experience, we all know that life is predominantly biased with pain and suffering rather than fun and happiness. Our earthly experience does not point to a benevolent creator; can this discrepancy with the evidence from near death experiences be reconciled? If not, something has to give. If killing is the way of nature, what is good and what is evil? Where is the demarcation line? This is why a little morality is also included in this chapter.

The final question is whether this creator is intrinsically good or evil? Is God a good God? Or do we just define God as good? Is whatever he does, good? So finally, this is why a little philosophy is also included in this chapter.

I simply did not want to end my book leaving too many loose ends. Consequently in this chapter, I shall, to all intents and purposes, assume the existence of a creator of the universe, whom we normally call "God", and try to answer some metaphysical, moral, and philosophical questions arising from the evidence we have of his creations. I shall also practically assume that the empirical evidence from accounts given in near death experiences and miracles regarding the soul, the nature of God, and the afterlife are evidence that is reliable enough—on average. Unfortunately, I lack the space to examine the details of every near-death-experience case or miracle closely and separately since there are so many of them. I admit that they might lack the rigorous nature of scientific evidence, but I really do believe there is enough of them to be convincing enough, especially if organized systematically—but that's a story for another entire book.

First Set of Questions

Following are several sets of questions intended to help the reader follow my footsteps as I discuss various topics.

The first set of questions is: assuming the Creator of the universe and life is an intelligent and perhaps a supernatural being, do we infer from his creations, that he is a benevolent or an evil creator? How do we define good and evil, anyway? Do we define anything that God does as good? Or is God really a good God?

Is God an Evil Creator?

God made life in such a way as to make it necessary for one organism to kill and feed on another in order to survive. Doesn't that make him an evil creator?

Could he not have created life without the need for food? Not really, because organisms need to grow; to do so they need energy, which can only come from food. Could he not, at least, have made all animals herbivores (plant eaters)? Plants have no nervous system whereas animals feel pain like us, I presume, while being killed. Or, even better, could he not have limited nourishment requirements to ingesting inanimate compounds and minerals, like water, salt, calcium, iron and so on? Why did he choose such a seemingly cruel food chain?

This is a very good question; in fact, there are many religions that believe that there are two creators: one good and one evil. Moreover, historically, many cultures believed that their gods were cruel and evil; such people are known to have sacrificed animals and even human beings (children) to appease their gods: so that the latter refrain from sending disasters along their way or leave them alone, at least.

In his treatise "The Problem of Pain", essayist and lay theologian Clive Staples (C.S.) Lewis tells us what he thought of God while he was still an atheist; that is, prior to converting to Christianity. He tells us that if someone had asked him why he did not believe in God, his reply would have been something along the lines of what follows. (I shall quote him significantly because it is so incisive.) [610]

> Look at the universe we live in. By far the greatest part of it consists of empty space, completely dark and unimaginably cold. The bodies which move in this space are so few and so small in comparison with the space itself that even if every one of them were known to be crowded as full as it could hold with perfectly happy creatures, it would still be difficult to believe that life and happiness were more than a byproduct to the power that made the universe. [611]

Lewis then references his contemporaneous scientific consensus, namely, that there are probably very few suns in space that have planets orbiting around them, that our Sun seems to be the only such star, and that in our solar system life probably only exists on Earth. There seems to have been no life on Earth for many millions of years, and the Earth will probably exist for many millions of years more after life on our planet has ended. In the meantime, while life is extant on Earth, its quality is nothing to write home about. [612]

> And what is it [life] like while it lasts? It is so arranged that all the forms of it can live by preying upon one another. In the lower forms this process entails only death, but in the higher there appears a new quality called consciousness which enables it to be attended with pain. The creatures

---

[610] Lewis, "The Problem of Pain", p. 551
[611] Lewis, "The Problem of Pain", p. 551
[612] Lewis, "The Problem of Pain", pp. 551–2.

cause pain by being born, and live by inflicting pain, and in pain they mostly die. In the most complex of all the creatures, Man, yet another quality appears which we call reason, whereby he is enabled to foresee his own pain which henceforth is preceded with acute mental suffering, and to foresee his own death while keenly desiring permanence. It also enables men by a hundred ingenious contrivances to inflict a great deal more pain than they otherwise would have done on one another and on the irrational creatures. This power they have exploited to the full. [613]

Lewis goes on to mention that the history of humanity is, to a great extent, a record of crime, littered by war and terror, coupled with outbreaks of disease. People occasionally experience short periods of happiness that are usually accompanied by the agonizing apprehension of losing this happiness, or the mental depression resulting from the memory of its loss. From time to time, various civilizations arise improving their lives for a while; while they survive, these civilizations have a tendency of causing their own characteristic sufferings, which are often greater than the pain they mitigate. There is no doubt whatsoever that this is the case with our own civilization. History also tells us that normally all civilizations come to an end after a while; our current civilization will also probably terminate, one day in the future: like all others have done before it. He continues by saying that even if it does not end, humanity is doomed to die and vanish, anyway; he explains: [614]

Every race that comes into being in any part of the universe is doomed; for the universe, they tell us, is running down, and will sometime be a uniform infinity of homogeneous matter at a low temperature. All stories will come to nothing; all life will turn out in the end to have been a transitory and senseless contortion upon the idiotic face

---

[613] Lewis, "The Problem of Pain", p. 552

[614] Lewis, "The Problem of Pain", p. 552

of infinite matter. If you ask me to believe that this is the work of a benevolent and omnipotent spirit, I reply that all the evidence points in the opposite direction. Either there is no spirit behind the universe, or else a spirit indifferent to good and evil, or else an evil spirit. [615]

Wow! Impressive! The obvious question, therefore, arises: Is God a good being or an evil being? Do we *define* as "good" whatever God decides to do? In other words, does God always do the right thing? How do we know what is right and what is wrong, benevolent or cruel, so that we can judge God's behavior?

The author of the above quote changed his views, eventually, and became one of the most authentic Christians that ever appeared in the public arena; that fact says something. Why did he change his mind? I surmise he realized that the above is how things appear from a creature's point of view; but it is not enough to view things in such a narrow-minded manner: one needs also to look at things from a designer's point of view. I am not trying to defend God at all; he does not need my defense— he can hold his own under any circumstances. However, looking at life from just one angle is an adolescent way of looking at things: in reality, life is much more complex, as we shall see presently.

Designer's Viewpoint

In developing the ideas in this last chapter, therefore, we shall try to see what things look like from a designer's point of view.

We have to keep in mind that a designer of an ecosystem has to set up a self-sustaining system: that is, a self-cleaning system where garbage, excrement, and carcasses are recycled. Moreover, the majority of the weak of every species need to be eliminated (killed) in order that a generally healthy offspring is born: hence the necessity of predators.

---

[615] Lewis, "The Problem of Pain", p. 552

At the top of the food chain, humans seem to be an exception to this general rule of survival of the fittest, because they also take care of their sick and their elderly. It also seems to be required of them by the way they generally think, reason, and empathize. We shall look into this concept more deeply later on in this chapter.

Overpopulation Threat

In order to consider things from the point of view of the designer, we can start by considering some mathematical realities and consequences of life.

In his most famous book *The Origin of Species*, naturalist and geologist Charles Darwin, reveals how easily any species, no matter how low its reproductive rate, can overpopulate the whole earth if it had no predators at all. He considers the case of the elephant, which is reckoned to be the lowest breeder of all known animals. He believed it would be conservative enough to assume that a typical (female) elephant reproduces from the age of thirty to the age of ninety, giving birth to one offspring every ten years (that is, 3 female and 3 male baby elephants in 60 years) and that elephants (male or female) live until they turn 100 years of age. If this be the case, calculations show that, after a period of between seven hundred and forty (740) and seven hundred and fifty (750) years, there would be about nineteen million (1.9×10^7) elephants alive from just an initial (male and female) pair of animals. [616] Now an adult elephant consumes between two hundred (200) and six hundred (600) pounds a day of grasses, tree foliage, bark, twigs, and other vegetation. Imagine the havoc nineteen million elephants can cause to an ecosystem.

I made some numerical calculations myself concerning a possible human population growth; take a look at the following numbers. I assumed we start with an equal male and female human population of one hundred people (that is, fifty men and fifty women) and that the population increases by just one percent (male or female) every year;

---

[616]   Darwin, *The Origin of Species*, pp. 91–2

of course, keeping the number of men roughly equal to the number of women, all the time. In other words, in any given year and in every group of one hundred people, the number of babies born exceeds the number of people who die by only one more person—male or female; but all the time the number of males remains equal to the number of females. Note that, in a given group, it is not required that someone dies every year.

After one thousand (1,000) years, the population would grow to about 2.1 million (2.1×10^6) people; after two thousand (2,000) years, the population would grow to about 4.4 billion (4.4×10^9) people; after three thousand (3,000) years, the population would grow to about 920 trillion (9.2×10^{14}) people; after four thousand (4,000) years, the population would grow to about 19 million trillion (1.9×10^{19}) people. Modern man seems to have existed for about two hundred thousand (200,000) years; I don't think I need to carry on for two hundred thousand (200,000) years: you get the drift. But the current world population is only about 7.4 billion (7.4×10^9) people; how come?

Incidentally, according to evolutionary principles, pleasure during human sexual intercourse is a survival instinct requiring a high reproductive rate to avoid extinction: ensuring that reproduction happens from time to time. But, my question is: is it really necessary in the case of humans? Or is it just a kind gesture on the part of our Creator? If every couple had just one intercourse every year, it would be ample opportunity to produce the above population explosion. It seems to me that sexual pleasure was designed to be conducive to building a loving relationship between a male and a female sharing their life together: to give them a break from the everyday problems of life.

Before I proceed further with our discussion, I would like to draw a more convincing picture of this threat of overpopulation; numbers are only impressive to a mathematician or a scientist, but a picture hits everyone to the core. In his book *The Selfish Gene*, evolutionary biologist Richard Dawkins writes about the threat of overpopulation as follows:

[T]he present population of Latin America is around 300 million, and already many of them are under-nourished. But if the population continued to increase at the present rate, it would take less than 500 years to form a solid human carpet over the whole area of the continent. ... In 1,000 years from now they would be standing on each other's shoulders more than a million deep. [617]

Now, imagine you were God anticipating these numbers.

Assuming you wanted to design a self-sustaining ecosystem, in which you do not have to interfere all the time; would you make life easier or more difficult for these people? If they procreate at a faster rate, more of them would starve to death. If you make life quite difficult for them, so that they dare not have more than two offspring, the population would possibly remain the same. How can God help a human carpet covering the whole continent? It would be too late. Now I ask the reader: what would *you* do? So much, then, for an evil Creator (or Designer) making our life impossible.

A solution, of course, would be not to allow organisms to procreate at all; but, obviously, that would be the end of life as we know it: life wouldn't last very long; unless, of course, living organisms did not change much and certainly not deteriorate at all: as, presumably, happens in the *afterlife* (see the corresponding section below). But then again, how can one prevent all of them from ever having an accident and getting hurt in this extremely long time they live? And, incidentally, would you like to work in the same job or doing housework for, say, a thousand years; how about a million years or a few billion years? Would you not complain about life being too long instead of complaining about death?

As the saying goes: "You're damned if you do, and you're damned if you don't." So why did God not forget the whole project? We shall also try to answer this question in the course of this chapter. Perhaps the reader is starting to realize that it's not so easy to be a designer; it almost makes one want to forget the whole thing. Let us now look

---

[617]   Dawkins, *The Selfish Gene*, p. 111

at some of the general experiences in life that we tend to blame God for or attribute to bad design on his part.

Good and Evil

It is true that sometimes human beings think of ingenious ways of inflicting physical pain on other human beings or animals; but most of us, including atheists, also believe that this is evil behavior. Why does God allow people to be evil? Why does he not just dispose of them?

Evil has to exist, because if it does not exist then neither would good exist; philosophically, they are part and parcel of one another: just like there cannot be a front without a back or a top without a bottom. Evil is the absence of good: like darkness is the absence of light, black is the absence of color, and a hole in the wall is simply the absence of wall material. Evil, even great evil, is basically various levels in the absence of good. Let me elaborate.

For example, most of us humans agree, deep down, that we should treat other human beings the same way we would like to be treated ourselves; so to us, it is good to behave accordingly, but evil to behave contrarily. So to kill someone and cannibalize one's body would be a very evil deed, because other human beings are not considered as part of our food chain; but to torture someone before killing and cannibalizing the person would be an even greater evil. On the other hand, if jungle law were our governing morality, then to kill others and cannibalize their bodies might not be considered evil at all! Perhaps killing without causing undue pain might still be considered to be the right thing to do; failure to comply with this norm then would be considered evil: but it would not be as evil as in the previous case. Consequently, evil is essentially only the absence of what one considers good or acceptable behavior: there are no hard and fast rules.

As we shall see more clearly later on in this chapter, God seems to have desired a personal relationship with every one of us;

335

consequently, he did not create us like robots: doing exactly what he desires all the time. It seems that he wanted to give us free will: that is, the freedom to choose whether to love him or hate him and whether to do good or evil to others whom he loves. God loves everyone, his creations, like a mother loves her baby, and like an engineer loves his own designs. Normally, a mother will still love her own child, even if the child is an evil person; she may not condone or justify her child's behavior, but she will still love her child. Similarly, God does not condone evil deeds, but he still loves evil people, because they are his creation—his children.

True, there are grey areas as to what human beings believe to be good or evil; but I believe that we all, including atheists, agree that we should treat others the same way that we would like to be treated ourselves, and that it is basically wrong to treat others differently. I also believe that we are hardwired, to some extent at least, to be able to feel, deep down, what is right and what is wrong: our conscience. What I propose here, for the time being, is that if one believes, deep down in oneself (not just in public), that something is wrong in this context, one should refrain from doing it; if one does not refrain from doing it, we shall *define* it as *evil*. Of course, there will still be different degrees of evil; for example, tripping someone on purpose is no big deal, but it is still wrong to do so to another person; on the other hand, maiming or killing a person is a much graver evil.

Normally we have to suffer the consequences of our actions; so if we are mean towards other people, people are not going to be very good to us in return: we are probably going to have a difficult life because of the way others will treat us. Most people will consider this fair enough in life; however, this is not always the case, as shown in the next section.

Pendulum Swings

Unfortunately, whenever people are really evil, it is usually not themselves that suffer; it is the next few generations that do.

Take, for example, the treatment of blacks by whites and the treatment of women by men one or two centuries ago; both were not even considered human or to have a soul; in fact, they were denied voting rights. So nowadays, whites and men have to be extra careful to be politically correct all the time; in addition, they have to suffer unfair discrimination. We may also mention the reputation of modern Germans, who are still considered Nazis and the perpetrators of the holocaust of the Jews in the Second World War; the current German generation did nothing wrong to the Jews. The consequences of the actions of a generation are not necessarily suffered by that same generation but by later generations; the pendulum has to swing far to the other side before it comes to the equilibrium (fair) position. Where is the justice in that?

When such things happen, instead of blaming previous generations for their evil doing, we tend to blame God for not stopping it. God's general policy is not to interfere, but to let us sort things out ourselves: this is the only way he can give us freedom of action—free will. We have to learn how to behave toward one another from our mistakes, our experience, and our foresight. It would also be a good idea, I think, to make these pendulum swings as minute as possible once evil has occurred.

Pain

It is probably easy to see that physical pain is a survival tool that protects us from getting seriously hurt.

Leprosy, for example, causes a type of inflammation of the nerves which results in a lack of the ability to feel pain. Consequently, this results in the loss of extremities (fingers and toes) due to repeated injuries caused by inadvertently hitting them.

I believe everyone feels pain to different degrees; natural selection probably opted for this much (average) pain for our own good: that is, so that we may be able to preserve a healthy body during our life span, and so be able to thrive in our environment. As evolutionary biologist Richard Dawkins points out in his book *The Greatest Show on Earth*, an "imperative admonition" or "a little red flag" in the brain would probably be insufficient to stop us from damaging our bodies at practically all costs. [618]

Evolution

Personally, I think that one of the most shocking facts showing the cruelty in nature is the case of the ichneumon wasp. The female wasp has an *ovipositor*, which is similar to a stinger; it uses this ovipositor to lay eggs inside the living bodies of caterpillars, so that when the larvae hatch they eat their host alive. How is it possible that a benevolent God would design something like this? As one would expect, books written by atheists never fail to mention this piece of information. [619] [620] [621]

However, although natural selection is probably God's general design, enabling the various species to survive and thrive in their natural environment, he did not dictate in detail what every species has to do: he left them alone to find their own way. I admit that natural selection is a somewhat selfish practice; in the process, a species may decide to be unkind and even cruel (by human standards) towards another species. When a particular species does so, it doesn't realize it is doing something wrong; we have to keep in mind that its behavioral standards are not as noble as those of human beings, nor does it possess human reasoning.

Given the situation, that is, after a species has taken advantage of another, God does not interfere in the prey-victim or parasite-host relationships so developed. We have already seen, in the section entitled "Overpopulation Threat", why predators are necessary in life; on the other hand however, a species may completely destroy

---

[618] Dawkins, *The Greatest Show on Earth*, p. 393

[619] Darwin, *The Origin of Species*, p. 255

[620] Stenger, *God and the Folly of Faith*, p. 107.

[621] Dawkins, *The Greatest Show on Earth*, p. 395

another species (to its own detriment) as a result of its assumed habits. For example, a very aggressive virus that kills its host in a very short time will not have good prospects of surviving for a long time itself.

I believe, however, that human beings are an exception to this general evolution process. I think God hardwired us a little differently; he gave us a level of intelligence and powers of reasoning that are much higher than all the other animals, coupled with a disposition to behave altruistically. Although atheists do not believe in God, they are still similarly hardwired: they still believe that it is a good thing to treat others the same way that they would like to be treated themselves.

To be fair, of course atheists do not attribute this universal human quality to God, but to some form of symbiotic evolution: where both sides gain. For example, as evolutionary biologist Richard Dawkins points out in his book *The Selfish Gene*, there is a lot to be said for a tit-for-tat policy in dealing with other people.

Briefly, a tit-for-tat policy starts by cooperating: that is, it trusts anyone it never met before. But after this first default state, it remembers the individual it dealt with and behaves the same way that individual behaved the last time they interacted. If one cooperated, then tit-for-tat will cooperate again; if one did not cooperate, then tit-for-tat will not cooperate. However, if the other organism decides, at any time, to start cooperating again, then tit-for-tat will cooperate again the next time around. It is benevolent, cooperative, fair, forgiving, and unenvious; it is therefore quite reasonable to conclude that it can easily evolve into a harmonious and stable society. [622]

As an aside, unfortunately, an individual can still end up dead or robbed by trusting a stranger the first time around, and even later. But we have to start somewhere, and I believe that tit-for-tat is a good start; one needs to be cautious, especially at the beginning, and trust strangers only where there is minimal risk.

---

[622] Dawkins, *The Selfish Gene*, pp. 210–33

Violence

Christians believe that God took *positive* action in this matter. They believe he conceived (possibly through a small miracle impregnating Mary) a human son, Jesus the Christ, who would show us how we ought to behave towards one another: God took such positive action to make sure we got this right!

Christ's teaching consisted of refraining from any kind of violence altogether. Violence has been the drug of choice of all of humanity throughout all the ages; for this reason, I believe, it is the sin par excellence. It is the "sin (singular) of the world" referred to by John the Baptist, in "The Gospel According to John", when he declared about Jesus: "There is the Lamb of God, who takes away the sin of the world!" [623]

Everyone knows there isn't just one sin in this world; but I believe that violence is *the* sin, and I think the Catholic Church is wrong in changing "sin" to "sins" like, for example, when it prays: "Lamb of God, you take away the sins of the world; have mercy on us". It shows a lack of respect for authenticity of text, misunderstanding, and possibly manipulation.

Jesus also reminded us that, unlike all the other animals, "man shall not live by bread alone". [624] There is more to human living than simple survival: full or real human living consists of loving others and building relationships with them.

Jesus also goes so far as to teach us to offer the other cheek when we are slapped in the face:

> "You have heard that it was said, 'An eye for an eye and a tooth for a tooth.' But now I tell you: do not take revenge on someone who wrongs you. If anyone slaps you on the right cheek, let him slap your left cheek too." [625]

---

[623] *New Testament and Psalms*; John: 1, 29.

[624] *The Holy Bible*; Matthew: 4, 4 & Deuteronomy: 8, 3.

[625] *New Testament and Psalms*; Matthew: 5, 38–9.

Apparently, this is not to be taken literally, because when Jesus was slapped in the face he did not turn the other cheek; in fact, he objected to it, saying:

> "If I have said anything wrong, tell everyone here what it was. But if I am right in what I have said, why do you hit me?" [626]

The point Jesus is making here, I think, is that we should only use words when objecting: never to retaliate by using violence. This resembles the teachings and actions of Mohandas Gandhi in dealing with the British rulers; unfortunately, he ended up badly by being assassinated, like Jesus.

Personally, I find it very reasonable to use the tit-for-tat policy, described above, and I think that most people do too; but I find it almost impossible to adopt an absolutely non-retaliatory policy. I can perhaps take it if only my person is being threatened (I hope) but if my family or friends are held hostage or hurt, I would probably lose it. I don't think I can measure up to this teaching, nor am I able to love my enemies as Jesus taught us:

> "You have heard that it was said, 'Love your friends, hate your enemies.' But now I tell to you, love your enemies, and pray for those who persecute you, so that you may become the sons of your Father in heaven. For he makes his sun to shine on bad and good people alike, and gives rain to those who do good and to those who do evil. Why should God reward if you love only the people who love you?" [627]

But, one might think, maybe this should not be taken literally, like the previous quote. Apparently, though, it seems that this should be taken literally; because while Jesus was dying on the cross he prayed for and excused his enemies, saying, "Father, forgive them; for they

---

[626] *New Testament and Psalms*; John: 18, 22–3.
[627] *New Testament and Psalms*; Matthew: 5, 43–6.

know not what they do." [628] It may be worth noticing also, that in the above quote, Jesus refers to God, our "Father in heaven", as being a *benevolent* God, over and above what one would expect him to be: he does not punish evil people, but still sends them the rain they need.

Motivational speaker and author Matthew Kelly, in his book *Rediscover Catholicism*, writes:

> In reference to a well-known fact that [Mohandas] Gandhi read the New Testament every day and often quoted the Christian Scriptures, a reporter once asked him why he had never become a Christian. He answered, "If I had ever met one, I would have become one." [629]

Unfortunately, I must admit that I too, personally, do not measure up; and Gandhi is right, most Christians don't either.

As a consequence most of us misinterpret God; so does the bible. I propose, therefore, that the bible is not infallible; it was written by people *trying* to find God: but they did not always get it right: because, being human, they did not always measure up. There are many passages in the Old Testament stating that God is a merciful and forgiving God; but in the greater majority of passages (unlike the New Testament) he is depicted as a vengeful and jealous God. In my opinion, this is why God conceived a human son: to set us straight, to tell us the truth about himself, and also to correct what was written about him in the bible (the Old Testament)—and we still got it wrong!

God conceived a human being in order that he may live among us in a tangible manner and clearly show us how we should treat one another. According to Christian teaching, Jesus is most like his Father (God), the Creator and Designer of the universe. Now, does Jesus strike you as a vindictive person from the above quotes and actions? By inference, then, would God be an evil Creator?

Incidentally, let me show you how wrong scriptures could get regarding violence. In the bible, looting and raping are considered to

---

[628]   *The Holy Bible*; Luke: 23, 34.
[629]   Kelly, *Rediscover Catholicism*, p. 24.

be the normal rewards for the victors of war; even bashing infants and children of defeated nations against rocks is considered desirable. Look at what is written in the famous psalm commonly known as "By the Rivers of Babylon":

> Babylon, you will be destroyed. Happy is the man who pays you back for what you have done to us—who takes your babies and smashes them against a rock. [630]

These are not God's words: they are the words of suffering human beings.

Suicide

It is so easy to go wrong in interpreting sacred scriptures. For example, in "The Gospel According to John", Jesus says:

> This is my commandment, that ye love one another, even as I have loved you. Greater love hath no man than this, that a man lay down his life for his friends. [631]

Should one, therefore, commit suicide if there is fear of being tortured by the enemy in war time, and there is a danger of betraying others? The *Catechism of the Catholic Church* says that suicide is never justified; however, it also points out that, in such cases, the responsibility of the person committing suicide is diminished. [632]

Suicide is probably not the meaning Jesus intended above; he was probably talking about dedicating one's whole life to one's family and friends: bearing all kinds of hardships for them. Moreover, there is a significant difference, of course, between being killed or offering to die for the benefit of others, like Maximilian Kolbe, and committing suicide. If one is not careful, one can go very wrong. One could go

---

[630]   *New Testament and Psalms*, Psalm 137: 8–9.
[631]   *The Holy Bible*; John: 15, 12–3.
[632]   *Catechism of the Catholic Church*: p. 467, para. 2282.

even more wrong than committing suicide in rare cases, as is clearly shown in the next section.

Terrorism

In the bible, "The Book of Judges" commends Samson for committing suicide by bringing down the idol Dagon's temple and killing a multitude of pagan Philistines (Palestinians?). [633] This is equivalent to terrorism; it is similar to driving a plane into a high-rise to kill thousands of believers of other faiths. The final words about Samson were: "the dead which he slew at his death were more than they which he slew in his life." [634] That's great! This is misguided teaching, and such stories should be censored for children. Again, this is not the word of God; it is the word of well-meaning but suffering human beings trying to find God.

Self Defense and War

Would the last quote from John's Gospel: "Greater love hath no man than this, that a man lay down his life for his friends." [635] justify a judicious use of violence? In other words, is violence justified if it is strictly limited to defending others? Is self-defense justified? Is there such a thing as a just war? Does killing people of other nations justify a nation's quest for what is called freedom?

I think the least we Christians can do, in practice, is to limit wars strictly to self-defense; that is, refrain from attacking, or colonizing, other nations. If every nation were to do this, wars will end in no time: or maybe even never happen. But then, some may reason, the best defense in war is attack: prior to the enemy's acquiring the opportunity to attack first. As one can see, it is all a labyrinth of reasons and counter-reasons; no wonder we do not have straight

---

[633]   *The Holy Bible*; Judges: 16, 23–30.
[634]   *The Holy Bible*, Judges: 16, 30.
[635]   *The Holy Bible*, John: 15, 12–3.

answers. It is not an easy task to stand opposed to such views. Currently, most nations probably think that war is unavoidable in certain cases; but what position will they take, in the future, if we end up destroying or sickening the greater majority of humanity through a nuclear world war?

God gave us freedom of action (free will) to sort things out among ourselves and learn from our mistakes, experience, and foresight; but he also sent us his son, Jesus, to clearly teach us to refrain from all violence, at a time when violence ruled the whole of the then known world—the Roman Empire. It was no trivial teaching. I wish I had more space to write about the Roman Empire and how the Jews hated the Romans; suffice it here to say that the Romans actually leveled Jerusalem and its temple in the year 70CE.

I do not think war is ever justified because it inevitably goes against one of the most basic commandments of human morality: "Thou shalt not kill". On the other hand, unless strictly blocking methods are used, self-defense goes directly against Jesus' prime teaching of using absolutely no violence. I think, the declaring of the Catholic Church that self-defense [636] and just war [637] are morally justified, boils down to watering down Jesus' teaching about the absolute non-use of violence. The Church should stick to the unadulterated teaching of Jesus; not to try to conform to the position taken by the State, or the opinion of the majority of its members.

As I previously pointed out, I personally do not measure up to this teaching, unfortunately, and I do not want to mislead the reader into thinking that I do. However, Gandhi and his followers were able to get it right when they used absolutely non-violent means; they succeeded in achieving what they wanted against overwhelming power and potential use of force. Jesus never used violence; not even when his life was threatened; [638] I believe this is the main lesson God sent his son Jesus to teach us. I don't think war is ever justified; but I suppose we shall never learn this lesson until we practically destroy

---

[636]   *Catechism of the Catholic Church*: p. 463, para. 2265.

[637]   *Catechism of the Catholic Church*: p. 472, para. 2309.

[638]   *New American Bible*; Luke: 22, 50–1.

humanity through nuclear war. A nation's arms race is a false sense of security: all it needs is a trigger.

Morality

I would now like to skim over the controversial subject of morality—that is, roughly speaking, what is right and what is wrong—and show that it is very hard to draw a demarcation line. It is not black and white at all; it is mostly shades of grey.

Regarding the general subject of killing someone else, everybody seems to agree, at first blush, that it is an evil act. In fact, even the law has no statute of limitation for murder: the file of an unsolved murder remains always open.

In the bible, however, even something as basic and obvious as "thou shalt not kill" did not actually mean that the Jews could not kill anyone; it only meant that they could not kill their fellow countrymen. They were, supposedly, allowed and even encouraged by God to exterminate entire neighboring nations. I don't believe God told them to do this; they just wanted to believe he did.

One would think that this particular commandment, at least, is something everyone should agree on; however even nowadays people kill each other when they are at war, without even giving it a second thought. Can one kill another person in self-defense? Regarding capital punishment, does the State have the right to kill a person? Regarding abortion, does a woman have a right to kill her unborn baby or a destroy fetus? What about euthanasia (mercy killing) and suicide? Even the Church's teaching, in my opinion, is probably wrong in half of these cases: particularly where it has to challenge the State.

Consequently, atheists propose that morality couldn't have been hardwired in us by God; that it is probably some sort of evolution of concepts and ideas, which self-declared atheist and evolutionary biologist Richard Dawkins calls *memes* in his book *The Selfish*

Gene. [639] I can't say I agree completely with atheists because most of us feel, deep down, that we should treat others the same way we would like to be treated, and I think that this is hardwired by God; all other animal species have no scruples stepping over other members of the same species. However, I must admit that atheists make a strong point here; it could be our reason, a realization, telling us that it is the best way to live harmoniously and peacefully together.

Moreover, in the same book Dawkins points to a couple of examples where things look good and humanitarian, but they may not be good at all. Realizing how easy it is to overpopulate a region, would you say that a welfare state is a good idea? It may be good for the elderly, but it may not be so good for those who can still reproduce offspring. Again, contraception seems to promote an anti-life policy; would you say, as the Catholic Church does, that contraception is bad? [640] I believe prevention is better than cure; unless the means are evil (see the section entitled "End and Means" below).

The Church says contraception is evil because the pill, for example, may have negative medical side effects on the woman: thus reducing her to an instrument of pleasure; but maybe she likes it and desires it too. It should be a medical and a personal decision, as in the case with most other drugs taken that have side effects; besides, I don't think condoms, for example, are detrimental to the female's health anyway; so why prohibit contraception across the board? To me, invoking the *natural law* to condemn contraception is ludicrous. The natural law is, by definition, "the use of reason to analyze both social and personal human nature to deduce binding rules of moral behavior". [641] I don't see reason being used here; it looks more like a clerical device to have it both ways. The Church is supposed to help people live a full life, not to overburden them with unreasonable regulations. I believe there *are* contraceptive methods that do not harm or hurt anyone and, when all is said and done, even if there

---

[639]   Dawkins, *The Selfish Gene*, p. 192

[640]   Dawkins, *The Selfish Gene*, p. 117

[641]   https://en.wikipedia.org/wiki/Natural_law (11 November 2016)

happen to be potential medical side effects, it is still a medical and a personal—not a religious—decision.

Again, would you say sweat shops are a bad thing? Do you think an employer that underpays his employees is evil? I think it is wrong if not evil. So, should God punish or destroy these people? But then half a loaf is better than no bread! If God destroys these people, the workers would end up with no bread at all. I am not trying to justify any behavior here; I'm only pointing out that it is difficult for God to know where to draw the line on evil people: because innocent people would end up suffering more. I'm only trying to explain why he might be leaving them alone, and that it does not follow that God is unjust, or that he does not care, or that he is cruel and enjoys seeing us suffer. What I believe is that he lets us sort things out among ourselves; something like a country's monetary system or the stock market does.

Everyone would probably instinctively agree that lying is at least wrong if not evil. But then, do you really think that telling someone that one is ugly or stupid, assuming one really believes it to be the case, is the Christian thing to do? Or would it be the right thing to say that one did not like the host's food after being cordially invited by her for dinner? Which, do you think would be the lesser of the two evils when you know that telling the truth is going to hurt someone? More seriously, should one betray a comrade in wartime in order not to tell a lie? I never bought into teachings like: using *mental reservation* does not constitute a lie, or withholding very relevant information. Both are still forms of lying to me: the *whole* truth constitutes the truth, nothing else—I like to call a spade a spade. I believe that choosing the lesser of two evils, assuming there is no other way, is the right thing to do.

Would you, for example, say that a single mother's lending her body to prostitution in order to be able to feed her children is evil? Or would you say that Yuri's (Dr. Zhivago) immensely loving both his wife (Tonya) and his mistress (Lara) was evil? Both the above actions might be normally wrong by our current moral standards, but I wouldn't call them evil under the circumstances. In fact, regarding the latter case, even in the bible's Old Testament polygamy (a man having multiple wives: that is, intimately loving and sexually interacting with more than one woman) was the norm.

Slavery is assumed to be wrong nowadays, but then again, even in the bible it was assumed to be the norm; only mistreating slaves was considered wrong. Allow me, now, to discuss a couple of currently controversial subjects, in order to make a salient point at the end.

In my opinion, abortion is evil because it is equivalent to killing a person. Once a zygote is conceived and aborted (whether in the zygote, embryo, or fetus stage) it cannot be re-conceived ever again: imagine aborting Einstein, Gandhi, Newton, or Shakespeare. Don't think there would have been someone else like them; you're wrong, there haven't been any—we just wouldn't know what we've lost. Prior to conception, a single male ejaculate consists of two hundred to five hundred million (2×10^8–5×10^8) sperms, every one of which has the possibility of becoming a human being—we have no control over that. But once the sperm penetrates the ovum, there is only one possibility: it becomes a *helpless* human being. I know that a lot of people think that a fetus does not constitute a person or human life; but I believe otherwise: I think new life starts at conception; because under normal circumstances, it becomes a baby and eventually a human being—unless it is deliberately destroyed in the meantime. Notice that this is not the case while the ovum and sperms are still separated. The phrase "prochoice" is meaningless to me; there should be no choice to do something wrong: one should not be allowed the freedom to do it if it is going to hurt others, period; and harming or killing a helpless being is wrong, in my books. If a helpless baby has the right to be protected, so does a helpless unborn child: fetus, embryo, or zygote. If anything, conception has to be prevented by abstaining from sexual intercourse or by contraceptive means, not by abortive means. Prior to conception there is no person, because under normal conditions an ovum separated from a sperm will not become a baby—that is the demarcation line. We cannot start killing people because of a threat of overpopulation or starvation; abortion is similar reasoning. Whether it creates hardship for a young single mother or its family is immaterial: it doesn't make it right; other alternatives like giving the baby up for adoption should be considered. I admit that cases like rape or unborn baby deformity (that may be revealed by ultrasound equipment) constitute a harder decision for the woman especially. Being a man, I admit, it is much easier for me to say this: in the former case, two wrongs do not make a right; while in the latter case, we cannot just be sailors of fine weather.

Again, I personally don't think homosexuality is evil if both (or multiple) partners agree to it; however, on the other hand, I don't think it is something to be flaunted, as some homosexuals do. It is a form of abnormality: similar to a man with an erectile dis-function or a bald woman. I suppose, nowadays, we give more importance to hormones rather than to body construction; but doesn't the wrong hormones in a given body structure give rise to deficiency, functionality, and performance issues? You know what I mean. Isn't that what one normally calls an abnormality or a defect? Some people insist on calling homosexuals normal; if so it must be to a wider specification of what constitutes a male or a female—one with a larger tolerance margin: just as we might still call certain handicapped or diseased people normal. I'm sorry, but again, I call a spade a spade. One might also mention here, in passing, the more obvious case of a sadist-masochist relationship; which may, *within certain limits*, still not be evil behavior if both parties agree: but it certainly is a sexual aberration. These abnormalities or defects do not make people less human, and such people should certainly not be mistreated or beaten because of them. But to *brag* about such things, as some homosexuals tend to demonstrate and parade in public rallies, makes absolutely no sense to me; I think these are private preferences that should be kept to oneself. Why do such people want to shove their marginal, and even eccentric, idiosyncrasies down the throat of the greater majority of the population? Maybe they need to prove that they are normal? Moreover, why does our society bend over backwards for the rights of the minority, at the same time neglecting the rights of the majority who are constantly being plagued by strange winds from every direction? Should we also, say, have parades for zoosexuals (people who have intercourse with animals) or pedophiles for that matter if, God forbid, it ever becomes legal? I can bet my bottom dollar that if ever pedophilia becomes legal, pedophiles will come out of the closet and go out parading in the streets; just like cannabis (marijuana) proponents are currently already in the process of creating general coalitions, now that it is about to become legal in Canada. Homosexuality was illegal too, at one time, like pedophilia is nowadays. I do agree with pedophilia's being illegal, because this way the law protects children who are practically helpless and incapable of an equal relationship with an adult; but then the law turns around and does not protect the even more helpless unborn child (fetus, embryo, or zygote) about to be aborted. Isn't that inconsistency in our laws? Reverting back to homosexuality, the current wisdom of our generation has culminated in legalizing same-sex marriages, to the chagrin of the children of such marriages: we all know how cruel other children can be towards them. In the past, I did not agree that homosexuals should be allowed to raise children, because children would then be exposed to an abnormality in their young age; that is, until I realized that I'd rather have a homosexual couple raise a child than that child's being aborted. There is a much higher probability of a same-sex couple's adopting an unwanted child than most heterosexual couples would, since the former cannot have children of their own: not belonging to both partners, anyway. So again, it is hard to define what is right and what is wrong. On the other side of the coin, the bible in the Old Testament condemns homosexuality outright, [642] and demands the death of such individuals: [643] utterly oblivious of the fact that killing someone might be much more of a heinous crime than a homosexual act! Not surprisingly, the Old Testament also demands stoning to death for anyone who commits adultery; [644] however, in the New Testament, Jesus showed tolerance and refrained from condemning

---

[642] *New American Bible*; Leviticus: 18, 22.

[643] *New American Bible*; Leviticus: 20, 13.

[644] *New American Bible*; Leviticus: 20, 10 & John: 8, 5.

an adulteress. [645] Did you ever think why Jesus changed things? Could it be because the bible is not *really* God's word?

I'm sure half of North America and Europe, on either side of my opinions, would like to pop out of this paper, right now, and give me a mouthful of what they think, or even grab me by the neck for writing the last two paragraphs. I apologize if I upset anybody; I wanted to discuss something controversial, and I confess it was my intention to get some people worked up for a little while, because I wanted to make a more important point than our disagreements. My conclusion on morality in general is that God hardwired in us reason and the golden (most basic) rule of morality: namely, "Do for others what you would like them to do for you."; [646] from there on, he expects us to behave as intelligent mature beings; but *most importantly* he expects us to be tolerant, and show love and understanding towards others who disagree with us. This is probably much more important: similar to tolerance to diverse religious and scientific beliefs. I have given my opinion above because it is *my* book, but if you don't agree with me let's just agree to disagree! When I was much younger, I used to believe that there is such a thing as a demarcation line between right and wrong: like black versus white; now that I got older, I'm not so sure anymore: it all seems like different shades of grey to me.

For example, some people may believe in the infallibility of the bible and therefore condemn homosexuality; [647] however, only a minority of humanity believes in the bible; so how are those people, who do not recognize the bible as being a holy book, going to conclude that homosexuality is so wrong? Therefore, how can the believer in the bible categorically declare that homosexuality is evil to others who never heard of such a thing? Moreover, according to the Catholic Church's teaching, if one does not know one is doing something wrong, one is not guilty of sin: [648] basically no harm is done; so what, really, makes an action definitely right or definitely wrong?

---

[645]   *New American Bible*; John: 8, 11.

[646]   *New Testament and Psalms*; Matthew: 7, 12 & Luke: 6, 31.

[647]   *New American Bible*; Leviticus: 18, 22.

[648]   *Catechism of the Catholic Church*: p. 391, paras. 1859–60

Allow me to give one more obvious example of what I mean. I'm no expert in Hinduism, but unless I am mistaken, Hindus believe that to kill a cow is like murder, maybe even worse. They believe that "there is no action more sinful in all of creation than cow killing." And they go on to say: "Cow killers and cow eaters are condemned to rot in hell for as many thousands of years as there are for each hair of the body of every cow they eat from." [649] When I read this I couldn't help shaking my head and rolling my eyes; yet there is over a billion people who believe this. Wow! Now, I ate beef in the past; I just hope it wasn't a cow: I don't want to end up in hell for thousands of years.

So, is the reader still convinced that there is a demarcation line between what is right and what is wrong?

End and Means

A good intention does not justify any means to achieve it. One of the most difficult things in morality is to use the right means to achieve a supposedly good goal.

For example, if one believes strongly in Christian principles, and that the whole world should become Christian, one cannot go about killing all the people of other religions so that all future offspring would be Christian: a good end does not justify evil means. Nor should one force someone else to convert to any religion under duress: with a gun or knife pointed at the latter   terrorism included. Nor can we start killing people because of a threat of overpopulation or starvation—abortion included. Such methods of an intentionally good action are *misled*.

The bottom line is that we have to use appropriate methods when aiming at something good: means that are not considered evil in isolation. Furthermore, we must respect one's freedom of action when we try to reach a supposedly good end. If it is so important to God that we tolerate and respect one another, why on earth do we

---

[649]   Varaha, "The Srimad Bhagavad-Gita and the Sacredness of All Cows", http://www.bhagavad-gita.org/Articles/holy-cow.html (29 April 2016)

affirm God's threatening us with eternal fire if we refuse to be in his company? Something must be wrong with this concept of God.

To recapitulate: God made life difficult because of physical *constraints* thus avoiding greater hardship; good and evil are *undefinable* except treating others as we would like to be treated; *tolerance* and *understanding* are the salient requirements.

Second Set of Questions

Why is there so much pain and suffering in life? Why are we here, anyway? What is the meaning of life? Why doesn't God tell us what it's all about?

Suffering

In this section I shall mainly deal with suffering, that is, psychological (mental) suffering, as opposed to physical pain; although it is not inconceivable that one may suffer mentally as a result of physical pain.

Why is it that we suffer psychological pain besides bodily pain? Isn't bodily pain enough to ensure our survival: to prevent us from getting seriously hurt? I can see fear of predators and of pain, and also physical (sexual) attraction being developed by evolutionary processes; perhaps also some level of love for one's helpless offspring, resulting in their protection. But why would evolution think we also need a complex suite of psychological feelings for our survival? What about empathy and altruism? We feel bad for someone who is obviously suffering pain or crying—someone we do not even know; we also often share hardship or help someone in need. Do they have any place in a struggle for survival? I can mention other psychological needs like companionship and heterosexual (or conjugal) love, without which the majority of us humans would feel that life is not worth living. Evolutionary biologists try to explain these contra-evolutionary emotions (strange by evolutionary standards) through convoluted arguments, which I generally find unconvincing. Unfortunately, I do not have the space to challenge these arguments individually here; let me just say that I am inclined to believe that these contra-evolutionary psychological feelings, even if they exist in some animals, are hardwired by God.

Do animals have feelings? It seems they do; some, at least, do. For example, dogs show it by wagging their tails, growling, whining, and tucking their tail between their legs. This is evidence of animal qualia; they are experienced only by the animal itself, and we will probably never know how much feeling they have: whether it is inferior or superior to ours. Dogs have been called "man's best friend" because of their loyalty to their masters; can they possibly exhibit such behavior without having any feelings for their master? Do dogs love their masters more than humans love their family members? Why is a dog ready to attack a human being or another animal, much larger than itself, in order to protect its master? That's not conducive to its survival! On the other hand, it may well be that animals do not fully appreciate what it means to die; they just live. Is this evidence of hardwiring or of an evolutionary process?

Most people will agree with me that life, in general, is very difficult: war and disasters, sickness and disease, bearing children, taking care of babies and a household from morning till night, going to work every day, unrecognition for effort and worth at work, holding a job in spite of an obnoxious boss, losing one's job and looking for another, setbacks in one's investments for the future, financial problems, unloving parents, ungrateful children, children bullied at school, difficult neighbors and co-workers, children or loved ones getting hurt, losing loved ones sometimes unexpectedly, people taking advantage of our trusting nature or naïveté, an unfaithful or unloving spouse, taking care of parents possibly coupled with financial difficulty, arthritis pain, dementia and inability to take care of oneself when one gets old, financial difficulties in sickness or choosing a retirement home, being abandoned at the end of one's life, awareness of the inescapability of death, and finally death itself.

I once heard it said that if a hundred people were to meet in a large room and dump all their problems in the middle of the room, intending to share all their problems equally among themselves, most people would again pick up their own original problems and walk back home with them. In other words, they would decide not to partake in the proposal after all: they would realize that most of the other people have worse problems than themselves. Once, when I was still a young lad, my mother said something to me that impressed me so much I still remember it to this very day; she said, "Life is a lick of sweetness and a barrel of bitterness." I'm sure most of my readers feel they have had their share of problems, too. So, the obvious question is: why all this pain and suffering?

Incidentally, in my opinion, the bible gives no satisfactory answer to the problem of pain and suffering in life; the best it could do in "The Book of Job" is, basically, God telling Job, "Shut up; I know better." Not so crudely, of course. Unfortunately, this answer did not impress me much; especially when one considers the fact that all the misfortunes Job suffered were the result of a challenge, a sort of bet, the devil made with God: namely, that the impeccable Job would denounce God if given enough hardship and mishaps. Job's story has a somewhat happy ending: in his old age, the faithful Job gets his health back and double the possessions he had lost (I doubt its possibility of happening in real life) but, of course, he never got his loved ones (children) back. A nice story perhaps, but it leaves a lot to be desired as an answer for all the pain and suffering we experience in life; it didn't do much for me, anyway: I must confess that I was quite disappointed.

In his treatise "The Problem of Pain", lay theologian C.S. Lewis tells us that he believes that pain and suffering are God's way of speaking to us loud and clear: through a loudspeaker or megaphone.

> God whispers to us in our pleasures, speaks in our conscience, but shouts in our pain: it is his megaphone to rouse a deaf world. [650]

---

[650] Lewis, "The Problem of Pain", p. 604

So why did God opt for so much suffering in our lives?

We have already encountered a practical physical constraint: the constant threat of overpopulation and subsequent starvation; consequently, God cannot make life too easy for us. However, one of the best answers I have for the problem of so much pain and suffering in our lives (which I shall elaborate even further later) is that, *in the long run*, it makes us better people: more compassionate, tolerant, and understanding of others: it tends to make us the best we can ever be. I think most people will agree with me that, if one has an easy life with lots of money at one's disposal, one becomes ruthless and insensitive of others. It is suffering that really connects individuals together: when they suffer together. There's nothing wrong with people having fun or pleasure together; however, there may be an element of selfishness in every person's sharing them. I always say: "It is funny how one can suffer for someone else; but no one can have fun for someone else." Doesn't that suck? Suffering for another is strictly altruistic: when one does not run away from it and decides to share someone else's burden. It shows that we care about someone if we are willing to suffer for or with that person. It has the tendency of drawing us closer together and thus knowing one another more intimately.

I do not wish the reader to misunderstand me here. God does not *enjoy* seeing us suffer, he empathizes; but he likes the end result it produces in us: it makes us better people. Love of God, here on earth, can only be shown through love of our neighbor—whoever he is. Behaving altruistically invariably draws us closer to God, enabling us to have a closer relationship with him. The concept of having a better relationship with God makes no sense unless there is another existence after our death where we can meet him: we cannot communicate directly with God here on earth. Is there enough evidence of an afterlife? I shall try to answer this question later in the corresponding section.

Hidden Reasons

The next obvious question is: If the reason for pain and suffering in life is to foster a personal relationship with God, why didn't God tell us his reasons?

I'll try to explain things by means of a parable or two. If a king wants to experience a real personal relationship with his subjects, he would wear like them and go among them; obviously, he wouldn't tell them that he was their king. It's no use telling such a king that you would have treated him, or his friends and family, better had he told you he was your king: that's the whole idea of a king mingling among his people incognito. God is present among us, all right; he has all his friends around us: our neighbor. The way we treat others is the way God will feel treated: God himself does not really need anything. God did not reveal everything to us. Why? There may be many reasons why a rich lover does not reveal how rich one is to one's significant other. These are some reasons why God may have opted to remain hidden from us: to find out what we are really made of.

On the other hand, I believe that he left enough evidence scattered around for us to be able to figure things out: if only we decided to take the time for it. However, I also believe that everyone will probably come to somewhat different conclusions about God, depending mainly on one's background and area of expertise; but that's exactly what a *personal* relationship is all about. What we learn about God is not really important to him—he is too complex anyway; what is most important to him, while we are here on earth, is that we care enough about him: enough so as to look for him and try to answer the questions everyone has about our existence, thereby building any kind of relationship with him.

Pain and suffering give us an opportunity to show love towards others—all of whom God loves: God loves people, his creation, like a mother loves her babies, an engineer loves one's own original designs, or an author loves one's own books. There is so much pain and suffering around everyone, in everyone's life, that there is no way one can have an excuse that one had no opportunity to help others: to show them love or to withhold it from them. All this pain and suffering on earth, is probably to provide ample opportunity for everyone to exercise kindness and tolerance towards others. Now, I just hope I can measure up, and be as good as my words: that is, taking all the pain and suffering that comes my way with a smile.

Another common observance among good people is that obnoxious people always seem to have an easy life and prosper more: never having money problems, things always coming their way, always getting promotions and salary raises, and always having a good time—very often at the expense of others. Why, God? Is God setting them up, leaving them no excuse for not helping others, so that he can punish them severely? This is what I used to think until a few years ago, and you may also find this concept written in several holy books; but I don't believe it any longer. As I said before, holy books were written by people trying to find God: they didn't always get it right. I lately realized that God loves everyone unconditionally; he's not setting them up, he's just leaving them alone: I guess he figures that's the best way of dealing with them. But that's not fair! That may be true, but look at the following examples.

Parents will probably understand what I'm going to say here better. Suppose you have two children, one of whom is reasonable while the other is not; you might ask the reasonable child to help you with the housework or maintenance, but not the other. Yet, if you think about it, you are making the reasonable child suffer and you are letting the unreasonable child go scot free, because it's too much of a hassle; however, at the same time, your relationship with the reasonable child is actually a deeper one. On the other hand, it doesn't mean that you don't love the mediocre, the unreasonable, or even the bad child. It is probably the only way God can keep such people as close to him as he can: otherwise they would sulk and drift away from him even further. But that's not fair! True! However, in the end, the one who suffers most will most probably have a closer relationship with God and will enjoy greater happiness in the afterlife.

Similarly, one might help a sibling who is always in debt, but never the other sibling who always balances one's budget. Is it fair?

No! But if one loves them both, and one is not very rich, the thrifty sibling does not need bailing out while the spendthrift sibling does. If the thrifty sibling needs bailing out in the future, one will seriously consider it, unless one cannot afford it any more at the time.

The bottom line is that God loves everyone unconditionally (as we love all our children unconditionally), and he does his utmost to keep them as close to himself as possible—in whatever way feasible. One may here refer to the parable of the "Prodigal Son", especially its ending when the faithful son complains about the father's extravagant treatment of the spendthrift son: [651] you will see what I mean.

We shall all, eventually, enjoy different degrees of intimacy and happiness in the presence of God; but every soul will be completely happy. Just as different sizes of glasses can be full to the brim (they cannot contain any more fluid—therefore they are all completely happy), and yet they all contain different amounts of fluid (obviously some have more than others). This is my concept of God's *justice*.

Sometimes God works in mysterious ways, too. Take, for example, the case of Saint Paul who loved God and Jesus indescribably. Yet, he was always in and out of jail as a result of his preaching about Jesus to the non-Jews (gentiles). Why? Well, in jail he had no choice but to write letters to the churches he had founded along the Mediterranean Sea: it was the only way he could communicate with them. These letters ended up to be the *only* authentic source of early Christianity remaining for posterity—later generations of Christians.

Supposedly, according to the Church's teaching, God never gives us more trouble than we can handle; but I don't know if this is just wishful thinking. When we consider the fact that so many people commit suicide, it doesn't seem that everyone is tested below one's breaking point. I just think it's more plausible that God will even everything out in the afterlife.

---

[651] *New American Bible*; Luke: 15, 11–32.

Meaning of Life

The usual complaint one hears from most people goes something like this: "Life is so short; most of it consists of hardship, pain, and suffering with only a little fun in between. Is this all we are here for? Why are we here anyway? What meaning is there to life? Why did God bother to create us? Why did he not forget the whole thing?" One of my friends, while pointing to his surroundings, lately told me: "What I cannot understand is what all this is about; if you could tell me anything that makes sense, I would greatly appreciate it." I don't pretend to know all the answers, but I shall outline my opinions as to the meaning of life: why we are here on earth.

Most modern (mainstream) scientists and philosophers are atheists; so obviously they do not believe in an intelligent Creator of the universe or of life. This normally leads to the inevitable philosophical conclusion that the universe is an accidental occurrence; therefore, there could be no past or future plan, or even logic, to the universe. As is clear from the previous chapters of this book, I do not agree with the majority of modern scientists and philosophers; I agree with a minority of scientists who believe the universe was designed by an intelligent being, not that it happened by chance.

We are certainly not at the center of the universe, and we may very well be situated in the backwaters of our universe; but we may still be the *reason* why this universe came into existence. I know, this may sound like wishful thinking and human pride; but let me explain why I think so.

When I was still rather young, right after graduating from university, I used to think that there must be many other life forms in the universe: the universe being so vast—the probability had to be high. However, over the years I found out that life is so improbable, that Earth seems to be the only planet containing life in the whole universe; moreover, we seem to be the only really intelligent species on earth, and therefore, by inference, in the whole universe. If this is actually the case, it would not be pretentious or preposterous to conclude that the universe was made for us. Maybe God really did have us, human beings, as an ultimate species in his mind with whom he could have and even wanted to have a real one-on-one relationship with—as equals. Am I the only one that thinks so? Well, James Trefil, one of the authors of the book *Are We Alone?* which investigates the possibility of extraterrestrial civilizations, concludes the book by saying:

> [T]he evidence we have at present clearly favors the conclusion that we are alone ... we are living on an insignificant speck of rock going around an undistinguished star in a low-rent section of the galaxy. We are not the center of the universe. Maybe so, but we are special ... If I were a religious man, I would say that everything we have learned about life in the past twenty years shows that we are unique, and therefore special in God's sight ... [652]

Bear with me for a minute longer, please. We have evidence of the existence of angels in the apparitions at Garabandal, seen in the previous chapter on "Miracles". Moreover, most Religions believe in the existence of angels. Angels are, supposedly, intelligent non-physical beings (spirits) who act as intermediaries between God and man: they are the messengers of God—from the Greek word "angellos" meaning "messenger". They are not human beings that went to heaven and grew their wings, as a lot of people believe; they are a completely different species—if you could call them that. Actually they are an utterly different creation from human beings: they have no body; they are intangible but intelligent spirits, very much like God himself.

Assuming the existence of angels, God seems to have created the angels in such a way as to know him for who he really is. Well, we are not angels, we are a completely different creation, so God planned something different for us; it seems he desired a relationship among equals from us, not an inferior-superior (or overwhelming type of) relationship—the type, I surmise, he seems to have with

---

[652] Rood & Trefil, *Are We Alone?*, pp. 251–2

his angels. By our being born on earth, we are not overwhelmed by God's greatness and our insignificance in comparison to him, as perhaps the angels might be; we really have to work hard to realize his greatness around us: it is so well hidden. The situation is somewhat similar to a grandparent desiring a personal relationship with a grandchild just reaching the age of reason. Is the grandparent going to learn something new from one's grandchild? Is such a relationship really important or necessary to the grandparent? Probably not; but a personal relationship with someone one loves is both desirable and satisfying.

It seems to me that the *main* reason for us being created, the real meaning of our lives, is that God desired us to *partake* in his role of *creation*: by our reproducing other human beings. He does not seem to have given the same opportunity to the angels, even though they are supposed to be overwhelmingly superior beings. If this is truly the case, I think it is truly an honor God is trying to bestow on us insignificant creatures; it may also explain why he opted for reproduction and life on earth, rather than creating us in a flash and placing us on earth or in heaven. I guess his plan was to place us here on earth first, allow us to educate ourselves, and *then* take us with him in heaven.

Personally, I also believe that what is most important in this life is to honestly look for God: to care about the spiritual aspect of life, not necessarily to find the truth. I wouldn't be surprised at all that a kind and charitable atheist, who is honestly convinced that God does not exist after researching the matter, will have a closer relationship with God than the churchgoer who despises others and thinks oneself superior to other people. Pastor Dwight Moody (1837-1899) once wrote:

> It has been said that there will be three things which will surprise us when we get to heaven—one, to find many there that we did not expect to find there; another, to find some not there whom we had expected; a third, and

363

perhaps the greatest wonder, will be to find ourselves there. [653]

In fact, in his article "Who's Afraid of Life after Death?", philosopher Neal Grossman summarizes the conclusions people with near death experiences arrived at regarding the meaning of life as follows:

> Research on the NDE has yielded the following unambiguous conclusion. NDErs confirm the basic values of the world's religions. The purpose of life, most NDErs agree, is divine knowledge and love. Studies on the transformative effect of the NDE show that the cultural values of wealth, status, and material possessions become much less important, and the perennial religious values of love, caring for others, and acquiring knowledge about the divine ascend to greater importance. [654]

The person who doesn't care at all about spiritual things is simply not investing in the right currency; at the same time, going to church is not as important as doing no harm and being good to others. So knowledge of God and love of neighbor are probably the things that give any real meaning to our lives. In "The Gospel According to Matthew", Jesus answers the law scholar's question as to which is the greatest commandment as follows:

> "Thou shalt love the Lord, thy God, with all thy heart, and with all thy soul, and with all thy mind. This is the greatest and the first commandment. And a second like *unto it* is this, Thou shalt love thy neighbor as thyself. On these two commandments hangeth the whole law, and the prophets." [655]

You cannot take possessions or fame with you after death. No matter how important and indispensable you presently are at your work, one

---

[653]   http://www.raptureready.com/resource/moody/m7.html (20 October 2015)

[654]   Grossman, "Who's Afraid of Life after Death?", p. 14

[655]   *The Holy Bible*; Matthew: 22, 37–40 (emphasis in original).

day you will have to retire. You will be unable to enjoy much of the food, or sex, once your body reaches a certain age. The beauty of the body fades; the body deteriorates and eventually dies. So, I think it is a good idea that, as one gets older, one gradually substitutes grace for beauty. The only things that endure are love of God and love of neighbor.

Eternal life may very well be every human being's wishful thinking; therefore, I admit, I could be wrong in what I am about to say. It is natural (normal) for the soul, the principle of life of us humans, to end with our body: that is, not to continue to exist after our body dies. However, the empirical (data based) evidence from many NDEs supports the view that the soul can exist apart from our body; therefore it is conceivable that it continues to exist after the body is dead. Jesus' resurrection seems to confirm that there is a different kind of life after death (refer to the section on the "Resurrection of the Body" in the previous chapter). Of course, God could choose to annihilate those souls that did not measure up; in other words, let them end their existence the natural (normal) manner. Unfortunately, it seems that we cannot be *sure* of anything in the afterlife until we get there.

Recall that, in his article "Who's Afraid of Life after Death?", philosopher Neal Grossman gives the following summary from NDEs in general:

> The evidence from the NDE ... suggests that God is not vengeful, does not judge us or condemn us and is not angry at us for our "sins"; there is judgement, to be sure, but the reports appear to be in agreement that all judgement comes from within the individual, and not from the Being of Light. It seems, in fact, that all God is capable of giving us is unconditional love. [656]

If there is actually eternal life after death, and if we are going to be in the company of a benevolent God who only has unconditional

---

[656] Grossman, "Who's Afraid of Life after Death?", p. 14

love for us, then there is nothing to fear in death. Furthermore, if bettering ourselves the best way we can be here on earth, will enable us to end up as close to God as our education in this life has taught us, I wouldn't mind my bettering myself here on earth and enjoying a better "time" with God for all of eternity. If this is really the case, then even a hundred years of the most difficult life is no comparison to eternity; even a life of suffering or disability, then, will make sense: only if this is the case, of course. If existence after death for humans, at least, is a reality, then life, pain, and suffering may make sense; if not …. In support of this we have the evidence of several NDEs if we decide to believe them.

The above NDE evidence also suggests that the only thing that may be capable of separating us from God seems to be our own judgement of ourselves. However, I wouldn't like to leave the reader with the wrong final impression regarding our personal judgement described above. Several NDErs reported a panoramic review of their whole life in which they also perceived how their actions affected others. [657] So I'd like to add that cardiologist and NDE researcher Pim van Lommel summarizes similar experiences as follows:

> All that has been done and thought seems to be significant and stored. Insight is obtained about whether love was given or on the contrary withheld. Because one is connected with the memories, emotions and consciousness of another person, you experience the consequences of your own thoughts, words and actions to that other person at the very moment in the past that they occurred. [658]

It seems, therefore, that the perpetrator experiences the same pain and anxiety one inflicted on one's victim; consequently, it appears to be a great disadvantage to one's own self to mistreat others. Possibly, in the afterlife, one remains in this state of pain and anxiety, experiencing it over and over again, until one really regrets what one has done to

---

[657] Beauregard & O'Leary, *The Spiritual Brain*, pp. 157–8
[658] Van Lommel, "About the Continuity of Our Consciousness", p. 121.

the victim. Credit is, of course, given to one's sincere repenting of one's deeds in this life—before dying.

This sounds like real justice to me, unlike the usual Church's teaching that Jesus "paid" for our sins. Jesus' contemporary Christians simply could not comprehend why Jesus, a person capable of performing startling deeds (miracles), allowed himself to become victim of Church and State; so they tried to explain it somehow. This is also probably what Judas banked on when he betrayed Jesus; but that's another story.

The effects of this life review are also summarized by philosopher Neal Grossman in his article "Who's Afraid of Life after Death?" He writes that the current NDE evidence tends to imply that:

> [W]e are reasonably certain that everyone will experience a life review, in which individuals experience not only every event and every emotion of their lives, but also, the effects their behavior, positive or negative, has had on others. The usual defense mechanisms with which we hide from ourselves our sometimes cruel and less than compassionate behavior towards others seem not to operate during the life review. ...
> A consequence of the life review is that it appears to be a great disadvantage to oneself to harm another person, either physically or psychologically, since whatever pain one inflicts on another is experienced as one's own in the life-review. [659]

Whatever pain and suffering we ever inflicted on others we shall experience ourselves; it sounds like perfect justice to me!

To recapitulate, my conclusion is that we are here on earth: first, to partake in God's plan of *creation* by reproducing other human beings; second, to *improve* ourselves and become the best beings we can be, by *loving* our neighbor, which is achieved mainly through *pain* and *suffering*; third, to *educate* ourselves in our knowledge of God; and finally, to form an uninhibited, free-willed, and personal

---

[659] Grossman, "Who's Afraid of Life after Death?", pp. 21–2

relationship with God. To me this is the real meaning of life, and the reason why we are here on earth rather than in heaven.

Third Set of Questions

Given that our lives, full of pain and suffering, make sense only if there is an afterlife, is there any evidence for an afterlife? Does the afterlife ensue at physical locations? Do we know of any physical locations that can possibly qualify as such? What does science say about the future of the earth and the universe?

Afterlife

Philosopher Neal Grossman starts the abstract of his article "Who's Afraid of Life after Death?" with the following statement:

> The evidence for an afterlife is sufficiently strong and compelling that an unbiased person ought to conclude that materialism [inexistence of mind or soul] is a false theory. Yet the academy [scientific community] refuses to examine the evidence, and clings to materialism as if it were *a priori* true, instead of *a posteriori* false. [660]

Modern scientists simply refuse to even look at any evidence in support of the possibility of our continuing existence after our death. Why? Because they think it is unscientific to even consider it. I hold that it is a question about reality and consequently a scientific question; it should therefore be treated with the same scientific rigor and principles. I agree that since our body is made of matter, it is the natural (default) thing to assume that its principle of life (its integrating function—the soul) ends at its death and so cease to exist thereafter. So there should be no afterlife for us; unless, of course, there is compelling evidence to the contrary. Do we have enough

---

[660] Grossman, "Who's Afraid of Life after Death?" p. 5 (emphasis in original).

evidence to conclude that the afterlife is a definite probability? Scientists do not even try to investigate the matter. This is what Grossman is concerned about in the above quote.

As we have seen in the chapter on "Miracles" there is some evidence that Jesus rose from the dead into another form of life. His contemporary followers strongly believed it; most importantly, even someone who previously hated him (a skeptic—Paul) came to believe it. They believed that Jesus rose from the dead and went to God the Father, and that God honored him and made him "sit at his right hand". They believed it so strongly that it resulted in the Christian movement and religion. This is all historic-scientific evidence; we cannot disregard it simply because it happened a long time ago.

It also implies that if one man really rose from the dead and is now in God's company, then life after death, the *afterlife*, must exist at least: it must be a reality. But then, does it necessarily follow that some other humans (at least a few) can rise from the dead? Could Jesus' followers, for example, also rise from the dead? The early Christians believed that their accepting Jesus as the son of God, being baptized, and living his moral teachings gave them the power to rise from the dead (as Jesus did) and to enjoy "eternal life" and "immortality" [661] with him in God's presence: it gave them a ticket to heaven, so to speak. Deplorably, early Christians did not care much about the rest of humanity; to them the latter were lost and they were going to "perish": [662] that is, God was going to punish them for their evil actions by an untimely death. Christians believed that there was "no resurrection to life" [663] for evil-doers and non-Christians: they were not going to rise at all after their death. Will Christians be the only ones to rise from the dead? I doubt that very much; in fact, modern Christianity believes that everyone will rise from the dead for their reward or punishment; that's quite a change from their original belief about the afterlife.

---

661   *New American Bible*; Romans: 2, 7.

662   *New American Bible*; First Corinthians: 1, 18.

663   *New American Bible*; Second Maccabees: 7, 14.

In any case, is Jesus' resurrection solid scientific evidence or just religious belief? I must admit it is rather shaky as scientific evidence: it happened long ago when documentation was very poor, the documents we have originated from sympathizers, and we do not have any outside corroborating documentation. We only have records of one contemporaneous skeptic that came to believe strongly that Jesus resurrected, and over five hundred of Jesus' followers who also witnessed it.

However, we also have modern compelling empirical (data-based) evidence from a great number of near-death-experience (NDE) accounts. Recall that several NDErs reported a panoramic review of their whole life in which they also perceived how their actions affected others. [664] Cardiologist Pim van Lommel summarizes them as follows:

> All that has been done and thought seems to be significant and stored. Insight is obtained about whether love was given or on the contrary withheld. Because one is connected with the memories, emotions and consciousness of another person, you experience the consequences of your own thoughts, words and actions to that other person at the very moment in the past that they occurred. [665]

Recall also several NDEs indicating a form of self-judgement, which philosopher Neal Grossman summarizes as follows:

> [T]here is judgement, to be sure, but the reports appear to be in agreement that all judgement comes from within the individual, and not from the Being of Light. It seems, in fact, that all God is capable of giving us is unconditional love. [666]

---

[664] Beauregard & O'Leary, *The Spiritual Brain*, pp. 157–8.

[665] Van Lommel, "About the Continuity of Our Consciousness", p. 121.

[666] Grossman, "Who's Afraid of Life after Death?", p. 14

So NDEs seem to show evidence of a judgement of some sort at the end of one's life; a judgement does not mean much if there isn't a time to experience the consequences of one's actions, at least for a while.

But is there any more direct evidence of another form of life after death—an afterlife? Yes, many NDErs reported seeing dead relatives leading them through a tunnel to a light at the end of it, and these dead relatives also led the NDErs back to their ordinary life again.

Other evidence we have of the existence of an afterlife comes from miracles. Our Lady's and Jesus' appearances in miracles, if we believe them, are direct evidence of their having existed for about two thousand years after their life here on earth.

As we shall see in more detail later, Our Lady apparently showed heaven, hell, and purgatory, filled with souls, to various seers during her apparitions to them, which the latter relayed on to us; from these revelations, the afterlife also seems to be unending: eternal.

Assuming there is an afterlife, there are several options God has for dealing with the soul of an individual after one dies. (1) He could annihilate the human soul (self) so that it ceases to exist; this is probably what happens to the souls of animals after they die. Maybe, this is how most people who failed to establish any sort of personal relationship with God will end up, too. (2) Alternately, God may opt to re-incarnate a soul and give it a second chance to live again; there is some evidence along these lines: that is, people having been re-incarnated. (3) The soul may end up with God living eternally with him. (4) The soul may opt to be permanently separated from him: I wonder if God gives a choice to an evil soul to be annihilated. (5) However, except in the case of annihilation, I don't believe any decision God, or possibly the soul, makes at one's death would be eternally irreversible.

Heaven, Hell, and Purgatory

I do not quite believe that the segregated heaven, hell, and purgatory, as described by the Christian Churches, exist. What is probably the case is that everyone will have a different personal relationship with God in the afterlife: some will be very close to him, others will be very far, and many will be in between. Humans are capable of having between one hundred and two hundred personal relationships; God, being infinite, is presumably capable of having billions of personal relationships.

My wife lately asked me what makes me think that I am right and all the Christian Churches are wrong on this subject. I answered her that it was a very good question and that the reason why I think so is because of inconsistencies in the Churches' own doctrines: things simply do not add up. My powers of reason, which I believe God hardwired into my existence (that is, my soul), tells me that they are wrong.

Regarding the subject of hell, for example, journalist Wayne Weible gives the following dialogue in his book *Medjugorje* between one of the seers, Mirjana, and Our Lady during one of her apparitions in Medjugorje (see the previous chapter):

> "I [Mirjana] asked her [Our Lady] how God can be so unmerciful as to throw people into Hell to suffer forever? … She told me that souls who go to Hell have ceased to think favorably of God—have cursed Him, more and more. So they have already become a part of Hell, and choose not to be delivered from it." [667]

Pay special attention to the last phrase "choose not to be delivered from it" in the quote: it seems that the damned *can* be delivered from hell if only they choose to. It looks like Our Lady purposely dropped us a subtle hint; we shall see later why Our Lady does not have much

---

[667] Weible, *Medjugorje*, p. 56.

choice, in practice, other than to conform to the traditional teachings of the Church. Mirjana continued:

> "Then she [Our Lady] told me that there are levels in Purgatory: levels close to Hell and higher and higher toward Heaven." [668]

Notice that this supports my hypothesis about heaven, hell, and purgatory not being segregated at all: there is a whole spectrum of relationship levels. They are relationship *states* everyone has with God: they are not physical locations. For simplicity, in the above sections, I was referring to heaven as if it were a place; it is not a place at all: it is a *mental* state.

Our concepts of heaven as being above, in the skies, and hell as being below, under the earth, are mythological concepts. For example, *Hades* (often translated into English as "Hell") was a Greek mythological concept of the afterlife: it was supposedly an underground place where the disembodied souls of the dead went after this life and simply existed there as zombies. Technically, above the earth has no real meaning, anyway, unless one believes the earth is flat. It depends on whether one is in the northern hemisphere or in the southern hemisphere of the earth—unless, of course, heaven is in the form of a large spherical shell surrounding the whole earth. Incidentally, which way is purgatory? In any case, I shall assume for the time being and for simplicity's sake that heaven, hell, and purgatory are outside our space. I shall deal with their possible "location" later in this chapter.

Mirjana continued to dialogue with Our Lady:

> "[I]f a person goes to Hell ... Don't people pray for their own salvation? Could God be so unmerciful as not to hear their prayers? Then the Madonna explained to me. People in Hell do not pray at all; instead they blame God for everything. In effect they become one with that Hell

---

[668] Weible, *Medjugorje*, p. 56.

> and they get used to it. They rage against God, and they
> suffer, but they always refuse to pray to God." [669]

It is interesting to note here that even the seer above, Mirjana, has the same gut feeling as I have, and probably also you, the reader: that something is wrong with the concept of God's punishing souls in hell for eternity.

Personally, I could never reconcile the concept of a benevolent and forgiving God with his condemning anyone to a hell of fire for all eternity and torturing one without reprieve, for whatever imaginable evil one could have done in a lifetime of a hundred odd years: the punishment does not fit the crime! Now God is infinitely just, much more just than I could ever dream to be; so how could he inflict a punishment that does not fit the crime? Moreover, would he not be a cruel and unforgiving God, rather than a "father" as Jesus described him? Obviously, it is impossible that I could be more compassionate than God himself: something does not add up, here. Not only that, but most people, if not everyone, will agree with me: they have the same gut feeling regarding this matter. Every single time I discussed this subject in some detail with anyone, that person agreed with me: in other words, our reason tells us so—we have all been hardwired this way by God himself.

Furthermore, God seems to have a soft spot for sinners. Jesus, whom Christians believe to be the son of God and most like him in character, spent most of his time with sinners (to the bewilderment of the religious people of his time) trying to bring them back to an amicable relationship with God—and he explains why: the ones that are forgiven most are the ones that love most. God is a fool for love: he does everything to gain someone's love. "God is love" [670] impersonated; he hardwired it into our souls, too.

I believe that the concept of a Christian hell is the wishful thinking of religious vengeful people who want to see their enemies punished unendingly and by the most painful means they can imagine, namely,

---

[669] Weible, *Medjugorje*, p. 57.
[670] *The Holy Bible*; 1 John: 4, 8.

fire: how dare you mistreat God's favorites; he's going to see to it that you pay for it dearly. As the saying goes, "Nice guys finish last"; being charitable to others, as Christians are supposed to be, has the tendency of leaving one open to being taken advantage of and ending up as the underdog; just as Jesus ended up the victim of Church and State and was crucified. But Jesus did not ask for revenge, as most Christians wish; on the cross, he asked God to forgive his tormentors. [671]

In its paragraph on Hell, the *Catechism of the Catholic Church* declares:

> To die in mortal sin without repenting and accepting God's merciful love means remaining separated from him for ever by our own free choice. This state of definitive self-exclusion from communion with God and the blessed is called "hell". [672]

I think the Church scored several good points in this paragraph, namely, that we go to hell by our own "free choice"; that hell is not a physical place, but a mental "state"; and that it is not God who forces us to be separated from him, but that it is a "self-exclusion". Recall from the above NDE accounts that, it seems, the soul judges itself, and it may decide to be separated from God, at least for a while.

However, and this is where I disagree with the Church's teachings, I also believe that a benevolent God would give people (souls) a second chance to repent of their sins or shortcomings even after their death; especially if they never had the opportunity of knowing anything about him. I believe God is the greatest gentleman, and he will give people a second chance after their death; everything is so fuzzy and uncertain here on earth, anyway. I also believe that the invitation to be reunited with him again at any time will always be open from his side; that is, if somehow we come to our senses— recall that NDE reports seem to indicate that God is only capable of

---

[671] *New American Bible*; Luke: 23, 34.
[672] *Catechism of the Catholic Church*: p. 221, para. 1033.

unconditional love. However, in the end, the soul can still decide to remain separated from God for a long time or "for ever"; God will not force the soul to join his company: as I said before, he is the greatest gentleman—but it would be a "self-exclusion" by the soul.

Moreover, I don't believe in the Church's traditions that there is physical pain (fire) or torture in hell; I believe the suffering we feel in hell is of a mental (psychological) nature resulting from our separation from God, which is not wise to underestimate: separation from a loved one, or from a place of fun and living in misery on top of that, can cause an awful feeling of depression. Jesus describes God as "our father"; do you really think that a father would prepare an eternal pit of fire as a punishment for his own children? This would be equivalent to *forcing* us to love him, and love cannot be forced. This is the main reason why I contest the Christian concept of hell.

The *Catechism of the Catholic Church* continues its paragraph on Hell as follows:

> Immediately after their death the souls of those who die in a state of mortal sin descend into hell, where they suffer the punishment of hell, "eternal fire". The chief punishment of hell is eternal separation from God, in whom alone man can possess the life and happiness for which he was created and for which he longs. [673]

It is hard to make any sense of this paragraph; it is typical of the contradictions in the Church's teachings (and in the bible, for that matter) that I was referring to in my discussions on this subject with my wife.

First, notice the mythological concept of hell: the souls are said to "descend" into hell; while the previous quote described hell as a "state" not a place. As already explained above, this is the Greek mythological concept of Hades, an underground place where the disembodied souls of the dead simply existed without doing anything. In fact, in the "Apostles' Creed", Christians profess that

---

[673] *Catechism of the Catholic Church*: p. 222, para. 1035.

Jesus "descended into hell" and that "on the third day he rose again"; all this simply means that he died and that he remained dead for three days. Had Jesus stayed in hell for three days his crucifixion ordeal would have paled compared to his suffering in those three days: and that's only three days, not an eternity in hell. Hell is not underground any more than heaven is in the skies; they are both mental states, not physical places.

Secondly, how can one burn a soul in physical fire, unless it is a *metaphorical* fire? Notice the parentheses in the above quote which are actually in the original. Why do you think "fire" is inside parentheses? The body cannot be suffering the pain of burning in fire "[i]mmediately after ... death" as the quote says, because we all know that it is still decomposing in the grave. Yet the Church insists it is a physical fire, and that the body will suffer, immediately after death—again, things don't add up.

If the Church were to admit that it is a metaphorical fire, not too many people would be scared enough by it; but the Church wants to have it both ways. She can't even wait to punish sinners until the last (universal) judgement at the end of the earth (or the universe) when supposedly all the bodies of the dead are resurrected (in which case it would make some sense); even though it still has an eternity to punish them in. Again, if the Church were to admit that people will be punished for their sins in the fires of hell after this earth (or the universe) passes away (in the universal judgement, that is, in a few billion years), the scare tactics wouldn't work too well with the ordinary individual—although it might still work with a well-educated person. Most people don't have a clue of what eternity really means, and billions of years from now is too far away for them to lose any sleep over. If you ask most people what they would do if they ended up in the fires of hell, they'll tell you that they'll be with other people, and that they'll cope with it the way other people do. That's the answer I once got from a man (while I still used to believe in a physical fiery hell) when I asked him the question.

Here is a third contradiction in the above-quoted paragraph. While I admit that being separated from a loved one can be quite painful, mentally and perhaps even physically to some extent, it doesn't compare to burning one's body in fire. You can feel free to ask any dedicated lover whether one would rather be thrown in fire eternally than be separated eternally from one's loved one; if a single person opts for eternal fire, I'm sure one will change one's mind after a few minutes of experiencing the fire. There are so many people who don't care about God, live without God, seem quite happy without God, and even actually hate God; why would they yearn for God so much if they were to be permanently separated from him? Having said this, I wouldn't recommend the reader's downplaying the painfulness of separation from God, either; we don't know much about the afterlife.

Finally, if, on the other hand, we accept as a fact that separation from God is so terribly painful, why would God add *another* terrible punishment: that of burning the soul (or body) in fire on top of it? Doesn't that make God cruel and vindictive?

I caution the reader not to take my word on this matter; there's too much at stake here. The testimonies given by the three little children of Fatima and a number of the seers in Medjugorje (please refer to the previous chapter on "Miracles") do affirm an eternal hell of fire. The only reasonable explanation I have for these revelations by Our Lady is that she probably *has to* stick to the Church's current traditional beliefs; otherwise the Church and its people would jump to the wrong conclusion: namely, that it is a devil's (satanic) apparition rather than a heavenly one. Even as it is, the Church has problems accepting and authenticating the apparitions experienced by these children and young adults. For example, Our Lady of Fatima showed the three children hell "under the earth". [674] As I said above, this is a Greek mythological concept. Hell is not underground any more than heaven is in the skies; they are both mental states, not physical places. But Our Lady, probably for the above reason, decides to act conventionally and not scientifically.

---

[674] De Marchi, *The True Story of Fatima*, p. 28

This question of whether there is fire in hell, and what hell is all about, is something I'm still working on and researching; when all is said and done, I do not want to take responsibility for any one's behavior: this is only my current personal opinion. As I pointed out above, the only reason I doubt my rational conclusions is Our Lady's revelations during her apparitions.

Is Hell a Physical Location?

So far, no underground mine has come across a physical location "under the earth" where the souls of the damned might be serving their time. However, the Catholic Church seems to insist that in hell there is a physical, not a metaphorical, fire; furthermore, Our Lady of Fatima seems to have opened the ground to show hell to the little children. So is there any physical evidence, at least, for the existence of such a place; albeit we may not be able to see the resident beings (devils and evil souls) in there. As far as current scientific evidence goes, the following is what geology, astronomy, and cosmology have to say.

(1) Geology

Seismological studies use shock waves, which are produced by explosions, in the earth's mass to investigate the earth's interior. These studies conclude that there does seem to be a rather small spherical volume at the center of the earth that is hot enough to qualify as the actual location for the Christian hell under the earth. According to Wikipedia, these seismological studies concluded that the innermost sphere of Earth, commonly known as its *inner core*, seems to consist of a solid ball of approximately one thousand two hundred (1200) kilometers (about 760 miles) radius—that is, about seventy percent (70%) of the moon's radius. It seems to be composed mainly of an iron-nickel alloy; its boundary temperature is about five thousand seven hundred degrees Kelvin (5700K or 5400C). [675]

The fact that it is solid does not help the Christian concept of hell much, it's supposed to be simply fire; however, it sure qualifies as hot enough (it's much hotter than a flame) and a spirit can, supposedly, move through matter like a knife through hot butter. (I can't imagine why a fallen angel, the head devil Satan, would be confined to such a small place, out of the whole vast universe, and so close to us—but anyway.)

What about the future of this place which might qualify as hell? It is supposed to last for eternity, right? This is the real problem with the concept of a physical hell. Everything physical changes according to the second law of thermodynamics: everything winds down, everything deteriorates.

(2) Astronomy

According to Wikipedia, after exhausting its core hydrogen in about 5.4 billion years, the sun will enter the *red giant* (elderly star) phase; during this phase its atmosphere will expand to over two hundred times its current size, and therefore it will (most probably) reach the earth's orbit. [676] If this happens, the earth will gradually lose speed, because of friction (drag) against the sun's atmosphere, and consequently it will spiral into the sun; then it will be absorbed by the sun—that is, incinerated inside the sun's core.

Our sun originally had about ten billion years of hydrogen fuel: fusing (burning) it into helium. This stage is the sun's main and most useful phase as a star: emitting heat and light—the sun is about half way there. Our sun is not large enough to eventually become a *supernova* (explosion of a star); from its mass scientists predict that, after its current phase of hydrogen fusion, it will become a *red giant*. When this happens, our sun will grow very much larger so

---

[675] https://en.wikipedia.org/wiki/Inner_core (10 April 2016)
[676] https://en.wikipedia.org/wiki/Sun (10 April 2016)

that its atmosphere will enshroud Mercury, Venus, and probably also Earth. [677] If the sun's atmosphere fails to reach our planet, Earth will get too hot, anyway (hotter than Venus is today—462C), and it will lose all its water.

NASA actually believes that the sun's atmosphere will drag down the earth's rotational speed around the sun: so the earth will eventually spiral into the sun's core.

> The radius of the red giant sun will be 100 times what it is now, lying just beyond the Earth's orbit, so that the Earth will plunge into the core of the red giant sun and be vaporized (source: NASA). [678]

The whole of our earth will be vaporized—so much for hell's original location under the earth.

But some people might still argue that it is hot enough in the sun's core, and that it can easily replace the former hell that was under the earth. I must admit they have a point. However, after the red giant phase, comes the helium-burning phase; during this phase our sun will shrink to about one-tenth of its current size. [679] Science writers Julia Layton and Craig Freudenrich describe the final fate of our sun as follows.

> When the helium fuel has exhausted, the core will expand and cool. ... Finally, the core will cool into a *white dwarf.* Eventually, it will further cool into a nearly invisible *black dwarf.* The entire process will take a few billion years. [680]

A *white dwarf* is a star's remnant consisting of a clump of matter devoid of their electrons. A white dwarf is very dense: it has a mass of

---

[677] https://en.wikipedia.org/wiki/Sun (10 April 2016)

[678] Layton & Freudenrich; http://science.howstuffworks.com/sun6.htm (10 April 2016)

[679] https://en.wikipedia.org/wiki/Sun (10 April 2016)

[680] Layton & Freudenrich; http://science.howstuffworks.com/sun6.htm (10 April 2016)

about that of our sun but a volume of about that of our earth. A white dwarf is only faintly luminous; its luminosity is the result of thermal (hot body) radiation, not fusion (like ordinary stars). [681]

A white dwarf continues to lose heat by radiating energy out into the universe; when it cools down enough, it eventually becomes a *black dwarf.* It is called black because it no longer emits significant light (or heat) and is therefore undetectable by any kind of telescope: it can only be detected through its gravitational influence. [682]

Cosmologists John Barrow and Frank Tipler estimated that it would take about a million billion (10^{15}) years for a white dwarf to reach five degrees Kelvin (5K or -268C); [683] since the universe has only existed for about fourteen billion (1.4×10^{10}) years, no black dwarfs exist yet: they are only theoretical. However, various white dwarfs, estimated to be between eleven and twelve billion (1.1×10^{10}–1.2×10^{10}) years old that cooled down below three thousand nine hundred degrees Kelvin (3900K), have been observed; this is why scientists theorize black dwarfs. [684]

(3) Cosmology

Of course three thousand nine hundred degrees Kelvin (3900K) is still hot enough to meet the requirements of hell, so to speak; but science says that it will cool down in a million billion (10^{15}) years. It will cool down to about two hundred and sixty-eight degrees Celsius below the freezing temperature of water (-268C): so literally hell, if that's where it was, will freeze over.

This may sound extremely long to the ordinary reader; however, any person with some concept of numbers knows that all this time pales compared to eternity: it is not even comparable, for example, to a drop of water in relation to all the oceans of the earth. Believe me,

---

[681]   https://en.wikipedia.org/wiki/White_dwarf (10 April 2016)
[682]   https://en.wikipedia.org/wiki/Black_dwarf (10 April 2016)
[683]   Barrow & Tipler, *The Anthropic Cosmological Principle*, (1988) Table 10.2
[684]   https://en.wikipedia.org/wiki/Black_dwarf (10 April 2016)

I'd hate to see anybody being punished in such fiery environments, without reprieve, any longer than this—for all the wicked deeds of one's whole life whatever they might have been; but still, it's no comparison to eternity! But Christianity still believes that God is capable of punishing like this quite a large number of souls that somehow end up there.

Science says that everything will cool down even further than five degrees Kelvin (5K or -268C); the current temperature of the universe, based on the cosmic microwave background radiation, is about three degrees Kelvin (3K or -270C); eventually everything will cool down asymptotically (tend) to zero degrees Kelvin (0K or -273C).

Some people might still argue that evil souls will still suffer for eternity at these extremely cold temperatures, just the same. Admittedly, this might be the case; however, I don't know of any Christian holy book that describes hell as a freezing place: one can't have it both ways. Or maybe one could try. For example, in Matthew's gospel Jesus describes hell as a place of "outer darkness"; [685] I could never figure out how one could have a dark (non-luminous) fire. Of course, bible scholars explain "darkness" as something else, like "loneliness"; but then it's metaphorical (not physical) to me.

On the other hand, God can break his own laws of physics through miracles; he can make the universe start contracting, thus keeping the fires of hell always kindled. But the above is as far as science goes. There does not seem to be (or rather, we don't know of) any laws of physics that can cause our universe to stop expanding and consequently (as we have seen) stop cooling further.

To recapitulate, near-death-experience accounts give ample data-based evidence of the *afterlife*, a benevolent God, and the soul's self-judgment; heaven, hell, and purgatory seem to be *mental* states reflecting our relationship with God rather than segregated physical locations; hell's supposed location and duration are *untenable*.

---

[685] *The Holy Bible*; Matthew: 8, 12; 22, 13; & 25, 30.

Fourth Set of Questions

Where is God? Is he everywhere? If so, is he big: the size of the universe? Is he far away? If so, how can he know everything about us or listen to our prayers in time? What does science have to say about all this?

Where Does God Reside?

Is God everywhere (omnipresent) in our universe, or is he outside our universe? God created space and time, so he is an external entity to spacetime, which is his creation: he is not a part or the whole of spacetime—that's pantheism.

Is God, a *spirit*, in any way associated with some limited amount of space, or does he permeate all of space? If he permeates all of space, then he must be very big and getting bigger since the universe is expanding. So how can the beings in heaven communicate with something so humongous? Where is his mind, everywhere? Ever since I was a young lad, I always had a problem accepting this; so I do not think one can associate a spirit with a location in our space.

God must also be independent of time because he created time. Recall the analogy of blowing up a balloon in the chapter on the "Big Bang Theory", time was analogous to the radius (size) of the balloon, and our three-dimensional space was analogous to the two-dimensional surface of the balloon. Time seems to be an illusion, anyway, something we human beings invented for our convenience: since everything around us is constantly changing, as we saw in the chapter on the "Second Law of Thermodynamics". We are trapped in time similar to the way we are trapped by gravity: this means that we cannot easily travel in time; we are confined to one time dimension—the present.

To understand what follows better, imagine a two-dimensional world: one confined to a very large flat horizontal sheet of paper, say. People and things drawn on this sheet of paper can go around one another, but none can go over or under one another by moving outside the plane of the paper. We, who are in a three-dimensional world, can intersect with (or cut through) this two-dimensional world, but they cannot really ask where we are; they can only ask where we intersect with them: where part of us meets their world.

If, for example we have a solid sphere passing through (intersecting with) their sheet world, they can only see a circular disk of it; depending on the sphere's relative position to their sheet world: they will see a large disk or a small disk. Similarly, if the sphere were hollow, they will only see a large or a small circle, depending on its relative position to the sheet. If the sphere is just sitting on their sheet world, in the extreme case, they will only see a point; and finally, if the sphere does not touch the sheet at all, they will not see any of it.

Notice that, in this last case, the sphere does not have to be very large in order to be outside the sheet—their space; and it has no meaning for them to ask where the sphere is because it is nowhere in their space. Notice, also, that the sphere can be close to anywhere in the sheet; at the same being outside the sheet all the time.

Do we have any evidence of this being the case in our universe; in other words, is there any evidence that the spirit world possesses the advantage of an additional dimension? In an article entitled, "4 Dimensional Afterlife", we are told that there have been "many well-documented" instances in support of this concept.

> There have been many well-documented cases of poltergeists [actions by ghosts] causing objects to appear and disappear or apparently "throwing" small objects, e.g., pebbles or coins. Frequently, the thrown object is not seen in trajectory but rather "appears" at its target which suggests that it does not use our time and space to get there. Imagine an entity in the afterlife collecting a 3D pebble with a 4D hand; what would we observe? Well, the pebble would vanish from our locality and reappear wherever the entity wished. In fact the pebble is withdrawn from

> 3D space via the 4th dimension and is then replaced, but
> would disappear during the process. [686]

God and spirits (angels and human souls) are independent of time, that is, they are not confined by time, as we are. So we cannot really ask where they are at any instant of time; just as the people confined to the sheet world cannot ask where we are: they can only ask where we interact with them, if we interact at all. If we do not interact with them at all, it does not mean that we do not exist, either; similarly with God and spirits interacting or not interacting with us at the present time.

I considered the possibility of God's residing in the black hole at the center of our galaxy (the Milky Way)—recall that no time elapses inside a black hole. But there are at least one hundred billion (10^{11}) galaxies, practically all of which have a black hole at their center; why would God choose ours? If we were the only living creatures in the universe, it may make some sense that he resides close by; and it would not be inconceivable that he would easily hear our prayers in time to make a difference. However, I rejected this idea because God also created our galaxy's black hole the same way he created time (or expansion of the universe): so he must be outside it, the same way he must be outside time, not confined to it.

Following successive layers of this and similar arguments, one has to conclude that God is outside space and time (spacetime). He can intersect with it from time to time; but he must be outside it: he is not confined to our universe. The universe is something he created, and he is not necessarily part of it; he can intersect with it whenever and wherever he chooses, but he does not have to be confined to it: consider the analogy of the two-dimensional flat-sheet world we pictured above. Moreover, he does not have to be big (humungous) either.

As I have mentioned before, I place considerable trust in the undisputed Pauline letters; because they are the earliest writings by

---

[686] http://4dimensionalafterlife.org/page6.htm (15 March 2016)

someone who was originally prejudiced against Christianity. As Paul himself declares in one of his undisputed letters "The Letter to the Galatians", he originally persecuted the Christian Church; then, he seems to have had an extraordinary supernatural experience from God himself revealing to him that Jesus was his son; finally, he seems to have done some real thinking about this supernatural experience for three years (quoted in full in the section on "Jesus' Resurrection").

> [Y]ou heard of my former way of life in Judaism, how I persecuted the church of God beyond measure and tried to destroy it, … But when (God) … was pleased to reveal his Son [Jesus] to me, … I did not immediately consult flesh and blood, nor did I go up to Jerusalem to those who were apostles before me; ….
> Then after three years I went up to Jerusalem to confer with Cephas [Peter] …. (As to what I am writing to you, behold, before God, I am not lying.) [687]

He also swears "before God" that he is not lying. I believe this man as I would believe Mohandas Gandhi.

In any case, Paul describes several other deep mystical (heavenly) experiences in another of these undisputed Pauline letters, "The Second Letter to the Corinthians":

> I must boast; not that it is profitable, but I will go to visions and revelations of the Lord. I know someone in Christ who, fourteen years ago (whether in body or out of body I do not know, God knows), was caught up to the third heaven. And I know that this person (whether in body or out of body I do not know, God knows) was caught up into Paradise and heard ineffable things which no one may utter. [688]

---

[687] *New American Bible*, Galatians: 1, 13, 15–8, 20.

[688] *New American Bible*, 2 Corinthians: 12, 1–4.

Believe it or not, Paul basically claims that his soul was caught up in ecstasy and transported into heaven where he heard privileged information. The *New American Bible* comments on this passage as follows:

> [H]e seemed no longer confined to bodily conditions, but he does not claim to understand the mechanics of the experience. ... [A]ncient cosmologies depicted a multilayered universe. Jewish intertestamental [420BCE to 50ACE] literature contains much speculation about the number of heavens. Seven is the number usually mentioned, but the Testament of Levi (2:7-10; 3:1-4) speaks of three; God himself dwelt in the third of these. Without giving us any clear picture of the cosmos, Paul indicates a mental journey to a nonearthly space, set apart by God, in which secrets were revealed to him ... i.e. privileged knowledge, which is not possible or permitted to divulge. [689]

In yet another of these undisputed Pauline letters, "The First Letter to the Corinthians", Paul says that we shall see God "face to face" when we go to heaven:

> [W]e know in part, and we prophesy in part: But when that which is perfect is come, that which is in part shall be done away. ... For now we see in a mirror, darkly; but then face to face: now we know in part; but then shall I know even as also I have been known [i.e., fully]. [690]

So for simplicity's sake, in our discussions, I shall henceforth assume that God resides outside our spacetime; that heaven, hell, and purgatory are also outside our spacetime; and that it is possible to see him and communicate with him "face to face" there—as a "Being of Light", perhaps, as he is described in NDEs.

---

[689] *New American Bible*, 2 Corinthians: 12, 1–4 footnote.
[690] *The Holy Bible*; 1 Corinthians: 13, 9–10, 12.

But now, if he is outside our space, at any moment in time he can be very near to us or very far away from us; so how can he hear our prayers in time to make any difference to us? A message (prayer) from us might take close to fourteen billion years to reach the end of our universe if it travelled at the speed of light.

However, *entangled* particles can communicate information *instantly*. Briefly, *entanglement* can be explained as follows. If two particles (or photons) A & B are *paired* like twins, and particle (or photon) B interacts with a third particle (or photon) C, then by observing the behavior of particle (or photon) A, which we assume to be very far-away from B, one can determine the kind of interaction B & C might have had *instantaneously*: there would be absolutely no time lag; therefore, particle C can practically be *teleported* to A's location instantly. This is not science fiction; it is another instance of the strangeness of quantum physics. [691]

God is a spirit, and he may not be in any way associable with space or time. But, when all is said and done, he needs some way of communicating with us in space and time, because we are in spacetime. Does God encompass all of our space? Or does he simply communicate with us (wherever we are in spacetime) in some way similar to the entanglement phenomenon described above? I opt for God being outside our spacetime; however, the universal wave function (the Holy Spirit) permeates all of our space and it gives him immediate access to everyone and everywhere in the universe. Christianity also holds that God knows everything we do at any time.

To recapitulate, God and heaven are probably *outside* spacetime; so are the souls in hell and purgatory; the souls in heaven can communicate with God face to face; it is possible for God to communicate with us *instantly* and to know everything about us.

---

[691] Greene, *The Fabric of the Cosmos*, pp. 442–6

Fifth (Final) Set of Questions

Does God, sometimes, break his own laws of nature? Does God ever grant our requests? What is prayer anyway?

Intervention

We saw in the previous chapter on "Miracles" that probably God does break the natural laws occasionally. Why does God sometimes break the laws he himself made? Maybe he has a plan for the universe, after all, and wants some control over what happens to the universe and to the earth; that is, he is not just a passive creator, as he has often been accused of being.

In fact, I believe, God begot a human son, Jesus, to tell us what life is all about and to show us how we should live; in particular, Jesus taught us to refrain from violence and to be ready to die for what we believe in. It is easy to end up victims of Church or State for what we believe in. Take, for example, a conscientious objector in war time, one who does not believe in killing anybody under any circumstances; if one refuses go to war, one might end up in jail. Similarly, a doctor who refuses to perform an abortion, because he believes it constitutes killing a human being, may lose his job.

We have also seen in the chapter on "Evolution" that probably God placed (or consistently places) new animal phyla (body plans) on earth from time to time. Why would he do something like that? This is a very good question, but maybe he wants to show us that he exists: that if we really look for him we will find him. Alternatively, it may be impossible to have an exclusively down-up type of evolution; maybe a down-up evolution only works effectively over the short run, as a survival or adaptation tool. It seems that his other natural law, the second law of thermodynamics, trumps a down-up style of evolution in the long run: either because the environment happens to change too drastically, or because subsequently an error catastrophe occurs.

If this is the case, and he wanted life to continue on earth, he would have to resort to this method: why not intervene to keep life going on?

From my experience in engineering, it is not possible to design something so versatile as to be able to cater for everything that might happen to it. It will be too costly and too cumbersome: design has to be selective. Engineering is not a dream wish; it is a trade-off of possibilities. One can only design for a few environmental conditions and for a few functions; this is another reason (constraint) why I believe the different animal body plans were designed separately—rather than one evolving from another.

God also seemed to have started the universe, anyway, and we don't know whether he intervened, from time to time during the Big Bang, to produce a starry universe: as occasionally causing a local or a temporary rapid expansion, or contraction, of space. The odds are so astronomically high against a starry universe; it appears that the natural laws alone are incapable of doing the job without any intervention.

Prayer

Of course, one cannot pray to God unless one believes in him in the first place. Prayer is, technically, the lifting up of one's mind and heart to God. It is supposed to be a one-on-one communication with God: getting to know him better. It is not supposed to be an asking of favors from God, as most people seem to think. God does listen to our pleading sometimes, but I would say it is rather infrequently, and only if he deems that granting us what we are asking for is going to make us better people altruistically or personally: more compassionate and tolerant towards others or more appreciative, happier, etc.

Jesus did teach us how to pray properly in what is known as "The Lord's Prayer", which starts: "Our father, who art in heaven, hallowed be thy name." This is not supposed to be an empty lip service to God; it is supposed to be something one really feels inside: as saying to one's wife or girlfriend, before going out on a dinner night when she is all dressed up, "how beautiful you look." God does not need to be glorified any more than he is by his angels; he prefers a personal relationship with us human beings: not someone who only shows holiness in church. When one takes the time to look for God in life, one's feelings for God change for better or for worse, and God would rather have that than total apathy about him.

The prayer continues: "Thy kingdom come, thy will be done on earth, as it is in heaven." God does not want to rule the world through any political or religious community; he wants to rule in our hearts: as a lover would like to rule in the heart of one's significant other. He does not want his will to be done across the board like a political tyrant; he would like his will to be done only because it means our greater happiness: his will is for us to behave altruistically and unselfishly, as Jesus showed us by example. His kingdom is not a political kingdom; the kingdom of God is a kingdom of altruism, justice, kindness, fairness, unselfishness, and total harmony: which is still a long way from coming to earth.

The prayer continues to ask for the necessities of life: "Give us this day our daily bread." Does God listen to prayers of need? Apparently he does listen to prayers but not all the time; perhaps even rarely. I think he would like everyone to carry one's own weight, and also to enjoy or suffer the consequences of one's actions. In one of the apparitions at Fatima, the nine-year-old seer, Lucia, asked Our Lady, "I have many [people's] favors to ask. Do you wish to grant them or not?" Our Lady replied, "Some I will. Others I will not! They must amend their lives ..." [692] To ask for our immediate indispensable needs is reasonable: food, shelter, income, health, etc.; but to be greedy and ask for riches so that we can satisfy our passions is not, and we should not expect God to grant such prayers.

Most atheists think that God does not exist because he does not grant such prayers; God only hears those prayers that are conducive for us to love others and our becoming better people, and it is hard for us to anticipate beforehand whether such requests are going to make

---

[692]   De Marchi, *The True Story of Fatima*, p. 53

us better people. A friend of mine lately asked me: "Why should God listen to my prayer, when there is so much misery and so many people suffering in third world countries?" It's true, that's why God seldom listens to prayers asking for favors; it's only fair that God lets people solve their own problems: God only helps those who help themselves. However, when people are in dire need he might, and he does at times; he usually does this in a very subtle way though, hardly detectable even to the person asking the favor.

Personally, I think there were probably half a dozen to a dozen times, in my whole life, when I believe God really helped me and my family. My friend above asked me how I knew it was God and not my own abilities that accomplished the desired ending. I answered that I didn't; I just felt that the circumstances indicated strongly that it was through his help that things worked out. I won't bore the reader with details trying to prove it; but if a loved one has a brain tumor, right on the artery where it has an ample supply of blood, and it doesn't grow ….

Probably the hardest part of the prayer follows: "And forgive us our trespasses, as we forgive those who trespass against us." Admitting our shortcomings, not retaliating, and perhaps even forgiving others for hurting us in any way, is pleasing conversation with someone who loves every one of his creations: the "father" of all human beings. Forgiving is probably the hardest thing in life; it sure goes against the grain of evolutionary principles: it has been shown through computers that a "tit-for-tat" policy is probably the best way both to thrive through cooperation and survive extinction.

I shall stop here: I don't want this chapter to end sounding like a sermon; I only wanted to convey the concept that prayer is a one-on-one natural conversation with God. Everyone has to create one's own private prayer, depending on what's most important to that person. My personal daily prayer is tailored on the very first part of the Lord's Prayer, it goes:

> Lord God almighty, father, I humbly ask you to help me in
> my search for the *truth*: help me to find *you* there. Help me
> to get to know you better, love you more, and fear you less.

I don't ask for favors except in very rare cases; I ask daily only to get to know God better. I don't pretend my prayer is the perfect prayer; but I try to follow Jesus' lead. God is not interested in showing off his greatness; but when his children truthfully appreciate it, he likes that. It means that his child is building a personal relationship with him: the child is interested in him.

When my son started university he also started to appreciate my knowledge of mathematics and physics; as I indicated before, I do not consider myself very knowledgeable scientifically, but my son's appreciation pleased me so much. When he was much younger he did not have a clue as to my knowledge in the subjects. At one time, when he was very young, he thought I knew less mathematics than his primary schoolteacher; even though I told him that I used to teach mathematics and physics at college level. [693] Did I need my son's appreciation? Not really. Most prophets are not recognized where they come from, anyway. [694] But the fact that he did, when he went to university, pleased me immensely, because it meant my having a personal relationship with him. I suppose this is how God feels when we really appreciate him: he feels a connection with us.

To recapitulate, God does seem to intervene by suspending his own natural laws from time to time; prayer is building a one-on-one relationship with this God.

---

[693] To be exact, I prepared Maltese students to sit for *Advanced Level* examinations issued by Oxford and London Universities in England, and the Associated Examining Board (AEB) in the United Kingdom.

[694] *New American Bible*; Mark: 6, 4.

CONCLUSION

Whenever I look at the complexity of the universe, or even empty space for that matter, I feel like someone looking at the operation of an automobile: its engine, its transmission, its couplings to the wheels, its controls, etc. The more I look at it, the more I understand how it works; but, at the same time, the more I appreciate its complexity and its ingenuity. Discovering or knowing how something works, however, does not explain how that something came about to be that way—far from it. Looking at an automobile engine, to continue with my analogy, we automatically conclude that it was designed by an *intelligent* person or a number of intelligent people: it never even enters our mind that it might have happened to come into existence by chance or as a result of the laws of nature.

Similarly, the fact that we humans are able to discover the background laws of physics does not explain how they came about to be that way. Contrary to what I might have implied in my introduction to this book, somehow, I do not feel that God is being pushed further and further into a corner with our scientific discoveries: on the contrary, I feel that God is being brought up more and more towards center stage by modern science. We know, for example, how the planets move under gravity; but we don't really know what causes gravity and how the laws governing gravity came about to be that way. Again, for some time, scientists thought they could explain the origin and diversity of life through abiotic evolution and descent from a common ancestor; nowadays, that we know more about the functionally specified information stored in deoxyribonucleic acid (DNA) and the evidence from the fossil record, a fair percentage of scientists have serious doubts about almost all of their original convictions. The number of things science can really explain pales compared to the number of things it cannot explain.

Scientific hypotheses should normally be provable or disprovable ideally experimentally in a laboratory but also through hard evidence observed in our universe. If they cannot be proved true or false, they enter into the realm of *pseudoscience* or religious *beliefs*. An arch-typical example of belief is the Roman Catholic concept of *transubstantiation* in the Holy Eucharist: in transubstantiation, none of the physical properties of consecrated bread or wine change, yet Jesus supposedly becomes present inside it. However, I believe that there are intangible things that can still be proved, beyond any reasonable doubt, through our powers of reason and inference. I admit it is a slippery terrain, but then neither is science always certain (100%) of its hypotheses: it has been wrong before, and it has also actually retracted hypotheses that were previously universally accepted by practically the whole of the scientific body.

Lately, I had a discussion on this matter with a very good friend who read my manuscript. He made a very good point; his point was that in order to invoke an *intelligent* or a *supernatural* origin in science we need to have much more stringent criteria than we have for ordinary science: otherwise we would be invoking a supernatural cause all the time.

For example, when declaring the discovery of the Higgs boson by the large hadron collider in Geneva, Switzerland, in July 2012, the scientists stated that the *significance* of their experimental measurement was of the order of five standard deviations (5 sigma, or 5σ). This means that the odds of their being wrong (that is, the observation being caused strictly by random chance alone) was 1 in 1,744,278. My friend insisted that such odds were not near enough sufficient to conclude an intelligent, or worse, a supernatural cause; and I agreed.

As cosmologist, astronomer, astrophysicist, and astrobiologist Carl Sagan used to say: "Extraordinary claims require extraordinary evidence." I responded to my friend by asking him to define any odds he would like, and the evidence I presented in this book would beat them hands down! Perhaps it is time for science to define an acceptable criterion (demarcation line) for the odds required to deem an intelligent or a supernatural cause to be a reasonable *scientific* conclusion. In other words, that a given phenomenon satisfies the criterion of being so beyond any reasonable doubt, and that it would be *unscientific* to think otherwise. My opinion is that science cannot say "never" to design or the supernatural any longer: there is too much evidence to the contrary; it cannot hide behind the concept of non-overlapping magisteria (NOMA) any longer.

My friend also thought that it would be best to steer away from design and the supernatural when dealing with scientific evidence, because there is a definite danger of its becoming a pseudoscience. I answered him by giving him a practical example (a parable) to deliver my point. Suppose we did not know anything about Newton's work and we happen to find it in a forlorn cave. We examine it, and we find it to be an intelligent creation comparable to genius, by our species' standards. Next we investigate whether it could have arisen by chance, the laws of physics, the laws of chemistry, natural selection, the weather, and any other natural law we can possibly think of. To every one of these questions the answer will most probably be "no"—overwhelmingly. The obvious conclusion would then be that someone with intelligence, at least comparable to that of our species, created it. But then someone stands up and says that if we do that we would have to find how its creator came about: who created the creator; and that can lead to an eternal regress argument where one creator creates another, ad infinitum. He continues that because of the possibility of this eternal regress argument, we must steer away from such a conclusion altogether: as scientists, we must decide not to look beyond the natural laws. So, as scientists, we conclude that we do not really know the origin of this ingenious creation, that we don't have enough knowledge yet, but we must *insist* that it is still the result of the natural laws (that is, definitely not created by an intelligent being): in other words, we should stay one step behind and leave it at that, and we should not go one step further and say what is probably the case.

But the bottom line is that our conclusion would still be *false*; we *know* of a cause that can create intelligent information comparable to Newton's work: our own intelligence, for a start—in our example above, it *was* actually designed. The *truth* is that we know more than we pretend to know when we publish such a report: we know deep down that it was probably created by someone with intelligence comparable to our species, or better; we just do not have the guts to come out and declare it openly. If, in the future, science discovers that it was wrong, science can take the decision back (as has, in fact, happened several times in the past); but we have to be honest with ourselves and with the general public and admit that that's what we think *today*: beyond any reasonable doubt. I had to disagree with my friend on this "scientific" attitude; and when I presented this scenario to him, he admitted that he saw my point; but I do not really know whether he agreed with me or not (I didn't push it).

Unfortunately, we were not at the scene of the crime when the universe and life started. In his book *Signature in the Cell*, science philosopher Stephen Meyer points out that if all the various competing natural explanations for the various phenomena observed in the universe are actually deemed inadequate, and there is actually just one plausible explanation left, then that plausible explanation, no matter how improbable it might seem, should be adopted as the only reasonable explanation at the time: the best explanation of reality, and the best approximation to the truth. This method of reasoning is called *abductive* reasoning: it consists of progressively eliminating hypotheses one by one until (ideally) only one hypothesis is left standing. [695]

This is what I tried to do throughout this entire book. The undeniable evidence of the universe around us and the sudden appearance of various animal phyla in the Cambrian explosion, the circumstantial evidence of the functionally specified information in

---

[695] Meyer, *Signature in the Cell*, pp. 153–5.

DNA, the empirical evidence from our consciousness and various near-death-experiences (NDEs), and the positive evidence of, at least, some of the miracle accounts leave no doubt at all, in my mind, that an intelligent superior being is the cause of all these observed phenomena. It is the only explanation that can encompass *all* these clues together. I hold that if we can prove, beyond any reasonable doubt, that there is no other competing satisfactory scientific natural explanation for some or all aspects of reality, and that there is actually strong evidence for a superior intelligent being; then one can resort to a supernatural first cause: I do not rule it out and exclude it from the realm of science—whether God exists or not is a scientific question! The scientists' calling the conclusion of such abductive reasoning a "god of the gaps" is somewhat of an *empty* boast when one considers that science has been unable, so far, to give any tangible explanations for the beginnings: that is, how the physical laws, matter, life, consciousness, etc., came about in our universe; all their hypotheses are so far-fetched and wishy-washy to the point of being unprovable. [696] I went along with this widespread scientific "slogan" in the introduction to this book simply because I did not want to let the cat out of the bag at the beginning of the book. I also wanted the reader to give fair consideration to mainstream scientific hypotheses because I have the utmost respect for science and its methods. But, I think it is time to question our paradigms.

In his book *The Trouble with Physics*, theoretical physicist Lee Smolin writes about an amusing conversation he once had with a group of *Creationists* on a plane: amusing, perhaps, because it sounds so unscientific. [697] His book, in my opinion, is a great book written by an authentic and honest scientist; so please don't get me wrong. But towards the end of his book he writes (just) one sentence which troubled me, since it was coming from a scientist of such balanced and relatively unbiased opinions: it really hit a nerve inside my being. In the chapter entitled "What Is Science?" he writes:

---

[696] Meyer, *Signature in the Cell*, pp. 324–6.
[697] Smolin, *The Trouble with Physics*, pp. 25–6

> [I]t becomes impossible to distinguish sciences like
> physics and biology from other belief systems—such as
> Marxism, witchcraft and intelligent design—that claim to
> be scientific. If no such distinction is made, the door is left
> open to a scary kind of relativism, in which all claims to
> truth and reality have equal footing. [698]

I am not sure whether Smolin, in the above quote, was simply not
distinguishing between Creationism (which bases its science on
the assumption that the bible is infallible) and Intelligent Design:
since they are often lumped together because they both believe
in an intelligent creator. But what he is actually implying in the
above quote is that intelligent design is not a science. I have to
disagree with Smolin: there's one exception to what he stands for.
I contend that if every natural explanation falls far too short of a
reasonable explanation; resorting to an intelligent (and perhaps even
a supernatural) explanation is not unscientific. As I proposed in the
introduction to this book: whether God exists or not is a scientific
question—as even atheists agree. It is a question about reality and it is
a question about truth: both of which are of direct interest to science.
Moreover, to be fair, if one wishes to investigate the existence of God
from scientific evidence, one cannot exclude the possibility, a priori.

There is a method of proof in mathematics known as *reductio ad
absurdum*, which is the Latin translation for "reduction to the absurd".
In this method of mathematical proof, one temporally assumes the
hypothesis in question to be *false*: that is, one initially stacks all the
odds against the hypothesis being true. Then, on the basis of this
assumption, one proves (mathematically) that an absurdity ensues;
such a case obviously implies that one's original assumption was
false, which therefore means that the opposite was true: in other
words, that the hypothesis in question was in fact true all the time,
after all. This is the kind of argument I claim to have laid down in
this book: I have taken the consensus of the mainstream scientific
community, the majority of which denies the existence of God, and

---

[698]   Smolin, *The Trouble with Physics*, p. 297

shot them down one by one. I may not have proved my original assumptions to be mathematically or philosophically absurd (I never had such intentions, as I stated in the introduction to the book); but I think I have proved them to be scientifically untenable.

So it is my final verdict that God exists: there is God's signature not only in the largest of things (the universe, the galaxies, the stars, and the planets) but even in the tiniest of things (in every atom and in every living cell). It seems as if God wrote his name in big letters all over the sky announcing his existence and ingenuity; not in English or in any other language, but in a universal language: *intelligence*. Intelligence is God's language, and this is how man was made in his image! God also deigned to make things in such a way that beings with our level of intelligence would be able to unravel the laws of nature (scientifically) and to be able to appreciate their ingenuity and complexity. If we take the time to examine the evidence around us, we can find God: that is, appreciate who he is and what he is all about; we may then be driven to form a personal relationship with him. The existence of the universe, life on earth, and our consciousness are the three main reasons why I believe in the existence of God.

In the last chapter of this book we looked at pain and suffering in life; we concluded that predators are necessary to curb the overpopulation of any species: they are conducive to a self-sufficient ecosystem. We also saw that survival of the fittest is a process which is not necessarily cruel; it is possibly also a requirement for a stable ecosystem. Furthermore, the universe seems to have been made especially for man; in fact, man is apparently supposed to be an exception to this general rule of evolution: he is expected to be altruistic towards other human beings.

Many near death experience (NDE) accounts, one of which is particularly compelling, seem to indicate quite clearly the existence of an afterlife, a personal self-judgement, and a benevolent God. A significant number of these NDE accounts (including the one just mentioned above) report seeing their long dead relatives, which implies there is probably a life after death; moreover, miracle seers describe apparitions of people who died about two thousand years ago, namely, Mary and Jesus; this afterlife seems to be eternal from the information given to us by these seers. Some of these miracles are supported by extraordinary cosmic phenomena witnessed by many people, others by extraordinary healings confirmed unexplainable by specially qualified doctors. I concede that these areas of NDEs and miracles require more rigid scientific scrutiny; however, it seems to me there is some solid evidence: but this is for future scientists to take up and research. Scientists should not always be motivated by grants; alternately, perhaps well to do scientists could adopt such projects. Aren't there any well-to-do scientists out there?

We also concluded that pain and suffering seem to be aimed at making us better individuals by helping others; this nourishes a personal relationship with God, whom we do not see, through love of our neighbor.

In undisputedly authentic letters by a former Pharisee by the name of Paul, who originally persecuted Christians, the author himself tells us that some three years after Jesus was crucified God revealed to him, personally, that Jesus was his son; he also swears before God (in whom he strongly believed) to the truth of this statement. Paul also tells us, in these undisputedly authentic letters, that although Jesus was known to have been killed, he personally saw Jesus alive again; so also did Jesus' apostles and five hundred (500) other Christians all at once. Historians have no doubt that contemporary Christians were convinced that Jesus rose from the dead, and that's what resulted in the Christian movement; otherwise Jesus would have been forgotten in in the annals history. I concluded by pointing out that Paul was a contemporary skeptic who decided to believe that Jesus resurrected. Unfortunately, we lack significant corroborating evidence (documents) from outsiders (non-Christians) testifying to Jesus' resurrection from the dead; on the other hand, there seems to be some outside evidence for Jesus' being a doer of miracles (startling deeds).

Jesus, being God's begotten son, is most like his Father in character. In the last chapter, by summarizing Jesus' teachings, I tried to show that, unlike the way God is described in the Old Testament,

God is not cruel, vindictive, and jealous but benevolent and capable only of unconditional love: "God is love", as the New Testament describes him. The evidence from a number of NDE accounts confirms this. Jesus has no problem challenging certain passages of the Old Testament where he believed its authors were off course, by saying: "You have heard that it was said, 'Love your friends, hate your enemies.' But now I tell you, love your enemies ...." [699] I have also shown several passages in the Old Testament where the authors went wrong and that therefore the bible is not infallible: in fact, I believe the Old Testament, especially, completely misrepresented God. Jesus tried to remedy that; through his teachings and his life of absolute non-violence he showed us what God is really like.

I do not wish the reader to get me wrong, however: I'm not trying to prove anything here. Personally, I wouldn't mind the non-existence of God or my total annihilation once I die: I'm not too keen on having to give account for every action of my life. When one is in front of a judge, it doesn't matter what everyone else has done, and one doesn't know how the judge is going to react to one's actions. (That is, of course, assuming God judges us, rather than we judge ourselves.) However, the overall evidence around us, including miracles and NDEs, seems to point in another direction, and I wouldn't like a nasty surprise at a time when I can't do anything about it.

Incidentally, one last note; in theory, I cannot agree that one should believe in God "just in case". If one is not convinced that God exists, deep down in his being, one still cannot help not believing in him.

My references to self-declared atheists were not intended to be derogatory; they were only intended to establish concepts about which most of humanity agrees; they were also used to indicate controversial issues and beliefs when it was the case. I have every respect for atheists because, in general, they are thinkers; they help us keep our feet on the ground: fending off our religions from becoming mere superstitions, perhaps even violent institutions.

---

[699] *New Testament and Psalms*; Matthew: 5, 43–4.

In his book *Subtle is the Lord*, physicist and science historian Abraham Pais quotes Albert Einstein as saying: "Science without religion is lame, religion without science is blind." [700] Pais' book's title comes from another one of Einstein's quotes: "God is subtle ...." I'm sure the reader appreciates, after reading this book, Einstein's convictions.

We read in Hindu sacred scripture *Bhagavad Gita* (Sacred Song): "There is no good without evil. There is no love without hate. There is no courage without fear. There is no faith without doubt." [701] I wrote this book for the benefit of those who are curious, but have serious scientific doubts about God's existence; I just hope it helped a little.

Bias

We all have our biases, and they may help or hinder our own progress in life. What is most important, however, is not to get carried away by them so as not to follow evidence and reason any longer. It is also important not to allow our biases to affect other people's lives negatively. Balance is the key; no doubt, this is easier said than done. Extreme bias normally leads to superstition on the one hand, or a radical form of skepticism on the other. I just hope my personal bias wasn't too sour for anybody's taste: if it was, please forgive me—I tried my best!

---

[700]   Pais, *Subtle Is the Lord*, p. 319.
[701]   Nithyananda, *Bhagavad Gita Demystified*, p. 420

ABOUT THE AUTHOR

Carmel Paul Attard was born in Gzira, Malta in 1947; he was educated at St. Aloysius' College, which is run by Jesuits; he became a member of the Society of Jesus (Jesuits) in 1963, staying for over six years; he graduated with a Bachelor of Science (BSc.) degree in Mathematics and Physics at the Royal University of Malta in 1970; his first occupation, in 1970, was problem solving in electronic component manufacturing; in 1974 he prepared students for advanced level mathematics and physics examinations organized by UK examining institutions; he emigrated to Canada in 1979, when he started calibrating and troubleshooting inertial navigation systems consisting of wheel gyros and accelerometers; in 1981 he was a member of the development and technical transfer team of a strap-down inertial navigation system using laser gyros; in 1985 he was promoted to sustaining engineering of accelerometers; in 1990 he became head of the engineering department for a central vacuum systems manufacturer, where he obtained a Canadian and US patent for a quiet canister vacuum cleaner in 1995; in 1994 he became head of the sustaining engineering department for a power supply manufacturer, solving nebulous problems and revisiting design shortcomings, where he is still working presently.

In 1972 he married Mary; they have three children: Lara, Omar and Kirk, and two grandchildren: Amelia and Diego; he is currently living with his wife of forty-four years' marriage in Brampton, in the Greater Toronto Area, Ontario, Canada. His pastimes are reading, writing, chess, bridge, and singing folk songs accompanied by guitar-playing.

APPENDIX I

(Mathematics)

Inevitably, in this book, I use some astronomically large and infinitesimally small numbers. For the benefit of those who are not mathematically inclined, following are some guidelines to help making sense of them.

A thousand, for example, is 1,000; that is, one (1) followed by three (3) zeros; mathematically this can be written as 10^3; notice the exponent "3" denoting the number of zeros. Consequently a million (1,000,000) can be written as 10^6 (that is, 1 followed by 6 zeros); a billion (1,000,000,000) is therefore 10^9 (that is, 1 followed by 9 zeros); and a trillion (1,000,000,000,000) is 10^{12} (that is, 1 followed by 12 zeros).

Similarly if one encounters a number like 10^{100}, for example, that means one (1) followed by one hundred (100) zeros: a humungous number indeed, probably too difficult to imagine and appreciate adequately. Although this last number is huge, it dwarfs compared to a number like $_{10}10^{12}$, for example, because this means one (1) followed by a trillion (10^{12}) zeros, a number that is almost impossible to imagine, or even write down (in full) on paper.

Just as they can be very large, numbers can also be very small. A millimeter, for example, is a meter divided equally into a thousand parts, or 1/1,000 (or 0.001) of a meter; this can be written down mathematically as 10^{-3}: notice the negative sign preceding the number "3". The exponent in this case is "-3". A millionth (1/1,000,000 or 0.000,001) would similarly be written as 10^{-6}; a billionth (1/1,000,000,000 or 0.000,000,001) as 10^{-9}; and so on.

APPENDIX II

(CHEMISTRY)

Atoms of chemical *elements* were for the longest time thought to be indivisible, because they cannot be divided or changed into other elements by chemical means alone. It was later discovered that atoms actually consist of three elementary particles: *protons*, *neutrons* and *electrons*. Electrons are very light; they weigh roughly 1/2,000 of a proton or a neutron, which weigh almost the same—the latter being very slightly heavier. A proton has a very small, single (indivisible) *positive* charge, while an electron has an equal *negative* charge; the neutron is *uncharged*—or electrically *neutral*.

For the longest time, a neutron was thought to be a combination of a proton and an electron, and for simplicity one can still think of it that way; but remember that modern particle physics discovered that this is not quite right (see "Appendix III" below).

An atom has a central *nucleus* consisting of protons and neutrons, bound together in a comparatively small space, and a number of electrons equal to the number of protons orbiting (circling) relatively far away around the nucleus. The reason why the electrons are not attracted onto the nucleus (since opposite charges attract) is because of their speed: behaving like a satellite or the moon around the earth. Since the number of electrons equals the number of protons, matter is normally electrically uncharged (neutral).

If an electron is extracted from an atom (*ionized*), it will have a net positive charge. An *ion* is an atom which has a net charge (positive or negative): that is, missing or having extra electrons.

In nature there are various basic chemical *elements* (roughly 100) determined by their number of protons or electrons. The number of proton and electron pairs of an element is called its *atomic number.* *Hydrogen* is the simplest element, having 1 proton and 1 electron (symbol: $_1$H); followed by *helium*, having 2 protons and 2 electrons ($_2$He). *Carbon* has 6 proton and electron pairs ($_6$C), *nitrogen* has 7 ($_7$N), *oxygen* has 8 ($_8$O), and *phosphorus* has 15 ($_{15}$P).

The *atomic mass number* is the sum of the number of protons and the number of neutrons in the nucleus of an atom. The atomic mass number of hydrogen is still 1 (^{1}H), helium's is 4 (^{4}He), carbon's is 12 (^{12}C), nitrogen's is 14 (^{14}N), oxygen's is 16 (^{16}O), and phosphorus' is 31 (^{31}P). There is no definite (exact) rule for the number of neutrons; as a general rule for the first few (lighter) elements, the number of neutrons is roughly equal to the number of protons; but the heavier elements tend to have a few more neutrons than protons.

APPENDIX III

(Physics)

Particle Physics

Scientists used to think that protons, neutrons and electrons were indivisible; but again they were mostly mistaken.

In the radioactive decay of Carbon-14 (^{14}C) to Nitrogen-14 (^{14}N), it seems like a neutron breaks down into a proton and an electron: the latter was known to be ejected from the nucleus. The uncharged neutron, in losing a negatively-charged electron, becomes a positively-charged proton; this type of radioactive decay is known as *beta-minus decay.* So the neutron seemed to consist of a proton and an electron; for decades that is how physics was taught in schools and universities. As expected, this phenomenon started to give the impression that the neutron was not really indivisible.

Cosmic rays, particle accelerators, and colliders were later instrumental in discovering many sub-atomic particles. The elementary particles of the *Standard Model* of particle physics (that is, the theory of matter and all the fundamental forces) include: (1) six flavors of *quarks*: up, down, bottom, top, strange, and charm; (2) six types of *leptons*: electron, electron neutrino, muon, muon neutrino, tau, and tau neutrino; (3) twelve *bosons*: the photon of electromagnetism, the three bosons W$^+$, W$^-$, and Z of the weak nuclear force, and eight gluons of the strong nuclear force; and (4) the *Higgs* boson.

There are 4 fundamental forces of nature: *electromagnetism*, the *weak nuclear force*, the *strong nuclear force* and *gravity* (see the corresponding section below).

Fermions make up matter, while *bosons* are force carriers. Fermions are subdivided into leptons and quarks (see below). *Baryons* are combinations of 3 quarks, held together by the strong nuclear force. For example, a proton is made of 2 up-quarks and a down-quark; while a neutron is made of 2 down-quarks and an up-quark. (So a neutron is not really made up of a proton and an electron, but for simplicity it is sometimes advantageous to consider it that way.)

Consequently, an atom of normal matter is made of leptons (electrons) and baryons (protons and neutrons). Beta-minus decay (see above) and the sun's thermonuclear process (*nuclear fusion*) are mediated by the weak nuclear force. The *photon* and *gluons* (see above) are massless (weightless); however the *W*, *Z* and *Higgs* bosons are very massive (heavy): the W-boson weighs about 80 times a proton, the Z-boson weighs about 90 times a proton, and the Higgs boson weighs about 125 times a proton. The Standard Model of particle physics does not include gravity, yet; however, various extensions of the theory predict the existence of a gravity boson, the *graviton*, and many other elementary particles.

Hadrons are composite particles held together by the strong force. They are characterized into two families: *baryons* (such as protons and neutrons, made of three quarks—see above) and *mesons* (such as *pions* and *kaons*, made of one quark and one antiquark).

Forces in Nature

Modern physics agrees that there are 4 main forces in our universe: *Gravity, Electromagnetism*, the *Strong Nuclear Force* and the *Weak Nuclear Force*.

We all know about the force of *Gravity*; we experience it every day, so we don't need to explain it further.

We are also familiar with the force of a magnet, and we probably have all heard of an electrostatic force, namely, that like charges repel and that unlike charges attract; although we may not have experienced it first-hand. *Electromagnetism* is the force associated with electric charges and magnetic fields; they are thought of as one force because we know that they are associated with one another, and we also know how they are interconnected with each other.

The forces of gravity and electromagnetism both extend to infinity obeying the inverse square law. This means that at a distance of two (2) meters they reduce to a quarter ($1/2^2=1/4=$one-fourth) of their strength at a distance of one (1) meter; at a distance of three (3) meters they reduce to one-ninth ($1/3^2=1/9$); at a distance of four (4) meters they reduce to one-sixteenth ($1/4^2=1/16$); and so on.

The *Strong Nuclear Force* is the force keeping protons, neutrons, and quarks, the basic building blocks inside the nucleus of an atom, together. Three quarks make up protons and neutrons. Protons, being positively charged, try to repel each other inside the nucleus of an atom, but the strong nuclear force keeps them bound together.

The *Weak Nuclear Force* is the force that acts between leptons (e.g., electrons); it is also the force involved in the decay of hadrons, which include baryons (e.g., protons and neutrons), and quarks. The weak nuclear force is also responsible for nuclear beta decay (more exactly, by changing the flavor of quarks) and for neutrino absorption and emission. The weak interaction happens through the emission or absorption of a W or a Z boson.

The strong force is about 137 times stronger than that of electromagnetism, a million (10^6) times stronger than the weak force, and many orders of magnitude stronger than gravity (about 10^{40} times).

The forces of gravity and electromagnetism extend to infinity.

The strong force is a very powerful attractive force at a distance of 10^{-15}m, but it reduces to virtually zero at 2.5×10^{-15}m; at 0.7×10^{-15}m it becomes repulsive, and this is what determines the size of the nuclei of atoms.

The weak force ranges between 10^{-17}m and 10^{-16}m; its strength in this range is about 10^{-5} that of electromagnetism; at 10^{-18}m it is comparable to electromagnetism, but at 3×10^{-17}m it is only about 10^{-4} that of electromagnetism.

Is this too complex? Indeed!

GLOSSARY

— **abiogenesis**: the original *evolution* of life or living organisms from inorganic or inanimate substances.

— **absolute temperature**: the temperature of an object on a scale where *absolute zero* is taken as its starting point. It is measured in Kelvin degrees (symbol "K"); i.e., 273.15 degrees higher than the *Celsius temperature*. E.g.: 0K (absolute zero) = -273.15°C; 0°C (freezing point of water) = 273.15K; 100°C (boiling point of water) = 373.15K.

— **absolute zero**: the lowest possible temperature, at which substances contain no heat energy.

— **accelerator, particle**: a machine that uses *electromagnetic fields* to propel charged particles to nearly light speed and to contain them in well-defined beams.

— **acid**: a *molecule* or *ion* capable of donating a *proton*, or hydrogen ion (H^+).

— **adenosine triphosphate (ATP)**: a *molecule* used in biological reactions requiring energy; hence it is often referred to as the "molecular unit of currency" of intracellular energy transfer. Most cellular functions need energy to be carried out; ATP transports chemical energy within cells for metabolism. When ATP releases energy to the cell, it loses one or two of its three phosphate groups, thus becoming adenosine diphosphate (ADP) or adenosine monophosphate (AMP), respectively. Both ADP and AMP may be considered as partially discharged batteries in the cell, which can be fully recharged to ATP.

— **afterlife (hereafter)**: life after death is the concept of a realm, or the realm itself whether physical or spiritual, in which an essential part of an individual's *identity* or *consciousness* continues to exist after the death of the body.

— **amino acid**: a biologically important organic *compound* containing amine ($-NH_2$) and carboxyl ($-CO_2H$) functional groups, along with a side-chain ("R" group) specific to each amino acid.

— **amphiphilic**: a chemical disposition possessing both *hydrophilic* (water-loving) and *hydrophobic* (water-fearing) properties.

— **annihilation**: the process that occurs when a subatomic particle collides with its respective *antiparticle*. Energy is conserved, and the annihilated particles are replaced by *photons*.

— **anthropic principle**: the philosophical consideration that observations of the universe must be compatible with the conscious and intelligent life that observes it. If life were impossible in our universe, no living entity would be there to observe it; thus it would not be known.

— **anticodon**: a unit made up of three *nucleotides* that correspond to the three *bases* of the *codon* on the *messenger RNA* (mRNA). Every *transfer RNA* (tRNA) contains a distinct *anticodon* triplet, a sequence that may base-pair to one or more codons for a particular (one) amino acid.

— **antimatter**: a material composed of antiparticles, which have the same *mass* as particles of ordinary matter but opposite charges and *fields*, as well as other particle properties such as *lepton* and *baryon* numbers.

— **arrow of time**: the one way direction of time.

— **atom**: the smallest constituent unit of ordinary matter that has the properties of a *chemical element*.

— **baryon**: a composite subatomic particle made up of three *quarks*; as distinct from *mesons*, which are composed of one quark and one antiquark. Baryons and mesons belong to the *hadron* family of particles, which are the quark-based particles.

— **base, organic**: an organic *compound* which acts as a base (the counterpart of an acid); it is usually, but not always, a *proton* acceptor containing nitrogen atoms, which can easily be protonated.

— **beta decay**: In beta minus (β^-) decay a *neutron* is converted to a *proton* and the process creates an *electron* and an electron antineutrino. In beta plus (β^+) decay a proton is converted to a neutron and the process creates a *positron* and an electron *neutrino*. Beta plus (β^+) decay is also known as positron emission.

— **big bang**: the rapid expansion of matter from a state of extremely high (possibly infinite) density and temperature that according to current *cosmological* theories marked the origin of the universe.

— **big crunch**: a possible scenario for the ultimate fate of the universe, in which the expansion of space eventually reverses and the universe re-collapses, ultimately ending as a *black hole* or causing a reformation of the universe starting with another *big bang*.

— **black body radiation**: the type of *electromagnetic* radiation emitted by a black body, an opaque and non-reflective body, assumed to be held at constant, uniform temperature.

— **black dwarf**: a theoretical *stellar* remnant, specifically a *white dwarf* that has cooled down to the temperature of the *cosmic microwave background*, and so is invisible. Because the time required for a white dwarf to reach this state is calculated to be longer than the current age of the universe (13.8 billion years), no black dwarfs are expected to exist in the universe yet.

— **black hole**: a region of *spacetime* exhibiting such strong *gravitational* effects that nothing—not even *electromagnetic* radiation such as light—can escape from inside it.

— **blueshift**: happens when visible light from an object is apparently decreased in wavelength, or shifted to the blue end of the *spectrum*—equivalent to a higher frequency. It is indicative of an *electromagnetic* source approaching the observer, or vice versa. The term applies to any decrease in wavelength and increase in frequency caused by relative motion, even outside the visible spectrum.

— **boson**: a subatomic particle, such as a *photon*, that has zero or integral *spin*; it is a force carrier.

— **Cambrian explosion**: was the relatively short (5 million year) *evolutionary* event, beginning around 530 million years ago in the Cambrian period, during which most major animal *phyla* appeared as indicated by the *fossil record*.

— **carbon-14 (^{14}C)**: spontaneously disintegrates to ordinary nitrogen (^{14}N) with a *half-life* of 5,730 years; it is normally used to date *fossils* of long dead organisms.

— **catalyst**: a substance that increases the rate of a chemical reaction without itself undergoing any permanent chemical change.

— **Celsius temperature**: the scale is based on 0°C for the freezing point of water and 100°C for the boiling point of water. The difference in the level of a mercury column, say, of a mercury thermometer is divided into 100 equal divisions giving a 1°C difference in temperature—also known as a "centigrade".

— **centriole**: a cylindrical cell structure composed mainly of a *protein* called "tubulin" that is found in most *eukaryotic* cells. An orthogonally arranged pair of centrioles, surrounded by a shapeless mass of dense material, makes up a compound structure called a *centrosome*.

— **centrosome**: an *organelle* that serves as the main *microtubule* organizing center of the animal cell during cell division, as well as a regulator of cell-cycle progression. The centrosome is copied only once per cell cycle so that each daughter cell inherits one centrosome, which contains two structures called *centrioles*. The centrosome replicates during the *S-phase* (the *DNA* replicating phase) of the cell cycle. During the *prophase* in the process of cell division called *mitosis*, the centrosomes migrate to opposite poles of the cell. The *mitotic spindle* then forms between the two centrosomes. Upon division, each daughter cell receives one centrosome. Aberrant numbers of centrosomes in a cell have been associated with cancer.

— **chlorophyll**: a green pigment found in the *chloroplasts* of plants. Chlorophyll is vital for *photosynthesis*, which allows plants to absorb energy from sunlight. It absorbs blue and red light, converting it into chemical energy, but reflects green light; hence it has a green appearance.

— **chloroplast**: an *organelle* found in plants; its main role is to conduct *photosynthesis*, where the pigment chlorophyll captures the energy from sunlight, converts it, and stores it in energy-storage molecules (such as *ATP*) while freeing oxygen from water.

— **chromosome**: a packaged and organized structure containing most of the *DNA* of a living organism. DNA is not usually found on its own, but rather is structured in long strands that are wrapped around protein complexes called "nucleosomes" that consist of proteins called "histones"—like thread around a spool. The DNA in chromosomes serves as the source for *transcription*. Most eukaryotic cells have a set of chromosomes (46 in humans) with the genetic material spread among them.

— **cladistics**: an approach to biological classification in which organisms are categorized based on shared derived characteristics that can be traced to a group's most recent common ancestor and are not present in more distant ancestors.

— **class**: a *taxonomic* rank of living organisms fitting between *phylum* and *order*.

— **classical physics**: physics that does not make use of *quantum mechanics* or the *relativity theory*. Many theories in classical physics break down when applied to extremely small objects such as *atoms* or to objects moving near the speed of light.

— **codon**: a sequence of three adjacent *nucleotides* on a strand of a *nucleic acid* (such as *DNA* or *RNA*) that constitutes the *genetic* code for a specific *amino acid* that is to be added to a *polypeptide* chain during *protein* synthesis.

— **collider, particle**: a type of particle accelerator involving directed beams of particles. Colliders may either be ring or linear accelerators, and may smash a single beam of particles against a stationary target or two beams head-on.

— **compound**: a *molecule* consisting of two or more atoms, at least two of which are from different elements, held together by chemical bonds.

— **consciousness**: the state or quality of awareness, or of being aware of an external object or something within oneself. It has been defined as: sentience, awareness, subjectivity, the ability to experience or to feel, wakefulness, having a sense of selfhood, and the executive control system of the mind.

— **cosmic microwave background (CMB)**: the thermal radiation left over from the time of recombination of electrons to atomic nuclei in *big bang cosmology.*

— **cosmological constant**: the value of the *energy density* of "empty" space (the "vacuum"). It was originally a mathematical device introduced by Albert Einstein as an addition to his general *relativity theory* to counteract the effect of *gravity* and achieve a *static universe*, which was the accepted view at the time.

— **cosmology**: the study of the origin, evolution, and eventual fate of the universe as a whole.

— **critical density**: that quantity of matter and energy per cubic meter of space that places the universe precisely on the edge between its continued expansion and collapse.

— **cytoplasm**: the material within a living cell, excluding the *nucleus.* It comprises the gel-like colorless substance enclosed within the cell membrane and the *organelles*, the cell's internal sub-structures. It comprises 90% of the volume of a cell. The cytoplasm itself is about 90% water.

— **cytoskeleton**: a microscopic network of *protein filaments* and *tubules* in the *cytoplasm* of many living cells, giving them shape and mechanical resistance to deformation; it also provides coherence, through association with extracellular connective *tissue* and other cells, thus stabilizing entire tissues.

— **dark energy**: an unknown form of energy which is hypothesized to permeate all of space, tending to accelerate the expansion of the universe. Dark energy is the most accepted hypothesis to explain the observations since the 1990s indicating that the universe is expanding at an accelerating rate.

— **dark matter**: an unidentified type of matter comprising approximately 27% of the mass and energy in the observable universe that is not accounted for by dark energy, *baryonic* matter (ordinary matter), and *neutrinos*.

— **density**: mass per unit volume of a substance; analogous to the weight of one cubic meter of a substance.

— **deoxyribonucleic acid (DNA)**: a *molecule* that carries the *genetic* instructions used in the growth, development, functioning, and reproduction of all known living organisms and many *viruses*. The four *bases* found in DNA are adenine (A), cytosine (C), guanine (G), and thymine (T). These four bases are attached to the sugar-phosphate to form the complete *nucleotide*.

— **determinism**: the doctrine that all events, including human action, are ultimately determined by causes external to the will. Some philosophers have taken determinism to imply that individual human beings have no free will and cannot be held morally responsible for their actions.

— **deuterium**: an *isotope* of hydrogen, having one *proton* and one *neutron*; its symbol is "^{2}H".

— **deuteron**: a *proton* and *neutron* bound together by the *strong nuclear force*; it is a nucleus of *deuterium*.

— **diproton**: two *protons* bound together by the *strong nuclear force*; it is the *nucleus* of a very unstable *isotope* of helium— symbol "^{2}He".

— **dogma**: an official system of principles or tenets of a church, or a strong belief that the ones adhering to it are not willing to discuss rationally.

— **domain**: the top *taxonomic* rank of living organisms followed by *kingdom*.

— **Doppler Effect (shift)**: the apparent change in frequency, or wavelength, of a wave to an observer moving relative to its source. When an observer approaches a source (or vice versa), the frequency emitted by the source apparently increases and the wavelength decreases. When an observer recedes from a source (or vice versa), the frequency emitted by the source apparently decreases and the wavelength increases.

— **duality, particle-wave**: the phenomenon of elementary particles behaving as waves and, vice versa, waves behaving as particles.

— **electromagnetism**: the fundamental force associated with electric and magnetic fields. The electromagnetic force is carried by the *photon* and is responsible for *atomic* structure, chemical reactions, the attractive and repulsive forces associated with electrical charge and magnetism, and all other electromagnetic phenomena. Michael Faraday showed that an electric current produces a magnetic *field*, and a changing magnetic field produces an electric current. James Maxwell completed the theory with a full mathematical description of the relationship between electric and magnetic fields.

— **electron**: a stable subatomic particle with a charge of elementary (indivisible) negative electricity, found in all atoms and acting as the primary carrier of electricity in solids.

— **element, chemical**: a substance that cannot be broken down into simpler substances by chemical means; an element is composed of *atoms* that have the same number of *protons* in the *nucleus*.

— **embryo**: a human or animal in the early stages of development before it is born, hatched, etc.

— **endoplasmic reticulum**: a type of *organelle* in *eukaryotic* cells that forms an interconnected network of flattened, membrane-enclosed sacs or tube-like structures. The membranes of the endoplasmic reticulum are continuous with the outer membrane of the *nucleus*. There are two types of endoplasmic reticulum: rough and smooth. The outer face of the rough endoplasmic reticulum is studded with *ribosomes*, which are the sites of *protein* synthesis. The smooth endoplasmic reticulum has no ribosomes; its functions are fat manufacture and metabolism, production of hormones, and detoxification.

— **energy density**: the amount of energy in one cubic meter of "empty" space (the "vacuum").

— **entropy**: a *thermodynamic* quantity representing the unavailability of a system's thermal energy for conversion into mechanical work; it is often described as the amount of disorder in the system.

— **enzyme**: a *protein molecule* that helps other organic molecules enter into chemical reactions with one another but is itself unaffected by these reactions. In other words, enzymes act as *catalysts* for organic biochemical reactions.

— **eukaryote**: an organism whose cells contain a *nucleus* and other *organelles* enclosed within membranes.

— **event horizon**: the point of no return of a *black hole*, beyond which neither matter nor radiation (not even light) can escape from it.

— **evolution**: the change in the heritable characteristics of biological populations over successive generations. The term is also used, generally, to mean progressive change.

— **exclusion principle, Pauli's**: the idea that two identical *fermions*, particles with half-integer ($\frac{1}{2}$) *spin*, cannot have, within the limits of the *uncertainty principle*, both the same position and the same velocity.

— **family**: a *taxonomic* rank of living organisms fitting between *order* and *genus*.

— **fermion**: can be an elementary particle, such as the *electron*, or it can be a composite particle, such as the *proton*; i.e., fermions make up matter. Particles with integer spin are *bosons*, while particles with half-integer spin are *fermions*.

— **fetus**: an unborn offspring of a mammal; in particular, an unborn human baby more than eight weeks after conception.

— **field**: something that exists throughout space and time, as opposed to a particle that exists at only one point at a time.

— **fission, nuclear**: either a *nuclear* reaction or a *radioactive* decay process in which the *nucleus* of an *atom* splits into smaller parts (lighter nuclei). The fission process often produces free *neutrons* and *gamma photons*, and releases a very large amount of energy.

— **flat earth hypothesis**: an archaic conception of the earth's shape as a plane or a disk.

— **fossil record**: the totality of preserved remains of organisms, both discovered and undiscovered, and their placement in fossil-containing rock formations and sedimentary layers (strata).

— **fusion, nuclear**: a reaction in which two or more *atomic nuclei* come close enough to form one or more different atomic nuclei and subatomic particles (*neutrons* and/or *protons*). The difference in *mass* between the products and reactants is manifested as the release of large amounts of energy.

— **galaxy**: a *gravitationally* bound system of *stars*, stellar remnants, interstellar gas, dust, and *dark matter*.

— **gamma ray**: penetrating *electromagnetic* radiation of a kind arising from the *radioactive* decay of *atomic nuclei* and consists of high-energy *photons*.

— **gene**: a region of *DNA* which is made up of *nucleotides* and is the *molecular* unit of heredity; the transmission of genes to an organism's offspring is the basis of the inheritance of traits. Most biological traits are under the influence of many different genes as well as the gene-environment interaction.

— **genome**: the *genetic* material of an organism; it consists of *DNA* (or *RNA* in some viruses). The genome includes the genes (the coding regions), the non-coding DNA, and the genomes of the *mitochondria* and *chloroplasts*. The human genome consists of 23 different pairs of like *chromosomes* (separate DNA complete strands).

— **genus**: a *taxonomic* rank of living organisms fitting between *family* and *species*.

— **geocentric hypothesis**: a superseded description of the universe with the earth at the center; under the geocentric model, the sun, moon, stars, and planets all circled earth. The geocentric model served as the predominant description of the cosmos in many ancient civilizations.

— **gluon**: an elementary particle that acts as the exchange particle for the *strong nuclear force* between *quarks*; analogous to the exchange of *photons* in the *electromagnetic* force between two charged particles. (In lay terms, they glue quarks together, forming protons and neutrons.)

— **god of the gaps**: a term used to describe observations of theological perspectives in which gaps in scientific knowledge are taken to be evidence or proof of God's existence.

— **Golgi apparatus**: an *organelle* found in eukaryotic *cells* consisting of a collection of fused, flattened membrane-enclosed disks. The Golgi apparatus packages *proteins* into membrane-bound enclosures inside the cell before they are sent to their destination. (In other words, it acts like a post office station.)

— **gravitational lensing**: a significant amount of concentrated matter, such as a cluster of galaxies or the sun, between a distant light source and an observer is capable of bending the light from the source as the light travels towards the observer. The amount of bending is one of the predictions of Albert Einstein's general theory of relativity; *classical physics* also predicts the bending of light, but only half that predicted by the general *relativity theory*.

— **graviton**: a hypothetical elementary particle that mediates the force of *gravity*. If it exists, the graviton is expected to be *massless* because the gravitational force appears to have unlimited range.

— **gravity**: a natural phenomenon by which all things with *mass* are attracted toward one another—including planets, stars, and galaxies.

— **hadron**: a composite particle made of *quarks* held together by the *strong nuclear force* in a similar way as *molecules* are held together by the *electromagnetic force*. Hadrons are categorized into two families: *baryons* made of three quarks, and *mesons*, made of one quark and one *antiquark*. *Protons* and *neutrons* are examples of baryons; *pions* are an example of a meson.

— **half-life**: the time required for a quantity to reduce to half its initial value; the term is commonly used in *nuclear* physics to describe how quickly unstable atoms undergo, or how long stable atoms survive, *radioactive* decay.

— **heat death**: a possible ultimate fate of the universe in which the universe has diminished to a state of no free energy and therefore can no longer sustain processes that increase entropy, such as life. Heat death does not necessarily imply that absolute zero temperature is reached; it only requires that temperature differences may no longer be exploited to provide any energy. This happens when the universe reaches thermodynamic equilibrium or maximum entropy.

— **heaven**: the location of the throne of God and the holy angels, as well as the abode for the righteous dead in the *afterlife*; though this is in varying degrees considered metaphorical. It is also considered a state or condition of existence, rather than a particular place somewhere in the cosmos, of the supreme fulfillment of transformation into the likeness of and union with God in the beatific (face to face) vision of God.

— **heliocentric theory**: the astronomical model in which the earth and planets revolve around the *sun* at the center of the solar system.

— **hell**: the place or state into which, by God's definitive judgment, unrepentant sinners pass either immediately after death (particular judgement) or in the general judgement. Its character is inferred from biblical texts interpreted literally. Theologians today generally see hell as the logical consequence of using free will to reject union with God and, because God will not force conformity, not incompatible with God's justice and mercy.

— **Higgs boson**: an elementary particle in the standard model of particle physics. It is the quantum excitation (embodiment) of the Higgs *field*, a fundamental field of crucial importance to particle physics *theory* first suspected to exist in the 1960s. Unlike other known fields such as the *electromagnetic* field, it takes a non-zero constant value almost everywhere. The question of the Higgs field's existence has been the last unverified part of the standard model of particle physics and, according to some, "the central problem in particle physics".

— **hydrophilic**: (water-loving) possessing an affinity for water or *soluble* in water. Such a substance tends to be polar (i.e., have an uneven charge distribution) so it tends to interact with water and other polar substances more favorably than with oil or other *hydrophobic* solvents.

— **hydrophobic**: (water-fearing/hating) not possessing an affinity for water. Such a substance tends to be nonpolar (i.e., its charge is evenly distributed) thus it prefers other neutral molecules and nonpolar solvents. Since water molecules are polar molecules, hydrophobic molecules do not dissolve well among them; hence a hydrophobic substance tends to aggregate in aqueous solutions, excluding water molecules.

— **hypothesis**: a proposed explanation for a phenomenon; for a hypothesis to be a scientific hypothesis, the scientific method requires that one can test it. Scientists generally base scientific hypotheses on previous observations that cannot satisfactorily be explained with the available scientific *theories*. Even though the words "hypothesis" and "theory" are often used synonymously, a scientific hypothesis is not the same as a scientific theory. A working hypothesis is a provisionally accepted hypothesis proposed for further research.

— **infrared**: an invisible *electromagnetic* radiation with longer wavelengths than those of visible light, extending from the nominal red edge of the visible spectrum at 700 nm to 1,000,000 nm, even though people can see infrared up to at least 1,050 nm. Most of the thermal radiation emitted by objects near room temperature is infrared.

— **intermediate filament**: a *cytoskeletal* component which has a diameter between that of a *microfilament* and that of a *microtubule*. While exhibiting considerable diversity in size, intermediate filaments act like steel girders: providing mechanical strength to cells.

— **inverse square law**: any physical law stating that a specified physical quantity or intensity is inversely proportional to the square of the distance from the source of that physical quantity.

The fundamental cause for this can be understood as geometric dilution corresponding to point-source radiation into three-dimensional space. The effect of the physical quantity decreases rapidly with distance, but theoretically it extends to infinity.

— **ion**: an *atom* or a *molecule* in which the total number of *electrons* is not equal to the total number of *protons*, giving the atom or molecule a net positive or negative electric charge. Ions can be created, by either chemical or physical means; the process is termed "ionization".

— **isotope**: variants of a particular chemical element which differ in *neutron* number. All isotopes of a given *element* have the same number of protons in every *atom*.

— **kaon**: is made of one *quark* and one *antiquark*; i.e., it is the bound states of a strange quark (or antiquark) and an up or down antiquark (or quark).

— **Kelvin temperature**: see *absolute temperature*.

— **kingdom**: a *taxonomic* rank of living organisms fitting between *domain* and *phylum*.

— **large hadron collider (LHC)**: the world's largest and most powerful *particle collider*, the largest, most complex experimental facility ever built, and the largest single machine in the world. It was built by the European Organization for Nuclear Research (CERN) between 1998 and 2008 in collaboration with over 10,000 scientists and engineers from over 100 countries, as well as hundreds of universities and laboratories. It lies in a tunnel 27 kilometers (17 miles) in circumference, as deep as 175 meters (574 feet) beneath the France-Switzerland border near Geneva, Switzerland.

— **latent heat**: the heat required to change the *phase* of a substance, i.e., from solid to liquid, or from liquid to gas.

— **lepton**: an elementary half-integer *spin* particle that does not undergo interactions of the *strong nuclear force*. Two main classes of leptons exist: charged leptons (also known as the *electron*-like leptons), and neutral leptons (better known as *neutrinos*). Charged leptons can combine with other particles to form various composite particles such as *atoms*, while neutrinos rarely interact with anything, and are consequently rarely observed. The best known of all leptons is the electron.

— **light year**: the distance travelled by light in one year, which is about 9.5×10^{12} km (6×10^{12} miles).

— **lipid bilayer**: a thin membrane made of two layers of *amphiphilic molecules*, called "phospholipids", that have a *hydrophilic* phosphate head and a *hydrophobic* tail consisting of two fatty acid chains. The space between the two layers is an oily (hydrophobic) environment; these membranes are flat sheets that form a continuous barrier around all cells. The cell membranes of almost all living organisms and many *viruses* are made of a lipid bilayer, as are the membranes surrounding the cell *nucleus* and other sub-cellular structures.

— **lysosome**: a membrane-bound *organelle* found in nearly all animal cells; they are spherical structures enclosed by a *lipid bilayer* which contain *enzymes* that can break down virtually all kinds of biological *molecules*. Lysosomes act as the waste disposal system of the cell by digesting unwanted materials in the *cytoplasm*, both from outside of the cell and obsolete components inside the cell.

— **macroevolution**: major evolutionary transition from one type of organism to another occurring at the level of the *species* and higher *taxa*.

— **mass**: the quantity of matter; its resistance to acceleration, i.e., changing its speed or direction.

— **matrix**: a collection of numbers ordered by rows and columns.

— **matrix algebra**: a system of mathematical operations involving *matrices*.

— **meiosis**: a type of cell division that reduces the number of *chromosomes* in the parent cell by half and produces four sex cells; this process is required to produce egg and sperm cells for sexual reproduction. In meiosis the chromosomes first duplicate, then the maternal and paternal chromosomes exchange genetic information. They then split into two daughter cells:

the chromatids (modified chromosomes) of one looking like the complement (like a photographic negative) of the other. The daughter cells then divide again separating the like chromatid pairs to form four sex cells. Male and female sex cells fuse during fertilization, creating a cell with a complete set of paired (maternal and paternal) chromosomes.

— **meme**: an idea, behavior, practice, style, or symbol that spreads from person to person within a culture; it can be transmitted from one mind to another through writing, speech, gestures, rituals, etc. Memes are cultural analogues to genes since they self-replicate, mutate, and are subjected to selective pressures.

— **meson**: a subatomic particle composed of one *quark* and one *antiquark*, bound together by the *strong nuclear force*. Mesons are a subfamily of *hadrons*.

— **messenger ribonucleic acid (mRNA)**: conveys genetic information from *DNA* to the *ribosome*, where it specifies the *amino acid* sequence of the *protein* products of *gene* expression. Following *transcription* the *mRNA* is *translated* into a *polymer* of amino acids—a protein.

— **metaphysics**: a branch of philosophy investigating the fundamental nature of being and the world that encompasses it. It tries to answer two basic questions. (1) What is there? (2) What is it like?

— **microevolution**: the change in *gene* frequencies that occurs over time within a population. This change is due to four different processes: mutation, selection (natural or artificial), gene migration, and genetic random drift. This change happens over a relatively short (in evolutionary terms) amount of time compared to the changes termed *macroevolution*, which is where greater differences in the population occur.

— **microfilament**: a filamentous structure in the *cytoplasm* of *eukaryotic* cells forming part of the *cytoskeleton*. Microfilaments are the thinnest filaments of the cytoskeleton; they keep the *organelles* in place inside the cell, and by pinching or folding the cell's membrane at the right places they give the cell the desired shape.

— **microtubule**: a component of the *cytoskeleton* found throughout the *cytoplasm*; they maintain the structure of the cell, provide tracks for intracellular transport, and also make up the internal structure of "cilia" (oar-like structures propelling the cell in its environment). They are also involved in *chromosome* separation during *meiosis* (production of sex cells) and *mitosis* and are the major constituents of *mitotic spindles*, which are used to pull apart *eukaryotic chromosomes*.

— **microwave radiation**: a form of *electromagnetic* radiation with wavelengths ranging from one meter (100 cm) to one millimeter (0.1 cm).

— **Milky Way**: the name of the galaxy in which our solar system happens to be situated.

— **mind**: a brain's integrating function; the set of faculties for acquiring knowledge and understanding, including: *consciousness*, perception, thinking, judgement, and memory.

— **miracle**: an event that violates the principles of science and is therefore normally considered to be the work of a divine agency. (Of course, tricks performed by entertaining illusionists are not miracles.)

— **mitochondria**: the powerhouse of the cell, generating most of the cell's supply of *adenosine triphosphate (ATP)*, which is used as a source of chemical energy.

— **mitosis**: a type of cell division that results in two daughter cells each having the same number and kind of *chromosomes* as the parent *nucleus*, typical of ordinary *tissue* growth. During mitosis, first the chromosomes are duplicated, and then they condense and attach to the *mitotic spindle* fibers that pull one copy of every maternal and paternal chromosome to opposite sides of the cell. When the cell divides, it ends up with two genetically identical daughter *nuclei*. Producing three or more daughter cells instead of the normal two is a mitotic error; certain types of cancer can arise from such mutations.

— **mitotic spindle**: the macromolecular machine that segregates *chromosomes* into two daughter cells during *mitosis*. The major structural elements of the spindle are *microtubule polymers*.

— **molecule**: an electrically neutral (uncharged) group of two or more *atoms* held together by chemical bonds. Molecules are distinguished from *ions* by their lack of electrical charge.

— **monomer**: a *molecule* that may bind chemically to other molecules to form a *polymer*. The process by which monomers combine to form a polymer is called "polymerization". Molecules made of a small number of monomer units (up to a few dozen) are called *oligomers*.

— **multiverse**: a hypothetical set of finite-sized and possibly infinite number of universes, including the universe in which we live. Together, these universes comprise everything that exists: the entirety of space, time, matter, energy, and the physical laws and constants that describe them.

— **natural law**: the use of reason to analyze both social and personal human nature to deduce binding rules of moral behavior.

— **natural laws**: the laws of physics, chemistry, and biology including *natural selection*.

— **natural selection**: the differential survival and reproduction of individuals resulting from an organism's particular traits. It is a key mechanism of *evolution*, the change in heritable traits of a population over time.

— **neuron**: an electrically excitable cell that processes and transmits information through electrical and chemical signals. These signals between neurons occur via *synapses*, specialized connections with other cells. Neurons are the core components of the brain, the spinal cord, and the nervous system; they can also connect to each other to form neural networks.

— **neutrino**: a fermion (an elementary particle with half-integer spin) that interacts only via *the weak nuclear force* and *gravity*. The *mass* of the *neutrino* is much smaller than that of the other known elementary particles (including the *electron*).

— **neutron**: a subatomic particle with no net electric charge and a *mass* slightly larger than that of a *proton*.

— **non-overlapping magisteria (NOMA)**: the view that science and religion each represent different areas of inquiry: facts versus values; so there is a difference between the areas over which they have a domain of teaching authority, and the two domains do not overlap.

— **nucleon**: one of the particles that make up the *atomic nucleus*; there are two known kinds of nucleons: the *neutron* and the *proton*.

— **nucleotide**: an organic *molecule* that serves as a *monomer*, or subunit, of *nucleic acids* like *DNA* and *RNA*. They are the building blocks of nucleic acids; nucleotides are composed of a nitrogenous *base*, a five-carbon sugar (ribose or deoxyribose), and at least one phosphate group.

— **nucleus, atomic**: the small, dense region consisting of *protons* and *neutrons* at the center of an *atom*.

— **nucleus, cell**: a membrane-enclosed *organelle* found in *eukaryotic* cells. Eukaryotic cells usually have a single nucleus containing most of the cell's *genetic* material, which is organized as multiple long linear *DNA molecules* called *chromosomes*.

— **oligomer**: a *molecular* complex that consists of a few *monomer* units, in contrast to a *polymer*, where the number of monomers is, in principle, not limited.

— **order**: a *taxonomic* rank of living organisms fitting between *phylum* and *family*.

— **organ**: a part of an organism that is typically self-contained and has a specific vital function; it is made up of a group of different *tissues* all working together to do that particular function: for example, heart, liver, lung, stomach, and brain.

— **organelle**: a specialized subunit within a cell that has a specific function; individual organelles are usually separately enclosed within their own *lipid bilayers*.

— **osmosis**: the spontaneous net movement of *solvent* (dissolving *molecules*) through a porous membrane into a region of higher *solute* (dissolved molecules) concentration, in the direction that tends to equalize the solute concentrations on the two sides.

— **osmotic pressure**: the minimum pressure which needs to be applied to a *solution* to prevent the inward flow of water across a porous membrane. It is also defined as the measure of the tendency of a solution to take in water by *osmosis*.

— **paranormal**: beyond normal explanation; i.e., impossible to explain by known natural forces or by science.

— **peptide**: biologically occurring short chains of *amino acid monomers* linked by *peptide bonds*.

— **peptide bond**: a chemical bond linking two consecutive *amino acid monomers* together. One amino acid loses a hydrogen *atom* and an oxygen atom from its carboxyl group (CO_2H) and the other loses a hydrogen atom from its amino group (NH_2). This reaction produces a loose *molecule* of water (H_2O) and two amino acids joined by a peptide bond (-CO-NH-).

— **phase, matter**: a state of matter, namely, solid, liquid, gas, and plasma.

— **phase, wave**: the position in the cycle of a wave at a specified time; it is a measure of how close it is to the crest or the trough of the wave.

— **photon**: an elementary particle, the *quantum* (energy packet) of all forms of *electromagnetic* radiation including light; it is the force carrier for the electromagnetic force. The photon has zero *mass* and as a result, the interactions of this force with matter are observable at long distance.

— **photosynthesis**: a process used by plants to convert light energy into chemical energy that can later be released to fuel the organisms' activities; it is the process by which green plants use red and blue light from sunlight to synthesize foods from carbon dioxide and water. Photosynthesis in plants generally involves the green pigment *chlorophyll* and generates oxygen as a byproduct. Photosynthesis is largely responsible for producing and maintaining the oxygen content of the earth's atmosphere;

it supplies all of the organic compounds and most of the energy necessary for life on earth.

— **phylum**: a *taxonomic* rank of living organisms fitting between *kingdom* and *order*.

— **pion**: a subatomic particle consisting of one *quark* and one *antiquark* and is therefore a *meson*.

— **Planck constant**: the proportionality constant relating the energy of an *electromagnetic photon* and its frequency, according to the equation E=hf, where "E" is the "energy" of the photon—or quantum of radiation—of frequency "f", and "h" is the Planck constant, 6.626×10^{-34} (a very small fixed number).

— **Planck length**: the smallest indivisible distance of space, 1.616×10^{-35} meter.

— **Planck time**: the smallest indivisible time span, 5.391×10^{-44} second.

— **planet**: an astronomical object orbiting a *star* or stellar remnant, is massive enough to be rounded by its own *gravity*, is not massive enough to cause *thermonuclear fusion*, and has cleared its neighboring region of circumstellar gas, dust, and debris.

— **polar**: having an unevenly distributed electric charge. Molecular polarity is dependent on the difference in charge between *atoms* in a *compound* and the asymmetry of the compound's structure. Polarity underlies a number of physical properties including solubility, surface tension, and boiling and melting points. The water molecule is a polar molecule and so tends to dissolve other polar molecules.

— **poltergeist**: a type of ghost or other supernatural entity which is responsible for physical disturbances, such as loud noises and objects being moved or destroyed.

— **polymer**: a large *molecule* composed of many repeated subunits called *monomers*. Polymers range from familiar synthetic plastics, such as polystyrene, to natural biological polymers such as *DNA* and *proteins* that are fundamental to biological structure and function.

— **polymerization**: the process of reacting *monomer molecules* together in a chemical reaction to form *polymer* chains or three-dimensional structures.

— **positron**: a positive *electron*; it is the antiparticle or the *antimatter* counterpart of the electron. The positron has a positive electric charge equal (in magnitude) to that of the electron; it also has the same *mass* and a *spin* of ½ like an electron.

— **prokaryote**: a single-celled organism that lacks a membrane-bound *nucleus, mitochondria,* or any other membrane-bound *organelle*.

— **prophase**: the first stage of cell division, during which the *chromosomes* become visible as double rod-shaped structures and the *nuclear* envelope disappears.

— **protein**: a large biological *molecule* consisting of one or more long chains of *amino acid monomers*. Proteins perform a vast array of functions in organisms including *catalyzing* metabolic reactions, *DNA* replication, responding to stimuli, and transporting molecules from one location to another. Proteins differ from one another primarily in their sequence of amino acids, which is dictated by the *nucleotide* sequence of their *genes*, and which usually results in the protein's folding into a specific three-dimensional structure that determines its activity.

— **proton**: a subatomic particle with a positive elementary (indivisible) electric charge and *mass* slightly less than that of a *neutron*.

— **purgatory**: an intermediate state after physical death, in the *afterlife*, in which those destined for *heaven* undergo purification, so as to achieve the holiness necessary to enter the joy of heaven. Only those who die with no un-repented (or unforgiven) grave sins, but have not yet fulfilled the temporal punishment due to their sin, can be in purgatory. Therefore no one in purgatory will remain forever in that state or go to *hell*. It is believed to be a place containing purifying fire similar to hell's, except that it is temporary rather than eternal; however it could also last for centuries, or millennia, and even to the end of the earth.

— **quantum**: the minimum amount of any physical entity involved in an interaction—an energy packet.

— **quantum mechanics**: a subset of physics explaining the nature and physical behavior of matter and energy at the *molecular*, *atomic*, and subatomic levels.

— **quantum physics**: see *quantum mechanics*.

— **quantum theory**: see *quantum mechanics*.

— **quantum tunnel**: a subatomic particle, with insufficient energy to surmount a potential barrier (hill), can tunnel to the other side, thus crossing the barrier, instead of rolling back down. Of course, the probability of this happening is very low; here, the particle borrows energy from its surroundings to tunnel through the wall or "roll over the hill".

— **quark**: an elementary particle and a fundamental constituent of matter. Quarks combine together to form composite particles called *hadrons*, the most stable of which are *protons* and *neutrons*, the components of *atomic nuclei*.

— **radioactive dating**: a technique used to date materials such as rocks or organisms, in which trace *radioactive* impurities were selectively incorporated when they were formed. The method compares the abundance of a naturally occurring radioactive *isotope* within the material to the abundance of its decay products, which form at a known constant rate of decay—the *half-life* of the isotope.

— **radioactivity**: the spontaneous breakdown of one type of *atomic nucleus* into another or more nuclei, emitting radiation in the process; therefore, the original chemical element changes into another or others.

— **red giant**: a luminous giant *star* of low or intermediate *mass* (about 0.3–8.0 solar masses) in a late phase of stellar evolution. The outer atmosphere becomes inflated and sparse, making the radius large and the surface temperature lower than 5,000K. The appearance of a red giant is from yellow-orange to red.

— **redshift**: happens when visible light from an object is apparently increased in wavelength, or shifted to the red end of the *spectrum*.

In general, whether or not the *electromagnetic* radiation is within the visible spectrum, "redder" means an increase in wavelength—equivalent to a lower frequency and lower *photon* energy. It is indicative of an electromagnetic source receding from the observer, or vice versa.

— **relativity theory**: encompasses two *theories* by Albert Einstein: special relativity (for non-accelerated motion) and general relativity (for accelerated motion). Concepts introduced by these two theories include *spacetime* as a unified entity of space and time, relativity of simultaneity, kinematic (motional) and *gravitational* time dilation and length contraction.

— **resurrection**: the concept of a living being coming back to life after one's death—never to die again.

— **reversible process**: a process in which absolutely no energy is converted into heat; in practice this is impossible, and is therefore only a theoretical concept.

— **ribonucleic acid (RNA)**: a *polymeric molecule* essential in various biological roles such as coding, decoding, and expression and regulation of *genes*. Both RNA and *DNA* are nucleic *acids*; like DNA, RNA is assembled as a chain of *nucleotides*; but unlike DNA, RNA is more often found in nature as a single-strand folded onto itself, rather than a paired double-strand. Cellular organisms use *messenger RNA (mRNA)* to convey genetic information (using the letters A, C, G, and U, to denote the nitrogenous *bases* adenine, cytosine, guanine, and uracil, respectively) that directs synthesis of specific *proteins*. Many *viruses* encode their genetic information using an RNA *genome*.

— **ribosome**: a complex *molecular* machine that serves as the site of *protein* synthesis (*translation*). Ribosomes link *amino acids* together in the order specified by *messenger RNA* (mRNA) molecules.

— **ribozyme**: *ribonucleic acid (RNA) molecules* that are capable of *catalyzing* specific biochemical reactions, similar to the action of *protein enzymes*.

— **RNA world hypothesis**: refers to the hypothesized self-replicating *ribonucleic acid (RNA) molecules* believed to have been the precursors to all current life on earth. The RNA world would have eventually been replaced by the *DNA*, RNA, and *protein* world of today.

— **S-phase**: (synthesis phase) is the period of the cell cycle in which *DNA* is replicated.

— **self (identity)**: an organism's principle of life, or the integrating function of its whole body—including, especially, the *mind*; an organism's essential being that distinguishes it from others.

— **solute**: the dissolved substance in a *solution*.

— **solution**: a homogeneous mixture composed of two or more substances. In such a mixture, a *solute* is the substance dissolved (e.g., sugar) in another substance, known as a *solvent* (e.g., water).

— **solvent**: the dissolving substance in a *solution*.

— **soul**: see *self*.

— **spacetime**: a mathematical model that combines space and time into a single interwoven continuous process.

— **species**: the bottom *taxonomic* rank of living organisms following *genus*.

— **spectrum**: the colors of light as seen in a rainbow, or of sunlight separated by a glass prism. Later it came to denote the component frequencies that make up the entire *electromagnetic* spectrum.

— **spin**: an internal property of elementary particles, related to, but not identical to the everyday concept of spin.

— **standard model of particle physics**: a theory concerning the *electromagnetic*, *weak*, and *strong* nuclear forces, as well as classifying all the subatomic particles known. (See "Appendix III (Physics)".)

— **star**: a luminous sphere of plasma (*ionized* gas consisting of positive *nuclei* and free *electrons*) held together by its own *gravity*. The nearest star to earth is the *sun*.

— **static universe**: a *cosmological* model in which the universe is both spatially infinite and temporally infinite, and space is neither expanding nor contracting.

— **statistics**: the science of collecting and analyzing numerical data in large quantities, especially for the purpose of inferring proportions in a whole from those in a representative sample.

— **string theory (hypothesis)**: a theoretical framework in which the point-like particles of particle physics are described as waves on (imaginary) one-dimensional objects called "strings". It describes how these strings propagate through space and interact with each other.

— **strong nuclear force**: holds ordinary matter together, confining *quarks* into *hadrons*, creating the *proton* and *neutron*, and the further binding of neutrons and protons creating *atomic nuclei*.

— **sugar code**: sequence-specific and information-rich structures comparable to the *DNA* code consisting of the various arrangements and orientations of several sugar building blocks outside cell membranes. The sugar code is what dictates the arrangements of different cell types during development.

— **sun**: the *star* at the center of our (earth's) solar system.

— **supernova**: an astronomical event that occurs during the last stellar evolutionary stages of a massive *star's* life, whose dramatic and catastrophic destruction is marked by one final titanic explosion. For a short time, this causes the sudden appearance of a "new" bright star, before slowly fading from sight over several weeks or months. During maximum brightness, the total equivalent radiant energies produced by supernovae may briefly outshine an entire output of a typical galaxy, consisting of billions of stars, and emit energies equal to that created over the lifetime of any solar-like star. Type 1a supernovae follow a characteristic (known) light curve—the graph of luminosity as a function of time—after the explosion.

— **survival of the fittest**: see *natural selection*.

— **synapse**: a structure that permits a *neuron* (or nerve cell) to pass an electrical or chemical signal to another neuron.

— **taxon**: a *taxonomic* rank of living organisms; that is, any one of the following: *domain, kingdom, phylum, order, family, genus,* and *species* (in descending order).

— **taxonomy**: a branch of science that encompasses the description, identification, nomenclature, and classification of organisms

— **thermodynamic equilibrium**: a state where there are no net macroscopic flows of matter or of energy, either within a system or between systems. In a system in its own state of internal thermodynamic equilibrium, no macroscopic change occurs. Systems in mutual thermodynamic equilibrium are simultaneously in mutual thermal, mechanical, chemical, and radiative equilibria.

— **thermodynamics**: a branch of physics concerning heat and temperature in relation to energy transfer.

— **thermonuclear**: the energy produced in *stars* as a result of *nuclear fusion*.

— **theory**: a well-confirmed type of explanation of nature, made in a way consistent with scientific methodology, and fulfilling the criteria required by modern science. Such theories are described in such a way that any scientist in the field is in a position to understand and either provide empirical support ("verify") or empirically contradict ("falsify") it. Scientific theories are the most reliable, rigorous, and comprehensive form of scientific knowledge, in contrast to more common uses of the word "theory" that imply that something is unproven or speculative, which is better characterized by the word "*hypothesis*".

— **tissue**: a cellular organizational level intermediate between cells and a complete *organ*. A tissue is an ensemble of similar cells from the same origin that together carry out a specific function; for example, blood, lining of the lung or stomach, and muscle. Organs are then formed by the functional grouping together of multiple tissues.

— **transcription**: the first step of *gene* expression, in which a particular segment of *DNA* is copied into *messenger RNA (mRNA)*.

— **transfer ribonucleic acid (tRNA)**: an adaptor *molecule* composed of *RNA* that serves as the physical link between the *mRNA* and the *amino acid* sequence of *proteins*. It does this by carrying an amino acid to the *ribosome*, the protein synthetic machinery of a cell, as directed by a *codon*, a three-nucleotide sequence in an mRNA. As such, tRNAs are a necessary component of *translation*, the biological synthesis of new proteins, in accordance with the *genetic* code.

— **translation**: the process in which cellular *ribosomes* create *proteins*. In translation, *mRNA*—produced by *transcription* from *DNA*—is decoded by a ribosome to produce a specific *amino acid* chain, or *polypeptide*.

— **tritium**: an *isotope* of hydrogen, having one *proton* and two *neutrons*; its symbol is "^{3}H".

— **ultraviolet**: an *electromagnetic* radiation with a wavelength from 10 nm to 400 nm; it is shorter than that of visible light but longer than *X-rays*. UV radiation is also present in sunlight.

— **uncertainty principle, Heisenberg's**: a fundamental limit to the precision with which certain pairs of physical properties of a particle, known as complimentary variables such as position and momentum, can be known; the more accurately one knows the one the less accurately one can know the other.

— **undisputed Pauline letters**: there is wide consensus in modern New Testament scholarship on a core group of letters whose authorship is rarely contested that they are authentically Paul's; they are, Romans, First and Second Corinthians, Galatians, Philippians, First Thessalonians, and Philemon.

— **universal common ancestor**: the hypothesis that all living organisms originated from a single organism.

— **universal wave function**: physicist and cosmologist Frank Tipler's idea of a physical *field*, encompassing all of space, connecting the natural with the supernatural; it converts inanimate material structures, like animal or plant bodies, to living organisms, giving animals consciousness and feelings; it is also identified as the Holy Spirit "the giver of life" in Christian traditions.

— **vacuole**: a membrane-enclosed *organelle* in plants serving as a reservoir for wastes, nutrients, and pigments. It occupies about 90% of the volume of some plant cells and is under high *osmotic* *pressure.* This pressure, pushing against the cell wall, stiffens the cell, thus also serving a structural role.

— **virus**: a small infectious agent that replicates only inside the living cells of other organisms; it hijacks the cell's *transcription* and *translation* mechanisms just to make more copies of itself, thus preventing it from performing its usual function.

— **W boson**: an elementary particle that mediates the *weak nuclear force*; it can have either a positive or negative (single) electric charge. The two oppositely charged *bosons* are each other's *antiparticles.*

— **weak nuclear force**: one of the four known fundamental forces of nature, alongside the *strong nuclear force, electromagnetism,* and *gravity.* The weak nuclear force is responsible for *radioactive* decay, which plays an essential role in *nuclear fission.* It is mediated by the exchange of the force carriers, the W and the Z *bosons.*

— **white dwarf**: a *stellar* remnant composed mostly of matter devoid of its *electrons.* A white dwarf is very dense; its mass is comparable to that of the sun, while its volume is comparable to that of the earth. A white dwarf's faint luminosity comes from the emission of stored thermal (heat) energy; no *fusion*, wherein mass is converted to energy, takes place in a white dwarf.

— **X-ray**: a form of *electromagnetic* radiation; X-rays have a wavelength ranging from 0.01nm to 10nm. X-ray wavelengths are shorter than those of *ultraviolet* rays and typically longer than those of *gamma rays.*

— **Z boson**: an elementary particle that mediates the *weak nuclear force.* It has no electric charge.

— **zygote**: the cell produced from the mating of a female egg and a male sperm.

BIBLIOGRAPHY

— Abelson, Philip H. "Chemical Events on the Primitive Earth." *Proceedings of the National Academy of Sciences USA*; vol. 55 (1966): pp. 1365–72.

— Ananthaswamy, Anil. *The Edge of Physics: A Journey to Earth's Extremes to Unlock the Secrets of the Universe*. New York, NY: Mariner Books, 2011; ISBN 9780547394527.

— *Arguing With Atheists: God of the Gaps*. http://www.arguingwithatheists.com/pages/Arguments.htm.

— Atkins, Peter William. *The Creation*; Oxford, U.K.: W.H. Freeman, 1981.

— Axe, Douglas D. "Estimating the Prevalence of Protein Sequences Adopting Functional Folds." *Journal of Molecular Biology*, vol. 341 (2004): pp. 1295–315.

— Ayala, Francisco Jose, Rzhetsky (Andrey), & Ayala (Francisco J.). "Origin of the Metazoan Phyla: Molecular Clocks Confirm Paleontological Estimates." *Proceedings of the National Academy of Sciences USA* vol. 95 (1998): pp. 606–11.

— Barrow, John David & Silk, Joseph Ivor. "The Structure of the Early Universe" *Scientific American*; vol. 242, no. 4 (1980): pp. 118–28.

— Barrow, John David & Tipler, Frank Jennings. *The Anthropic Cosmological Principle*; New York, NY: Oxford University Press, 1986.

— Barrow, John David & Tipler, Frank Jennings. *The Anthropic Cosmological Principle*; New York, NY: Oxford University Press, 1988.

— Beauregard, Mario & O'Leary, Denyse. *The Spiritual Brain: A Neurologist's Case for the Existence of the Soul.* Toronto, ON: Harper Perennial, 2008; ISBN 9781554682188.

— Behe, Michael J. *Darwin's Black Box: The Biochemical Challenge to Evolution.* New York, NY: Free Press, 2006; ISBN 0743290313, 9780743290319.

— Brachet, Jean, Denis (Helene), & De Vitry (F.) "The Effects of Actinomycin D and Puromycin on Morphogenesis in Amphibian Eggs and Acetabularia Mediterranea" *Developmental Biology*; vol. 9, no. 3 (1964): pp. 398–434.

— *British Society for Cell Biology: Endoplasmic Reticulum.* http://bscb.org/learning-resources/softcell-e-learning/endoplasmic-reticulum-rough-and-smooth/.

— *British Society for Cell Biology: Ribosome.* http://bscb.org/learning-resources/softcell-e-learning/ribosome/.

— Britten, Roy J. & Davidson, Eric H. "Gene Regulation for Higher Cells: A Theory" *Science* vol. 165 (1969): pp. 349–57.

— Bronham, Lindell, Rambaut (Andrew), Fortey (Richard), Cooper (Alan), & Penny (David). "Testing the Cambrian Explosion Hypothesis by Using a Molecular Dating Technique" *Proceedings of the National Academy of Sciences USA* vol. 95 (1998): pp. 12386–9.

— Brooks, Jim. *The Origins of Life*; Belleville, MI: Lion Publishing Inc., 1985.

— Brysse, Keynyn. "From Weird Wonders to Stem Lineages: The Second Reclassification of the Burgess Shale Fauna." *Studies in the History and Philosophy of Science Part C: Studies in the History and Philosophy of Biological and Biomedical Sciences* vol. 39, no. 3 (2008):pp. 298–313.

— Carr, Bernard J. "On the Origin, Evolution and Purpose of the Physical Universe." *Irish Astronomical Journal* vol. 15, no. 3 (1982): pp. 237–53.

— Carr, Bernard J. & Rees, Martin John. "The Anthropic Principle and the Structure of the Physical World" *Nature* vol. 278 (1979): pp. 605–612.

— Carter, Brandon. "Large Number Coincidences and the Anthropic Principle" *Confrontation of Cosmological Theories with Observational Data: International Astronomical Union Symposia*. eds. Longair, Malcolm S. Dordrecht, Holland: D. Reidel Publishing Company, 1974: pp. 290–8.

— Carter, Brandon. "Understanding the Fundamental Constants." in *Atomic Masses and Fundamental Constants 5*. New York, NY: Plenum Press, 1976; eds. Sanders, J. H. & Wapstra, A. H.: pp. 650–4.

— *Catechism of the Catholic Church*. Vatican City, Italy: Libreria Editrice Vaticana (Latin original), 1992; trans. Canadian Conference of Catholic Bishops; Ottawa, ON: Publications Service, 1994; ISBN 0889972818.

— *Christian Apologetics and Research Ministry: Taxonomic Rank*. https://carm.org/dictionary-taxonomic-rank.

— Chung, Wen-Yu, Wadhawan (Samir), Szklarczyk (Radek), Pond (Sergei Kosakovsky), & Nekrutenko (Anton). "A First Look at the ARFome: Dual-coding Genes in Mammalian Genomes" *PLoS Computational Biology*; vol. 3, no. 5 (2007): e91.

— Close, Frank. *Antimatter*; Oxford, U.K.: Oxford University Press, 2010; ISBN 9780199578870.

— Craig, William Lane. "Cosmos and Creator" *Origins and Design* vol. 17, no. 2 (1996): pp.18-28.

— Crump, Thomas. *A Brief History of Science: As Seen Through the Development of Scientific Instruments*. London, U.K.: Robinson, 2002; ISBN 9781841195220.

— *Crystalinks: Quantum Mechanics*. http://www.crystalinks.com/quantumechanics.html.

— Darwin, Charles. *The Origin of Species: By Means of Natural Selection or The Preservation of Favored Races in the Struggle for Life*. New York, NY: Modern Library, 2009; ISBN 9780375751462.

— Davies, Paul Charles William. *Other Worlds*; New York, NY: Touchstone/Simon & Schuster, 1980.

— Davies, Paul Charles William. *Superforce*; New York, NY: Touchstone, 1984.

— Davies, Paul Charles William. "The Anthropic Principle" *Progress in Particle and Nuclear Physics* vol. 10(C) (1983): pp. 1–38.

— Dawkins, Richard. *The God Delusion*; New York, NY: Mariner Books, 2008; ISBN 0618918248, 9780618918249.

— Dawkins, Richard. *The Greatest Show on Earth: The Evidence for Evolution.* New York, NY: Free Press, 2010; ISBN 9781416594796.

— Dawkins, Richard. *The Selfish Gene (30th anniversary edition).* New York, NY: Oxford University Press, 2006; ISBN 9780199291151.

— De Marchi, John. *The Immaculate Heart: The True Story of Our Lady of Fatima.* ed. Fay, William; New York, NY: Farrar, Straus & Young, 1952.

— De Marchi, John. *The True Story of Fatima: A Complete Account of the Fatima Apparitions.*

— Demaret, Jacques & Barbier, Christian. "Le Principe Anthropique en Cosmologie" *Revue des Questions Scientifiques* vol. 152, no. 4 (1981): pp. 461–509.

— Dembski, William Albert. *The Design Inference: Eliminating Chance Through Small Probabilities.* Cambridge, UK: Cambridge University Press, 1998.

— Dennett, Daniel Clement III. *Consciousness Explained.* Boston, MA: Little, Brown and Company, 1991.

— Dicke, Robert Henry. "Gravitation without a Principle of Equivalence" *Reviews of Modern Physics* vol. 29, no. 3 (1957): pp. 363–76.

— Dicke, Robert Henry & Peebles, Philip James Erwin. *General Relativity; an Einstein Centenary Survey*; eds. Hawking, Stephen William & Israel, Werner; Cambridge, U.K.: Cambridge University Press Archive, 1979.

— Distefano, Matthew. *Biology: Homework Helpers*, Franklin Lakes, NJ: Career Press, 2004; ISBN 1564147207.

— Dyson, Freeman. "Energy in the Universe" *Scientific American* vol. 225, no. 3 (1971): pp. 50–9.

— Ehrman, Bart D. *Did Jesus Exist?—The Historical Argument for Jesus of Nazareth* ; New York, NY: Harper One, 2012; ISBN 9780062204608.

— Ehrman, Bart D. *How Jesus Became God: The Exaltation of a Jewish Preacher from Galilee.* New York, NY: Harper One, 2014; ISBN 9780061778186.

— Erwin, Douglas H., Laflamme (Marc), Tweedt (Sarah M.), Sperling (Eric A.), Pisani (Davide), & Peterson (Kevin J.). "The Cambrian Conundrum: Early Divergence and Later Ecological Success in the Early History of Animals." *Science* vol. 334 (2011): pp. 1091–7.

— Etinger, Judah. *Foolish Faith: What 21ˢᵗ Century Man Says about God.* Green Forest, AR: Master Books, 2003; ISBN 0890513996.

— *Fatima, Crusader: Miracle of the Sun.* http://www.fatima.org/crusader/crintro/crintropg16.asp.

— *Four Dimensional Afterlife*; http://4dimensionalafterlife.org/page6.htm.

— Gabius, Hans-Joachim. "Biological Information Transfer beyond the Genetic Code: The Sugar Code." *Naturwissenschaften* vol. 87 (2000): pp. 108–21.

— Gabius, Hans-Joachim, Siebert (Hans-Christian), Andre (Sabine), Jimenez-Barbero (Jesus), & Rudiger (Harold). "Chemical Biology of the Sugar Code" *Chembiochem* vol. 5, no. 6 (2004): pp. 740–64.

— Gale, George. "The Anthropic Principle" *Scientific American* vol. 245, no. 6 (1981): pp. 154–71.

— Gil, Rosario, Silva (Francisco J.), Pereto (Juli), & Moyal (Andres). "Determination of the Core of a Minimal Bacterial Gene Set" *Microbiology and Molecular Biology Reviews* vol. 68 (2004): pp. 518–37.

— Gilbert, Walter. "Origin of Life: The RNA World." *Nature* vol. 319 (1986): p.618.

— Graham, Loren R. *Science and Philosophy in the Soviet Union.* London, U.K.: Allen Lane, 1973.

— Graur, Dan & Martin, William. "Reading the Entrails of Chickens: Molecular Timescales of Evolution and the Illusion of Precision" *Trends in Genetics* vol. 20, no. 2 (2004): pp. 80–6.

— Greene, Brian. *The Fabric of the Cosmos: Space, Time, and the Texture of Reality*. New York, NY: Vintage Books, 2005; ISBN 0375727205.

— Grossman, Neal. "Who's Afraid of Life after Death?" *Journal of Near-Death Studies* vol. 21, no.1 (2002): pp. 5–24.

— Halvorson, Hans. "The Measure of All Things: Quantum Mechanics and the Soul." in *The Soul Hypothesis: Investigations into the Existence of the Soul*. eds. Baker, Mark C. & Goetz, Stewart; New York, NY: The Continuum International Publishing Group Inc., 2011; ISBN 9781441152244: pp. 138–67.

— Harvey, Ethel Browne. "A Comparison of the Development of Nucleate and Non-nucleate Eggs of Arbacia Punctulata" *Biological Bulletin* vol. 79 (1940): pp. 166–87.

— Harvey, Ethel Browne; "Parthenogenetic Merogony or Cleavage without Nuclei in Arbacia Punctulata." *Biological Bulletin* vol. 71 (1936): pp. 101–21.

— Hawking, Stephen William. "The Anisotropy of the Universe at Larger Times" *Confrontation of Cosmological Theories with Observational Data: International Astronomical Union Symposia*. ed. Longair, Malcolm S. Dordrecht, Holland: D. Reidel Publishing Company, 1974: pp. 283–9.

— Hawking, Stephen William. *The Universe in a Nutshell*; New York, NY: Bantam Books, 2001; ISBN 055380202X.

— Hoekstra, Hopi E. & Coyne, Jerry A. "The Locus of Evolution: Evo Devo and the Genetics of Adaptation." *Evolution* vol. 61 (2007): pp. 995–1016.

— Holland, Heinrich D. *The Chemical Evolution of the Atmosphere and Oceans*, Princeton, NJ: Princeton University Press, 1984.

— Hood, Bruce. *The Self Illusion: How the Social Brain Creates Identity*. Toronto, ON: HarperCollins Publishers Ltd, 2012; ISBN 9781443405225.

— Hoyle, Fred. "On Nuclear Reactions Occurring in Very Hot Stars: I. The Synthesis of Elements from Carbon to Nickel." *Astrophysical Journal Supplement* vol. 1 (1954): pp. 121–46.

— Hoyle, Fred. *The Intelligent Universe*; London, UK; Michael Joseph, 1983

— Hume, David. *An Enquiry Concerning Human Understanding*, 1758

— Illangasekare, Mali, Sanchez (Giselle), Nickles (Tim), & Yarus (Michael). "Aminoacyl-RNA Synthesis Catalyzed by an RNA." *Science* vol. 267 (1995): pp. 643–7.

— Isham, Christopher J., Penrose (Roger), & Sciama (Dennis William), eds. *Quantum Gravity 2: A Second Oxford Symposium.* Oxford, U.K.: Clarendon Press, 1981.

— Jaki, Stanley L. *God and the Sun at Fatima.* Fraser, MI: Real View Books, 1999.

— Johnston, Wendy K., Unrau (Peter J.), Lawrence (Michael S.), Glasner (Margaret E.), & Bartel (David P.) "RNA-Catalyzed RNA Polymerization: Accurate and General RNA-Templated Primer Extension." *Science* vol. 292 (2001): pp. 1319–25.

— Josephus, Flavius. "From the Banishment of Archelus to the Departure from Babylon" *Antiquities of the Jews: Book XVIII.* Rome, Italy: (Greek original), 93CE; trans. Whiston, William (1737).

— Josephus, Flavius. "From Fadus the Procurator to Florus" *Antiquities of the Jews: Book XX.* Rome, Italy: (Greek original), 93CE; trans. Whiston, William (1737).

— Joyce, Gerald & Orgel, Leslie. "Progress toward Understanding the Origin of the RNA World." in *The RNA World: The Nature of the Modern RNA Suggests a Prebiotic RNA World*; eds. Gesteland (Raymond F.) Cech (Thomas), & Atkins (John F.); Woodbury, NY: Cold Spring Harbor Laboratory Press, 2006: pp. 23–56.

— Joyce, Gerald F. & Orgel, Leslie. "Prospects for Understanding the Origin of the RNA World." in *The RNA World*; eds. Gesteland, Raymond F. & Atkins, John J.; Cold Spring Harbor, NY: Cold Spring Harbor Laboratory Press, 1993: pp. 1–25.

— Kamminga, Harmke. "Studies in the History of Ideas on the Origin of Life" (Ph.D. dissertation) London, U.K.: University of London, 1980.

— Kelly, Matthew. *Rediscover Catholicism: a Spiritual Guide to Living with Passion & Purpose*. Boston, MA: Beacon Publishing, 2010; ISBN 9781937509675.

— Kenyon, Dean H. Forward in Thaxton, Charles B., Bradley, Walter L. & Olsen, Roger L. *The Mystery of Life's Origin: Reassessing Current Theories*. New York, NY: Philosophical Library, 1984: pp. v–viii.

— *Kimball's Biology Pages: Genome Sizes.* http://users.rcn.com/jkimball.ma.ultranet/BiologyPages/G/GenomeSizes.html.

— Lakatos, Imre. *Proofs and Refutations*; Cambridge, U.K.: Cambridge University Press, 1976.

— Lawrence, Peter A. & Struhl, Gary. "Morphogens, Compartments and Pattern: Lessons from Drosophila?" *Cell* vol. 85 (1996): pp. 951–61.

— Layton, Julia & Freudenrich, Craig. http://science.howstuffworks.com/sun6.htm.

— Leibniz, Gottfried. *The Principles of Nature and of Grace, Based on Reason*. Vienna, Austria: (French original), 1714; trans. Latta, Robert (1898).

— Leslie, John Andrew. "Anthropic Principle, World Ensemble, Design." *American Philosophical Quarterly* vol. 19 (1982): pp. 141–51.

— Leslie, John Andrew. "The Prerequisites of Life in Our Universe" *Newton and the New Direction of Science*; eds. Coyne, George V., Heller, Michal, & Zycinski, Jozef; Vatican City (Italy): Specola Vaticana, (1988) pp. 229–58.

— Levy, Matthew & Miller, Stanley Lloyd. "The Stability of the RNA Bases: Implications for the Origin of Life." *Proceedings of the National Academy of Science USA* vol. 95 (1998): pp. 7933–8.

— Lewis, Clive Staples. "The Problem of Pain." in *The Complete C. S. Lewis Signature Classics*; designed by Smelser, Claudia; New York, NY: HarperOne, 2007; ISBN 0061208493, 9780061208492.

— *Liquisearch: Miracle of the Sun, Critical Evaluation*; http://www.liquisearch.com/miracle of the sun/ critical evaluation of the event.

— Martin, Robert A. *Missing Links: Evolutionary Concepts and Transitions through Time.* Boston, MA: Jones & Bartlett, 2004.

— Masui, Y., Forer, A., & Zimmerman, A.M. "Induction of Cleavage in Nucleated and Enucleated Frog Eggs by Injection of Isolated Sea-Urchin Mitotic Apparatus." *Journal of Cell Science* vol. 31 (1978): pp. 117–35.

— Matzke, Nicholas. "Meyer's Hopeless Monster, Part II" *Panda's Thumb* (19 June 2013): http://pandasthumb.org/archives/2013/06/meyers-hopeless-2.html.

— McClure, Kevin. *The Evidence for Visions of the Virgin Mary*; Wellingborough, U.K.: Aquarian Press, 1983.

— McDonald, John F. "The Molecular Basis of Adaptation: A Critical Review of Relevant Ideas and Observations." *Annual Review of Ecology and Systematics* vol. 14 (1983): pp. 77–102.

— Meyer, Stephen C. *Darwin's Doubt: The Explosive Origin of Animal Life and the Case for Intelligent Design.* New York, NY: HarperOne, 2014; ISBN 9780062071484.

— Meyer, Stephen C. "Evidence for Design in Physics and Biology: From the Origin of the Universe to the Origin of Life." in *Science and Evidence for Design in the Universe*; eds. Behe (Michael J.), Dembski (William A.), & Meyer (Stephen C.); San Francisco, California: Ignatius Press, 2000: pp. 53–111.

— Meyer, Stephen C. *Signature in the Cell: DNA and the Evidence for Intelligent Design.* New York, NY: Harper One, 2009; ISBN: 9780061472794.

— Miller, Stanley Lloyd. "A Production of Amino Acids under Possible Primitive Earth Conditions" *Science*, vol. 117, no. 3046 (1953): pp. 528–9.

— Nagel, Thomas. "What Is It Like to Be a Bat?" *The Philosophical Review*, vol. 85, no. 4 (1974) pp. 435–50.

— *Nature, Scitable: Protein Function.* http://www.nature.com/scitable/topicpage/protein-function-14123348.

— Nelson, Paul. Contemporaneous notes taken while attending Weischaus' (Eric) Nobel Lecture: "From Molecular Patterns to Morphogenesis: The Lessons from Drosophila." 8 December 1995: pp. 314–26.

— *New American Bible: Revised Edition*; translated from original languages, authorized by Confraternity of Christian Doctrine; Totowa, NJ: Catholic Book Publishing Corp., 2010; ISBN 9780899429519.

— *New Testament and Psalms: Fourth Edition*; American Bible Society, New York, NY: Canadian Bible Society, 1976; ISBN 0888346395.

— Nithyananda, *Bhagavad Gita Demystified: Third Edition*. Volume 2, Chapters 7–12; Nithyananda Mission, Karnataka, India (2011)

— Noller, Harry F., Hoffarth (Vernita), & Zimniak (Ludwika). "Unusual Resistance of Peptidyl Transferase to Protein Extraction Procedures" *Science* vol. 256 (1992): pp. 1416–9.

— Nusslein-Volhard, Christiane & Weischaus, Eric. "Mutations Affecting Segment Number and Polarity in Drosophila" *Nature* vol. 287 (1980): pp. 795–801.

— Oparin, Aleksandr Ivanovich. *Genesis and Evolutionary Development of Life*; New York, NY: Academic: trans. Maass, Eleonor; 1968.

— Oparin, Aleksandr Ivanovich. *The Origin of Life*, trans. Morgulis, Sergius. New York, NY: Macmillan, 1938.

— Pais, Abraham. *Subtle Is the Lord—The Science and the Life of Albert Einstein*; New York, NY: Oxford University Press (2005).

— Parfit, Derek. *Reason and Persons*; Oxford, U.K.: Oxford University Press, 1986.

— Pelletier, Joseph A. *The Queen of Peace Visits Medugorje*; Worchester, MA: Assumption Publications, 1988.

— Penrose, Roger. *The Emperor's New Mind*. New York, New York: Oxford University Press, 1989.

— Phillips, Helen. "The Ten Biggest Mysteries of Life"; *New Scientist*, (2004): pp. 24–34.

— *Piercedhearts: Garabandal.* http://www.piercedhearts.org/treasures/shrines/garabandal.htm.
— Plutarch, Lucius Mestrius. *Theseus.* Chaeronea, Greece, 75CE, trans. Dryden, John http://classics.mit.edu/Plutarch/theseus.html.
— Press, William H., Lightman (Alan Paige), Peierls (Rudolf), & Gold (T.). "Dependence of Macrophysical Phenomena on the Values of the Fundamental Constants" *Philosophical Transactions of the Royal Society of London A.* vol. 310 (1983): pp. 323–36.
— *Raptureready: Moody.* http://www.raptureready.com/resource/moody/m7.html.
— Rees, Martin John. "Our Universe and Others" *Quarterly Journal of the Royal Astronomical Society* vol. 22 (1981): pp. 109–24.
— Rees, Martin John & McCrea, William Hunter. "Large Numbers and Ratios in Astrophysics and Cosmology" *Philosophical Transactions of the Royal Society of London A.* Vol. 310 (1983): pp. 311–22.
— Ridley, Matt. *Genome: The Autobiography of a Species in 23 Chapters.* New York, NY: Harper Perennial, 2006; ISBN 0060894083, 9780060894085.
— Rood, Robert T. & Trefil, James S. *Are We Alone?: The Possibility of Extraterrestrial Civilizations.* New York, NY: Scribner, 1982.
— Rosenberg, Alex. *The Atheist's Guide to Reality: Enjoying Life without Illusions.* New York, NY: W. W. Norton & Company Inc., 2012; ISBN 9780393344110.
— Rozental, Iosif Leonidovich. *Elementary Particles and the Structure of the Universe;* Moscow, Russia: Izdatel'stvo Nauka (Russian original), 1984.
— Rozental, Iosif Leonidovich. *On Numerical Values of Fundamental Constants;* Moscow, Russia: (NASA STI/Recon Technical Report) 1980.
— Rozental, Iosif Leonidovich. "Physical Laws and the Numerical Values of Fundamental Constants" *Soviet Physics: Uspekhi.* vol. 23, no.6 (1980): pp. 296–305.
— Ryden, Barbara. "The First Three Minutes." http://www.astronomy.ohio-state.edu/~ryden/ast162_10/notes44.html

— Ryle, Gilbert. "Descartes' Myth." in *The Concept of Mind.* Hutchinson, London, UK, 1949

— Sabom, Michael. *Life and Death: One doctor's Fascinating Account of Near-Neath Experiences.* Grand Rapids, MI: Zondervan, 1998.

— Salpeter, Edwin Ernest. "Accretion of Interstellar Matter by Massive Objects" *The Astrophysical Journal*; vol. 140 (1964): pp. 796–800.

— Salpeter, Edwin Ernest. "Nuclear Reactions in Stars: Buildup from Helium." *Physical Review* vol. 107, no. 2 (1957): pp. 516–24.

— Schnaar, Ronald L. "The Membrane is the Message." *The Sciences* May–June (1985): pp. 34–40.

— Schrödinger, Erwin. *What is Life? : The Physical Aspects of the Living Cell* (repr. ed.) Cambridge, U.K.: Cambridge University Press, 2001.

— Schwartz, Alan W. "Intractable Mixtures and the Origin of Life" *Chemistry and Biodiversity.* vol. 4, no. 4 (2007): pp. 656–64.

— Searle, John Rogers. "The Problem of Consciousness" *Consciousness and Language*; Cambridge, U.K.: Cambridge University Press, 2002: pp. 7–17.

— Shapiro, Robert. "Prebiotic Cytosine Synthesis: A Critical Analysis and Implications for the Origin of Life." *Proceedings of the National Academy of Science USA* vol. 96 (1999): pp. 4396–401.

— Shapiro, Robert. "Prebiotic Ribose Synthesis: A Critical Analysis." *Origins of Life and Evolution of the Biosphere* vol.18 (1988): pp. 71–85.

— Singh, Simon. *Big Bang: The Most Important Scientific Discovery of All Time and Why You Need to Know About It.* London, U.K.: Harper Perennial, 2005; ISBN 0007152523.

— Smolin, Lee. *The Trouble with Physics: The Rise of String Theory, The Fall of a Science, and What Comes Next.* Boston, New York: Haughton Mifflin Company, 2006; ISBN 0618551050, 9780618551057.

— Spemann, Hans & Mangold, Hilde. "Induction of Embryonic Primordia by Implantation of Organizers from a Different Species." trans. Hamburger, Viktor; ed. Sander, Klaus; *International Journal of Developmental Biology* vol. 45 (2001): pp. 13–38.

— Sperry, Roger Wolcott. "Brain Mechanisms in Behavior" *Engineering and Science*; vol. 20, no. 8, (1957) pp. 24–9

— Steinhardt, Paul. "Theories of Anything" *2014 : What Scientific Idea Is Ready for Retirement* https://www.edge.org/response-detail/25405.

— Stenger, Victor J. *God and the Folly of Faith: The Incompatibility of Science and Religion.* Amherst, New York: Prometheus Books, 2012; ISBN 9781616145996.

— Sternberg, Richard. "Discovering Signs in the Genome by Thinking Outside the BioLogos Box." in *Signature of Controversy: Responses to Critics of Signature in the Cell.* ed. Klinghoffer, David; Seattle, WA: Discovery Institute Press, 2010; ISBN 0979014182, 9780979014185: pp.77–81.

— Strobel, Lee. *The Case for a Creator: A Journalist Investigates Scientific Evidence That Points Toward God.* Grand Rapids, Michigan: Zondervan, 2004; ISBN 0310259134.

— Swisher, Carl C. III, Wang (Yuan-qing), Wang (Xiao-lin), Xu (Xing), & Wang (Yuan). "Cretaceous Age for the Feathered Dinosaurs of Liaoning, China" *Nature* vol. 400 (1999): pp. 58–61.

— Szostak, Jack W., Bartel (David P.), & Luisi, (Luigi P.). "Synthesizing Life" *Nature* vol. 209 (2001): pp.387–90.

— Taliaferro, Charles. "The Soul of the Matter." in *The Soul Hypothesis: Investigations into the Existence of the Soul.* eds. Baker, Mark C. & Goetz, Stewart; New York, NY: The Continuum International Publishing Group Inc., 2011; ISBN 9781441152244: pp. 26–40.

— Thaxton, Charles B. & Bradley, Walter L. "Information and the Origin of Life." in *The Creation Hypothesis: Scientific Evidence for an Intelligent Designer.* ed. Moreland James Porter; Downers Grove, IL: InterVarsity, 1994: pp. 173–210.

— Thaxton (Charles B.), Bradley (Walter L.) & Olsen, (Roger L.). *The Mystery of Life's Origin: Reassessing Current Theories.* New York, NY: Philosophical Library, 1984.

— *The Holy Bible*: translated from the original languages, Octopus Books; Chancellor Press, London, UK, 1981; ISBN 0907486258.

— *Time, Health Land: Junk DNA.* http://healthland.time. com/2012/09/06/junk-dna-not-so-useless-after-all/.

— Tipler, Frank Jennings. *The Physics of Christianity*, New York, NY: Doubleday, 2007.

— Tipler, Frank Jennings. *The Physics of Immortality: Modern Cosmology, God and the Resurrection of the Dead.* New York, NY: Anchor Books, 1995; ISBN 0385467990.

— Towe, Kenneth M. "Environmental Oxygen Conditions during the Origin and Early Evolution of Life." *Advances in Space Research*, vol. 18 (1996): pp. 7–15.

— Trimble, Virginia Louise. "Cosmology: Man's Place in the Universe." *American Scientist* vol. 65 (1977): pp. 76–86.

— *University of California, Museum of Paleontology, Berkeley: Phylogenetics.* http://www.ucmp.berkeley.edu/glossary/gloss1phylo.html.

— *University of Cincinnati, Clermont College, Biology and Chemistry: Photosynthesis.* http://biology.clc.uc.edu/courses/bio104/photosyn.htm.

— *University of Wisconsin, Madison: Chemistry.* http://www.chem.wisc.edu/deptfiles/genchem/sstutorial/Text3/Tx33/tx33.html.

— Valentine, James W., Jablonski (David), & Erwin (Douglas H.). "Fossils, Molecules and Embryos: New Perspectives on the Cambrian Explosion." *Development* vol. 126 (1999): pp. 851–9.

— Van Lommel, Pim. "About the Continuity of Our Consciousness" *Advances in Experimental Medicine and Biology* vol. 550 (2004): pp. 115–32.

— Varaha, Swami Bhakti Vedanta; "The Srimad Bhagavad-Gita and the Sacredness of All Cows." http://www.bhagavad-gita.org/Articles/holy-cow.html.

— Von Baer, Karl Ernst Ritter. *Uber Entwickelungsgeschichte der Thiere: Beobachtung und Reflexion.* (*On the Development of Animals*); Bei den Gebrugdern Borntrager, Konigsberg, Germany (1828)

— Wallace, Alan B. *The Taboo of Subjectivity: Toward a New Science of Consciousness*, Oxford, U.K.: Oxford University Press, 2000.

— Wallace, Arthur. *The Origin of Animal Body Plans: A Study in Evolutionary Developmental Biology.* Cambridge, U.K.: Cambridge University Press, 1997.

— Wallace, Bruce. "Adaptation, Neo-Darwinian Tautology, and Population Fitness: A Reply." *Evolutionary Biology* vol. 17 (1984): pp. 59–71.

— *Wayne's Word: Kingdoms.* http://waynesword.palomar.edu/trfeb98.htm.

— Weible, Wayne. *Medjugorje: The Message.* Brewster, MA: Paraclete Press, 1989; ISBN 155725009X.

— Weidner, Richard T. & Sells, Robert L. *Elementary Classical Physics (Volume 1).* Boston, MA; Allyn and Bacon Inc., 1965 (1971 printing); LOCCCN: 65-10891.

— Weinberg, Steven. "A Designer Universe?" *The New York Review of Books*, 21 October 1999.

— Weischaus, Eric "From Molecular Patterns to Morphogenesis: The Lessons from Drosophila." (Nobel Lecture.) 8 December 1995: pp. 314–26.

— *Wikipedia: Abiogenesis.* https://en.'/wiki/Abiogenesis.

— *Wikipedia: Archaeopteryx.* https://en.wikipedia.org/wiki/Archaeopteryx.

— *Wikipedia: Black Dwarf.* https://en.wikipedia.org/wiki/Black_dwarf.

— *Wikipedia: Black Hole.* https://en.wikipedia.org/wiki/Black_hole.

— *Wikipedia: Centrosome.* https://en.wikipedia.org/wiki/Centrosome.

— *Wikipedia: Chinese Room.* https://en.wikipedia.org/wiki/Chinese_room.

— *Wikipedia: Cis-Regulatory Element.* https://en.wikipedia.org/wiki/Cis-Regulatory_element.
— *Wikipedia: Cladistics.* https://en.wikipedia.org/wiki/Cladistics.
— *Wikipedia: Color Blindness.* https://en.wikipedia.org/wiki/Color_blindness.
— *Wikipedia: Cosmic Microwave Background.* https://en.wikipedia.org/wiki/Cosmic_microwave_background.
— *Wikipedia: Cosmological Constant.* https://en.wikipedia.org/wiki/Cosmological_constant.
— *Wikipedia: Entropy.* https://en.wikipedia.org/wiki/Entropy.
— *Wikipedia: Entropy (Order and Disorder).* https://en.wikipedia.org/wiki/Entropy_(order_and_disorder).
— *Wikipedia: Escherichia Coli Long-Term Evolution Experiment.* https://en.wikipedia.org/wiki/E._coli_long-term_evolution_experiment.
— *Wikipedia: Fundamental Interaction.* https://en.wikipedia.org/wiki/Fundamental_interaction.
— *Wikipedia: Future of the Earth.* https://en.wikipedia.org/wiki/Future_of_the_Earth.
— *Wikipedia: Garabandal Apparitions.* https://en.wikipedia.org/wiki/Garabandal_apparitions.
— *Wikipedia: Gravitational Lens.* https://en.wikipedia.org/wiki/Gravitational_lens.
— *Wikipedia: Inner Core.* https://en.wikipedia.org/wiki/Inner_core.
— *Wikipedia: Margaret Mary Alacoque.* https://en.wikipedia.org/wiki/Margaret_Mary_Alacoque.
— *Wikipedia: Miracle of the Sun.* https://en.wikipedia.org/wiki/Miracle_of_the_Sun.
— *Wikipedia: Multiverse.* https://en.wikipedia.org/wiki/Multiverse.
— *Wikipedia: Mutation Rate.* https://en.wikipedia.org/wiki/Mutation_rate.
— *Wikipedia: Natural Law.* https://en.wikipedia.org/wiki/Natural_law.
— *Wikipedia: Neuron.* https://en.wikipedia.org/wiki/Neuron.

— *Wikipedia: Our Lady of Guadeloupe.* https://en.wikipedia.org/wiki/Our_Lady_of_Guadalupe.

— *Wikipedia: Our Lady of Lourdes.* https://en.wikipedia.org/wiki/Our_Lady_of_Lourdes.

— *Wikipedia: Our Lady of Medjugorje.* https://en.wikipedia.org/wiki/Our_Lady_of_Medjugorje.

— *Wikipedia: Phantom Pain.* https://en.wikipedia.org/wiki/Phantom_pain.

— *Wikipedia: Phylum.* https://en.wikipedia.org/wiki/Phylum.

— *Wikipedia: Qualia.* https://en.wikipedia.org/wiki/Qualia.

— *Wikipedia: Quantum Mechanics.* https://en.wikipedia.org/wiki/Quantum_mechanics.

— *Wikipedia: Red Giant.* https://en.wikipedia.org/wiki/Red_giant.

— *Wikipedia: Ship of Theseus.* https://en.wikipedia.org/wiki/Ship_of_Theseus.

— *Wikipedia: Sun.* https://en.wikipedia.org/wiki/Sun.

— *Wikipedia: Supernova.* https://en.wikipedia.org/wiki/Supernova.

— *Wikipedia: Thermodynamic Equilibrium.* https://en.wikipedia.org/wiki/Thermodynamic_Equilibrium.

— *Wikipedia: Turing Test.* https://en.wikipedia.org/wiki/Turing_test.

— *Wikipedia: White Dwarf.* https://en.wikipedia.org/wiki/White_dwarf.

— Wolf, Yuri I. & Koonin, Eugene V. "On the Origin of the Translation System and the Genetic Code in the RNA World by Means of Natural Selection, Exaptation, and Subfunctionalization." *Biology Direct* vol. 2 (2007): pp. 1–25.

— Wray, Gregory A., Levinton (Jeffrey S.), & Shapiro (Leo H.). "Molecular Evidence for Deep Precambrian Divergences among Metazoan Phyla" *Science* vol. 274 (1996): pp. 568–73.

— Zhang, Biliang & Cech, Thomas R. "Peptide Bond Formation by In Vitro selected Ribozymes." *Nature* vol. 390 (1997): pp. 470–5.

INDEX

www.ingramcontent.com/pod-product-compliance
Lightning Source LLC
Chambersburg PA
CBHW020719180526
45163CB00001B/35